BEGINNING ALGEBRA

EIGHTH EDITION

PRINTED TEST BANK

JON BECKER

Indiana University Northwest

Margaret L. Lial
American River College

John Hornsby
University of New Orleans

▲▲ **ADDISON-WESLEY**

An imprint of Addison Wesley Longman, Inc.

Reading, Massachusetts • Menlo Park, California • New York • Harlow, England
Don Mills, Ontario • Sydney • Mexico City • Madrid • Amsterdam

Reproduced by Addison Wesley Longman from camera-ready copy supplied by the author.

Copyright © 2000 Addison Wesley Longman, Inc.

ISBN: 0-321-06192-6

1 2 3 4 5 6 7 8 9 10 PHC 02 01 00 99

PREFACE

This book provides several resources for instructors using *Beginning Algebra*, Eighth Edition, by Margaret L. Lial and John Hornsby.

- Conceptual, writing, and other special exercises in the textbook are listed to enable instructors to select and use these exercises for assignments and projects.

- One open-response and one multiple-choice form of a pretest are supplied to assist faculty with student placement.

- Four open-response and two multiple-choice forms are provided for each chapter.

- One open-response and one multiple-choice form of a final exam are provided.

- Additional exercises are included for each chapter. These exercises can be used as a source of problems for tests, quizzes, and additional practice.

The following people have made valuable contributions to the production of this Printed Test Bank: Jennifer Crum, editor; Adam Hamel and Kari Heen, Assistant Editors; Kathyrn Becker, typist; Darryl Nester, artist; Joe Vetere, technical support specialist; and Hope Martin, Tracy Fugate, and Anthony Gagliardi, accuracy checkers. Without the able and invaluable assistance of these fine people, this manual could *not* have been completed.

Jon Becker
Indiana University Northwest

CONTENTS

of Special Exercises for *Beginning Algebra*, 8e Lial/Hornsby

ion	Journal	Writing	Challenge	Scientific Calculator	Graphing Calculator	Relating Concepts	Tech Insights	Theme
	49, 50							67, 68, 71, 75, 76
	21, 39	22, 70, 77						79, 80, 82
		7-12, 57, 58	83-86	25, 26, 41, 42, 83-86				83-86
	13, 79	14	73-78					15, 16, 20, 65-68
	5, 6							94
	41, 42	65, 108		37, 38				105, 106
	33, 34	53, 54				83-86		
		35, 36, 69				63-70		
iew		119, 120, 139, 140						78
st		8						22
	3, 21, 22	4, 41	75-80					
	1	19, 21, 22, 36-38				35-40		55, 56
	3							11-13, 19, 20, 23, 52
		1, 2		23-26, 33-44		65-68		37
	11	12		27, 28, 37-42, 49-54, 59		55-58		35, 36
		39, 44, 47		7-16, 58		40-43		7-10, 15, 16, 54, 56, 58, 59
		1, 25-27, 76				71-76, 87-98		81, 82

Section	Journal	Writing	Challenge	Scientific Calculator	Graphing Calculator	Relating Concepts	Tech Insights	T...
2 Review		34, 59, 81, 91		32, 47, 52, 56, 85				19, 48, 7
2 Test				11				14,
Cum Rev				23				
3.1	18, 31	14, 33, 34, 43, 77						11-
3.2	20	19, 41		43(a), 44(a), 45, 46			37-42	43, 48,
3.3	14	1, 9, 15, 40	24, 25	57(a)		51-56	61-66	5
3 Review		30, 45		34				19,
3 Test		6		2				1,
Cum Rev		25		17, 19				
4.1	58	57	79-86	45, 46, 81, 82		43-46	41, 42	
4.2	89	13, 14, 23, 24, 37, 38, 81, 82, 90		91-94			39-42	
4.3	39, 40, 63	64	41-54			55-62		
4.4	2, 18	36, 53, 54				31-40		
4.5		73, 74				53-56		
4.6		5, 31, 40, 75, 76, 84	33-36, 73, 74, 77-80			37-40, 81-84		
4.7	13, 14		59, 60				45-50	55
4 Review	40	12, 41, 63, 88				83-88		8
4 Test								
Cum Rev								
5.1	6, 62	4, 61, 89(b), 90				83-88		

ion	Journal	Writing	Challenge	Scientific Calculator	Graphing Calculator	Relating Concepts	Tech Insights	Theme
2	3, 10	4, 63, 64	71-76			23-30		
3	16	15, 65, 66	67-74			75-82		
4	63	27, 28, 66, 78	68-71, 79-82			64-67, 72-77		
nary						81-88		
5	12	9, 67, 68	57-66				69-72	
6		26, 30		35-38, 45-47		49-52		20, 45-47
7		38	31-34				27-30	
iew		57(c), 79(b), 86		57, 79				57, 79, 83, 85
est		15		22				22
Rev								
1	3, 4	21, 22	53, 54, 61-66				67-72	
2	19, 20		45-50, 52			15-18		
3	61, 62	22				19-22, 39-44		
	17, 18	5-7	47-62			1-8		
	5, 6	3, 4, 42	35, 36, 43-48			39-42		
	15	16, 63, 64, 83, 84	59-62			79-84	85-90	
ary								
	73	74	25, 26, 37-40	11-16, 25, 26, 58			67-70	7-10, 71, 72
ew		4, 37(c), 67		52		59-68		56, 77
st								
Rev								26

Section	Journal	Writing	Challenge	Scientific Calculator	Graphing Calculator	Relating Concepts	Tech Insights	Th
7.1	36	5, 49(c), 50(c)	35, 43, 44, 49, 50	45, 46, 49, 50		31-34, 55-62	51-54	45,
7.2	17	18, 31, 32	37-40				33-36, 41-44	2-4, 4
7.3	59, 60	45, 58(b), 58(c)		39, 40		29-32	51-57	41
7 Review		28(c), 37		28, 45, 46		38-44		45
7 Test		12, 16						13
Cum Rev								
8.1	12, 31, 32	2, 30, 33			55, 56	57-62	51-54	34
8.2		1, 2, 21, 22,31,41,42		29, 30	35-42	31-34		29
8.3	23, 24	33-36, 48	37-42	45, 47		43-48		4:
8.4	6		33, 34	19, 20				9, 1 16,
8.5	20	24	25-28				29-32	
8 Review	23	5, 6, 11, 12, 63				41-52		25,
8 Test								14
Cum Rev								9
9.1	76	41, 42, 75	49-54	33-40, 43-60, 65, 66, 73, 74		93-98	89-92	
9.2	35, 36	18, 75(c), 81, 82	80	75, 76		83-86		
9.3	45	26, 46, 50	51-54	26, 51-54, 57, 58		47-50		5
9.4				26, 53, 54				
9.5		35, 36, 74	55-68, 76	75, 76		69-74		
9.6	27	25, 26, 28, 45,46,51(b)	47-50	47-52			53-57	

ion	Journal	Writing	Challenge	Scientific Calculator	Graphing Calculator	Relating Concepts	Tech Insights	Theme
7		83, 84	81, 82	69-76, 81, 82		63-68	77-80	
iew		66, 91(d)		17-20, 43, 91		116-120		91
st		25		2(b), 14(b)				
Rev								10, 34
1	12	11, 43	39-42	45-48, 55-58		49-54		55-58
2	33		34	29-32, 38, 42				37-40
3		15, 16, 53	55, 56	41-44, 57-62, 67, 76		70-76		57-65, 67
ary		43						
4	65	40, 59				35-40	21-28	
5	49, 50	43-48	50		43-48		19-26	41, 42
iew		18(d), 25, 35		64		65-70	49, 50	15
st				16, 24(c)				15, 16
Rev		40(c)						31

DIAGNOSTIC PRETESTS

DIAGNOSTIC PRETEST, FORM A   BEGINNING ALGEBRA

NAME_____   SECTION_____

For Exercises 1-3, write your answer in lowest terms.

1. Write $\dfrac{108}{144}$ in lowest terms.   1. _____

2. Add: $\dfrac{1}{12} + \dfrac{3}{4} + \dfrac{5}{8}$.   2. _____

3. Divide: $\dfrac{25}{32} \div \dfrac{24}{35}$.   3. _____

For Exercises 4-7, simplify the given expression.

4. $3(4+5) + 8 \cdot 5 - 6^2$ 4. _____

5. $\dfrac{7(4+3) + 11}{2(3) + 4}$ 5. _____

6. $|-9| - |-8|$ 6. _____

7. $-|-23|$ 7. _____

For Exercises 8-11, simplify the given expression.

8. $-12 + (-5)$ 8. _____

9. $-7 - (-19)$ 9. _____

10. $-3 + 12 + 16 + (-3 - 6)$ 10. _____

11. $5 - [-8 - (-3)]$ 11. _____

For Exercises 12-13, solve the equation.

12. $8x - 18 - 2x + 2 + 3 + 7x = 11x - 6 - 3$ 12. _____

13. $4(y - 3) - y = y - 4$ 13. _____

14.   The perimeter of a rectangle is 68 centimeters. The length of the rectangle is 7 centimeters more than the width. Find the width of the rectangle.

14. _____

For Exercises 15-16, solve the inequality.

15. $5y + 4 - 6 > -2y + 7 + 4y$

15. _____

16. $-\dfrac{3}{4}a \leq 18$

16. _____

17.   Evaluate $(-5)^4$

17.   Value: _____
      Base: _____
      Exponent: _____

For Exercises 18-19, simplify the expression. Write the answer with positive exponents.

18. $(3z^6)(-2z^3)$

18. _____

19. $\left(\dfrac{2^3 x^{-3}}{5^{-1} x^2}\right)^{-2}$

19. _____

For Exercises 20-22, perform the indicated operation.

20.   Subtract $6x^3 - 4x^2 + 2$ from $11x^3 + 2x^2 - 8$.

20. _____

21.   Multiply $(5y + 3)$ and $(2y - 6)$.

21. _____

22.   Divide $12z^3 - 16z^2 - z + 10$ by $3z + 2$.

22. _____

For Exercises 23-24, write each number in prime factored form.

23.   252

23. _____

24.   480

24. _____

For Exercises 25-27, factor each polynomial completely.

25. $m^2 - 4m - 21$

25. _____

26. $9b^2 - 24b + 16$

26. _____

27. $8x^2 - 2xy - 15y^2$

27. _____

28.   Solve the quadratic equation

$$y^2 + 7y = -12$$

28. _____

29.   The product of two consecutive odd integers is
1 less than five times their sum. Find the integers.

29. _____

For Exercises 30-31, perform the indicated operation.

30. $\dfrac{y^2 - 3y}{y^2 - 2y - 3} \cdot \dfrac{y^2 + 6y + 5}{y^2 + 9y + 20}$

30. _____

31. $\dfrac{q}{q^2 - 1} + \dfrac{q}{q + 1}$

31. _____

32.   Solve the equation

$$\frac{2}{t^2 - t} = \frac{1}{t^2 - 1}$$

32. _____

For Exercises 33-35, graph the equation or inequality.

33. $5x - 2y = -10$

33.

34. $y = 2$ 34.

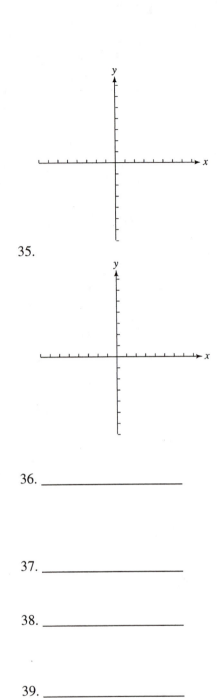

35. $4x - 3y < -12$ 35.

36.   Find the slope of the line through the
         points (-5, 8) and (9, -4).   36. _____

For Exercises 37-38, solve the system of equations.

37. $3x + 4y = -6$
         $2x - 3y = 13$ 37. _____

38. $-a + 3b = -2$
         $4a - 12b = 8$ 38. _____

For Exercises 39-40, find the roots.

39. $\sqrt[3]{-125}$ 39. _____

40. $-\sqrt{576}$ 40. _____

For Exercises 41-44, simplify the given expression.

41. $\sqrt{75}$ 41. _____

42. $\sqrt[3]{54}$ 42. _____

43. $3\sqrt{18} + 4\sqrt{50}$ 43. _____

44. $\sqrt{20} \cdot \sqrt{45}$ 44. _____

For Exercises 45-46, rationalize the denominator.

45. $\dfrac{12}{\sqrt{6}}$ 45. _____

46. $\dfrac{-2}{3 - \sqrt{2}}$ 46. _____

For Exercises 47-49, solve the equation.

47. $3\sqrt{x} = \sqrt{x + 8}$ 47. _____

48. $y^2 = -3y + 4$ 48. _____

49. $2z^2 = -\dfrac{5}{2}z + \dfrac{25}{4}$ 49. _____

50. Graph $y = 2 - x^2$ 50.

DIAGNOSTIC PRETEST, FORM B   BEGINNING ALGEBRA

NAME_____   SECTION_____

__For Exercises 1-3, choose the best answer.__

1. Write $\dfrac{154}{385}$ in lowest terms.

    (a) $\dfrac{3}{5}$ (b) $\dfrac{2}{3}$ (c) $\dfrac{6}{15}$ (d) $\dfrac{2}{5}$ 1. _____

2.   Add: $\dfrac{1}{2}+\dfrac{3}{5}+\dfrac{7}{8}$.

    (a) $\dfrac{11}{15}$ (b) $\dfrac{3}{8}$ (c) $1\dfrac{39}{40}$ (d) $1\dfrac{29}{40}$ 2. _____

3.   Divide: $\dfrac{11}{16}\div\dfrac{9}{8}$.

    (a) $\dfrac{11}{18}$ (b) $\dfrac{99}{128}$ (c) $\dfrac{22}{9}$ (d) $\dfrac{88}{36}$ 3. _____

__For Exercises 4-7, simplify the expression.__

4. $4(13-7)+14\cdot 2-5^2$

    (a) 150   (b) 21   (c) 27   (d) –21   4. _____

5. $\dfrac{6(3+7)+4}{5\cdot 7-3}$

    (a) $2\dfrac{1}{16}$ (b) 2   (c) –2   (d) $3\dfrac{1}{5}$ 5. _____

6. $|-9|+|6|$

    (a) 15   (b) –15   (c) 3   (d) –3   6. _____

7. $-\left|-14\right|-\left|-14\right|$

    (a) –28   (b) 28   (c) 0   (d) 1   7. _____

For Exercises 8-11, perform the indicated operation.

8. $(-5)+(-33)$

    (a) –38   (b) 38   (c) –28   (d) 28   8. _____

9. $(-20)-(-36)$

    (a) –56   (b) 56   (c) –16   (d) 16   9. _____

10. $(-7)+12+(-3)+(-13+6)$

    (a) –5   (b) 5   (c) 1   (d) 9   10. _____

11. $6-\left[-16-(-3)\right]$

    (a) –7   (b) 7   (c) 19   (d) –19   11. _____

For Exercises 12-13, solve the equation.

12. $7y-9+4y-6=-8y+13+12y$

    (a) –4   (b) 4   (c) $\dfrac{2}{7}$ (d) $-\dfrac{2}{7}$ 12. _____

13. $5(2a-7)-a=3(a-9)$

    (a) $\dfrac{3}{4}$ (b) $-\dfrac{3}{4}$ (c) $\dfrac{4}{3}$ (d) $-\dfrac{4}{3}$ 13. _____

14.   The perimeter of a rectangle is 48 feet. The length of
the rectangle is 4 feet longer than three times the width.
Find the length of the rectangle.

    (a) 5 feet   (b) 24 feet   (c) 19 feet   (d) 15 feet   14. _____

For Exercises 15-16, solve the inequality.

15. $4(t+3)-6 > -2t-6$

    (a) $t > 2$ (b) $t > -2$ (c) $t < 2$ (d) $t < -2$ 15. _____

16. $-\dfrac{4}{7}z \le 28$

    (a) $z \ge -49$ (b) $z \le -49$ (c) $z \ge -16$ (d) $z \le -16$ 16. _____

17. Evaluate $(-2)^5$ and name the base.

    (a) 32, base is –2   (b) –32, base is 5

    (c) –32, base is –2   (d) –32, base is 2   17. _____

For Exercises 18-19, simplify the expression. Write the answer with only positive exponents.

18. $(-4a^3b^5)(-5a^4b^8)$

    (a) $20a^{12}b^{40}$ (b) $20a^7b^{13}$ (c) $-9a^{12}b^{40}$ (d) $-20a^7b^{13}$ 18. _____

19. $\left(\dfrac{3^3x^{-2}}{3^{-1}x^2}\right)^{-2}$

    (a) $\dfrac{x^8}{3^8}$ (b) $\dfrac{3^8}{x^8}$ (c) $\dfrac{3^6}{x^6}$ (d) $\dfrac{x^6}{3^6}$ 19. _____

For Exercises 20-22, perform the indicated operation.

20. Subtract $4y^3-8y^2-3$ from $-12y^3+2y^2+y-1$.

    (a) $16y^3-10y^2-y-2$ (b) $-16y^3+10y^2+y+2$

    (c) $-8y^3-16y^2+y-2$ (d) $8y^3+16y^2-y+2$ 20. _____

21. Multiply $(4x-3)$ and $(2x+5)$.

    (a) $8x^2+26x-15$ (b) $8x^2+26x+15$

    (c) $8x^2+14x-15$ (d) $8x^2-14x-15$ 21. _____

22. Divide $6y^3 - 17y^2 + 15y - 2$ by $3y - 4$.

(a) $2y^2 - 3y + 1$ (b) $2y^2 + 3y + 1$

(c) $2y^2 - 3y + 1 + \dfrac{2}{3y-4}$ (d) $2y^2 - 3y + 1 + \dfrac{-2}{3y-4}$ 22. _____

For Exercises 23-24, write the given number in prime factored form.

23. 450

(a) $2 \cdot 3^2 \cdot 5^2$ (b) $2 \cdot 3 \cdot 5^2$ (c) $2 \cdot 3 \cdot 5$ (d) $2^2 \cdot 3^2 \cdot 5^2$ 23. _____

24. 504

(a) $18 \cdot 28$ (b) $2^2 \cdot 3^2 \cdot 7$ (c) $2 \cdot 3 \cdot 7$ (d) $2^3 \cdot 3^2 \cdot 7$ 24. _____

For Exercises 25-27, factor the polynomial completely.

25. $36a^2 - 64b^4$

(a) $(6a - 8b^2)^2$ (b) $4(9a^2 - 16b^4)$

(c) $4(3a - 4b^2)(3a + 4b^2)$ (d) $(6a - 8b^2)(6a + 8b^2)$ 25. _____

26. $16a^2 - 20ac + 12ab - 15bc$

(a) $(4a + 5b)(4a - 3c)$ (b) $(4a - 3b)(4a + 5c)$

(c) $(4a + 3b)(4a - 5c)$ (d) $(8a + 3b)(2a - 5c)$ 26. _____

27. $100x^2 - 60xy + 9y^2$

(a) $(10x + 3y)^2$ (b) $(10x + 3y)(10x - 3y)$

(c) $(10x - 3y)^2$ (d) $(25x - 3y)(4x + 3y)$ 27. _____

28. Solve the quadratic equation

$$x^2 - 9x = -20$$

(a) $4, -5$ (b) $4, 5$ (c) $-4, -5$ (d) $-4, 5$ 28. _____

29. A rectangle with an area of 36 square yards has a length that is one foot longer than twice its width. Find the length.

(a) 12 yards   (b) 6 yards   (c) 4 yards   (d) 9 yards   29. _____

For Exercises 30-31, perform the indicated operation.

30. $\dfrac{x^2 - 7x + 12}{x^2 - 4} \cdot \dfrac{x^2 + 4x - 12}{x^2 + 3x - 18}$

(a) $\dfrac{x-4}{x+2}$ (b) $x - 2$

(c) $\dfrac{x+2}{x-4}$ (d) $\dfrac{(x-3)(x+4)}{(x+6)}$ 30. _____

31. $\dfrac{5}{b-3} - \dfrac{2}{b+3}$

(a) $\dfrac{3}{(b-3)(b+3)}$ (b) $\dfrac{3}{b-3}$

(c) $\dfrac{3(b+7)}{(b-3)(b+3)}$ (d) $\dfrac{3}{2x}$ 31. _____

32. Solve the equation: $\dfrac{8}{y^2} + \dfrac{2}{y} = 1$

(a) –4, 2   (b) –2, 4
(c) 4 only   (d) No solution   32. _____

For Exercises 33-35, graph the equation or inequality.

33. $5x - 2y = 10$ 33. _____

(a)   (b)   (c)   (d)

34. $x = 2$ 34. _____

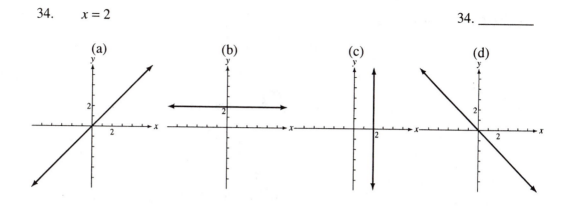

35. $3x - 4y > 12$ 35. _____

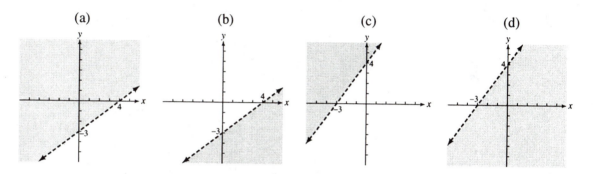

36.   Find the slope of the line through the points (-5, -1) and (5, -4).

(a) $\dfrac{1}{6}$ (b) $-\dfrac{4}{9}$ (c) $-\dfrac{3}{10}$ (d) undefined   36. _____

For Exercises 37-38, solve the system of equations.

37. $2x + 4y = 8$
           $-6x - 12y = -24$

(a) (4, 0)   (b) (0, 2)
(c) Infinite number of solutions   (d) No solution   37. _____

38. $6x + y = 2$
           $3x + 5y = -17$

(a) (1, –4)   (b) (0, 2)
(c) Infinite number of solutions   (d) No solution   38. _____

For Exercises 39-40, find the root.

39. $-\sqrt{256}$

　　　　(a) 16 　　　　　　　　　　(b) –16
　　　　(c) 16, –16 　　　　　　　　(d) Not a real number 　　39. _____

40. $\sqrt{-256}$

　　　　(a) 16 　　　　　　　　　　(b) –16
　　　　(c) 16, –16 　　　　　　　　(d) Not a real number 　　40. _____

For Exercises 41-44, simplify the radical equation.

41. $\sqrt{45}$

　　　　(a) $3\sqrt{5}$ 　　　　　　　　(b) $\sqrt{3}\cdot\sqrt{5}$
　　　　(c) $5\sqrt{3}$ 　　　　　　　　(d) $5\sqrt{9}$ 　　　　　41. _____

42. $\sqrt[3]{48}$

　　　　(a) $2\sqrt[3]{6}$ 　　　　　　　(b) $\sqrt[3]{16}\cdot\sqrt[3]{3}$
　　　　(c) Cannot be simplified 　　　　(d) $6\sqrt[3]{2}$ 　　　42. _____

43. $2\sqrt{32}-3\sqrt{50}$

　　　　(a) $7\sqrt{2}$ 　　　　　　　　(b) $-7\sqrt{2}$
　　　　(c) Cannot be simplified 　　　　(d) $-\sqrt{18}$ 　　　43. _____

44. $\sqrt{72}\cdot\sqrt{162}$

　　　　(a) $54\sqrt{2}$ 　　　　　　　(b) $2\sqrt{54}$
　　　　(c) Cannot be simplified 　　　　(d) 108 　　　　　44. _____

For Exercises 45-46, rationalize the denominator.

45. $\dfrac{14}{\sqrt{7}}$

　　　　(a) $\sqrt{2}$ 　　　　　　　　(b) $2\sqrt{7}$
　　　　(c) $\sqrt{7}$ 　　　　　　　　(d) 2 　　　　　　45. _____

46. $\dfrac{4}{3+\sqrt{5}}$

(a) $3-\sqrt{5}$ (b) $4(3-\sqrt{5})$

(c) $\dfrac{3-\sqrt{5}}{4}$ (d) $\dfrac{12-\sqrt{5}}{4}$

46. _____

For Exercises 47-49, solve the equation.

47. $j-3=\sqrt{j-1}$

(a) 2 only   (b) 5 only
(c) 2, 5   (d) No solution

47. _____

48. $2x^2=5x+3$

(a) $-1,\ -\dfrac{3}{2}$ (b) $1,\ \dfrac{3}{2}$

(c) $-\dfrac{1}{2},\ 3$ (d) $\dfrac{1}{2},\ -3$

48. _____

49. $\dfrac{1}{4}z=\dfrac{3}{8}z^2-\dfrac{1}{2}$

(a) $\dfrac{1+2\sqrt{13}}{3},\ \dfrac{1-2\sqrt{13}}{3}$ (b) No real number solution

(c) $\dfrac{-1+\sqrt{13}}{3},\ \dfrac{-1-\sqrt{13}}{3}$ (d) $\dfrac{1+\sqrt{13}}{3},\ \dfrac{1-\sqrt{13}}{3}$

49. _____

50.   Graph $y=-2-x^2$

50. _____

(a)   (b)   (c)   (d)

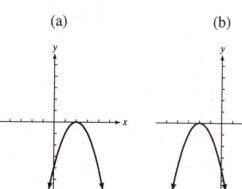

**Answers to Diagnostic Pretest,
Form A**

1. $\dfrac{3}{4}$

2. $\dfrac{35}{24}$ or $1\dfrac{11}{24}$

3. $\dfrac{875}{768}$ or $1\dfrac{107}{768}$

4. 31

5. 6

6. 1

7. −23

8. −17

9. 12

10. 16

11. 10

12. $x = 2$

13. $y = 4$

14. 13.5 cm

15. $y > 3$

16. $a \geq -24$

17. Value: 625;

    base: -5; exponent:

    4

18. $-6z^9$

19. $\dfrac{x^{10}}{5^2 \cdot 2^6}$

20. $5x^3 + 6x^2 - 10$

21. $10y^2 - 24y - 18$

22. $4z^2 - 8z + 5$

23. $2^2 \cdot 3^2 \cdot 7$

24. $2^5 \cdot 3 \cdot 5$

25. $(m - 7)(m + 3)$

26. $(3b - 4)^2$

27. $(2x - 3y)(4x + 5y)$

28. $y = -3, -4$

29. 9,11 or −1,1

30. $\dfrac{y}{y + 4}$

31. $\dfrac{q^2}{q^2 - 1}$

32. $t = -2$

33.

34.

35.

36. $-\dfrac{6}{7}$

37. $(2, -3)$

38. Infinite solutions

39. −5

40. −24

41. $5\sqrt{3}$

42. $3\sqrt[3]{2}$

43. $29\sqrt{2}$

44. 30

45. $2\sqrt{6}$

46. $\dfrac{-2\left(3 + \sqrt{2}\right)}{7}$

47. $x = 1$

48. $y = 1, -4$

49. $x = \dfrac{5}{4}, -\dfrac{5}{2}$

50.

**Diagnostic Pretest,
Form B**

1. D
2. C
3. A
4. C
5. B
6. A
7. A
8. A
9. D
10. A
11. C
12. B
13. C
14. C
15. B
16. A
17. C
18. B
19. A
20. B
21. C
22. C
23. A
24. D
25. C
26. C
27. C
28. B
29. D
30. A
31. C
32. B
33. A
34. C
35. B
36. C
37. C
38. A
39. B
40. D
41. A
42. A
43. B
44. D
45. B
46. A
47. B
48. C
49. D
50. D

CHAPTER TEST
FORMS

CHAPTER ONE, FORM A

BEGINNING ALGEBRA

NAME_____   SECTION_____

For Exercises 1-3, write your answer in lowest terms.

1.   Write $\frac{1296}{702}$ in lowest terms.   1._____

2.   Add: $\frac{2}{3} + \frac{3}{4} + \frac{5}{12}$.   2._____

3.   Divide: $\frac{12}{70} \div \frac{6}{50}$.   3._____

4.   Decide whether $4\left[-20 + 7(-2)\right] \geq 135$
      is true or false.   4._____

5.   Graph the group of numbers
      $\left|-4\right|, -\left|2\right|, -3\frac{3}{8}, -\left|-1\right|$
      on the given number line.   5.

6.   Select the smaller number: $-9, \ -\left|-8\right|$ 6._____

7.   Write the following in symbols;
      then simplify the expression.

            The product of 5 and –6, divided
            by the sum of –3 and –7.   7._____

8.   In the expression $\frac{r-s}{r \cdot s}$, if the value of
      r is negative and the value of s is positive,
      is the value of the expression
      positive or negative?   8._____

For Exercises 9-15, perform the indicated operations whenever possible. Leave all fractional answers in lowest terms.

9. $5-(6-18)+(-10)$ 9. _____

10. $-12\frac{3}{4}+5\frac{2}{3}$ 10. _____

11. $-9-\left[-4-(7-12)\right]$ 11. _____

12. $3^3+(-2)^2-\left(5^2-4\right)$ 12. _____

13. $(-6)(-9)+7(-2)+(-3)^2$ 13. _____

14. $\dfrac{14(-3-2)}{12(-3)+(-8-1)(3-7)}$ 14. _____

15. $\dfrac{-9\left[3+(-2+6)\right]}{-7[3-(-4)]-8(-5)}$ 15. _____

For Exercises 16-17, find the solution for each equation from the set $\{-5, -4, -3, -2, -1, 0, 1, 2, 3, 4, 5\}$ by guessing or by trial and error.

16. $\dfrac{16}{x}=-4$ 16. _____

17. $5a+3=-12$ 17. _____

For Exercises 18-19, evaluate the expressions, given $m = 3$ and $p = -5$.

18. $7m-2p^2$ 18. _____

19. $\dfrac{4m+8p}{-2(m+p)}$ 19. _____

20.   Cara lives in Marietta, Georgia, where the altitude is 1118 feet. Her brother Dan lives in Calipatria, California, where the altitude is 184 feet below sea level. Find the difference in altitude between the two cities.

20. _____

For Exercises 21-25, match the property in Column I with its example in Column II.

	COLUMN I		COLUMN II	

21. Commutative   A. $1 \cdot x = x$ 21. _____

22. Associative   B. $6x + 3 = 3 + 6x$ 22. _____

23. Identity   C. $40 + 4y = 4(10 + y)$ 23. _____

24. Inverse   D. $3x + (3y + 6) = (3x + 3y) + 6$ 24. _____

25. Distributive   E. $-\frac{1}{2}(-2) = 1$ 25. _____

In Exercises 26-27, simplify by combining like terms.

26. $8x + 4x - 6x + x + 14x$ 26. _____

27. $5(2x - 1) - (x - 12) + 2(3x - 5)$ 27. _____

28. Use the distributive property to rewrite
the expression. Simplify if possible.

$$-2(3a - 4b + 2c)$$

28. _____

29. The following bar graph shows the number of units,
in thousands, produced by Gem Electronics during
the years 1987 through 1991. Use a signed number
to represent the change from 1989 to 1990.

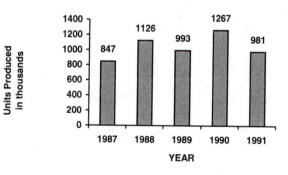

29. _____

30.   One country's wheat prices and production figures
       for a recent ten-year period are given below.

Year	Price ($ Per Bushel)	Production (Millions Of Bushels)
1982	2.70	2200
1983	2.30	2000
1984	2.95	1750
1985	3.80	2200
1986	3.90	2400
1987	3.60	2800
1988	3.55	2800
1989	3.50	2450
1990	3.35	2600
1991	3.20	2750

Determine the change in production
from 1988 to 1989 by subtraction.   30. _____

For Exercises 31-32, refer to the following pie chart.

Sales Income Allocation

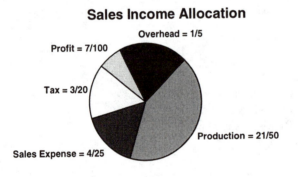

Overhead = 1/5
Profit = 7/100
Tax = 3/20
Production = 21/50
Sales Expense = 4/25

31.   What fractional part of total sales income
       went to areas other than overhead?   31. _____

32.   If 20.5 million dollars was the total sales
       income, how much of it went for overhead?   32. _____

33.   Give an example which shows that division
       is not a commutative operation.   33. _____

CHAPTER ONE, FORM B   **BEGINNING ALGEBRA**

NAME_____   SECTION_____

For Exercises 1-3, write your answer in lowest terms.

1.   Write $\frac{450}{252}$ in lowest terms.   1._____

2.   Add: $\frac{3}{5} + \frac{1}{3} + \frac{7}{15}$.   2._____

3.   Divide: $\frac{21}{56} \div \frac{7}{40}$.   3._____

4.   Decide whether $-7[-16 + 7(-3)] \le 256$
      is true or false.   4._____

5.   Graph the group of numbers

      $-|3|, -|-1|, -4\frac{1}{8}, |-2|$

      on the given number line.   5.

6.   Select the smaller number: $-8, \ -|-9|$ 6._____

7.   Write the following in symbols;
      then simplify the expression.

             The product of 2 and –14, divided
             by the sum of 5 and –7.   7._____

8.   In the expression $\frac{a-b}{a \cdot b}$, if the value of
      a is positive and the value of b is negative,
      is the value of the expression
      positive or negative?   8._____

For Exercises 9-15, perform the indicated operations whenever possible. Leave all fractional answers in lowest terms.

9. $3 - (7 - 15) + (-11)$ 9. _____

10. $-9\frac{4}{5} + 4\frac{1}{4}$ 10. _____

11. $-9 - [-6 - (12 - 7)]$ 11. _____

12. $3^2 + (-2)^3 + (4^2 - 5)$ 12. _____

13. $(-16)(-4) + 2(-7) + (-5)^2$ 13. _____

14. $\dfrac{10(-3 - 2)}{3(-13) - (-5 - 4)(1 - 5)}$ 14. _____

15. $\dfrac{-4[5 + (-2 + 11)]}{-5[4 - (-3)] - 4(-7)}$ 15. _____

For Exercises 16-17, find the solution for each equation from the set {-5, -4, -3, -2, -1, 0, 1, 2, 3, 4, 5} by guessing or by trial and error.

16. $\dfrac{20}{y} = -4$ 16. _____

17. $5a + 7 = -8$ 17. _____

For Exercises 18-19, evaluate the expressions, given m = -3 and p = 5.

18. $9m - 4p^2$ 18. _____

19. $\dfrac{-4m + 6p}{2(p - m)}$ 19. _____

20.   On March 28, the temperature in Chicago
       at 6:00 A.M. was $-21°$ F. By 2:00 P.M.,
       the temperature had risen $17°$. What was
       the temperature at 2:00 P.M. on March 28?   20. _____

For Exercises 21-25, match the property in Column I with its example in Column II.

COLUMN I   COLUMN II

21. Commutative   A. $\left(-\dfrac{2}{3}\right)\left(-\dfrac{3}{2}\right) = 1$ 21. _____

22. Associative   B. $54 + 6w = 6(9 + w)$ 22. _____

23. Identity   C. $3a + 4 = 4 + 3a$ 23. _____

24. Inverse   D. $2a + (3b + 6) = (2a + 3b) + 6$ 24. _____

25. Distributive   E. $1 \cdot c = c$ 25. _____

In Exercises 26-27, simplify by combining like terms.

26. $7y - 4y + 13y - y + 5y$ 26. _____

27. $4(5x - 3) - (x - 9) + 5(2x - 7)$ 27. _____

28. Use the distributive property to rewrite the expression. Simplify if possible.

$$-3(2r - 5s + 7t)$$

     28. _____

29. The following bar graph shows the number of units, in thousands, produced by Gem Electronics during the years 1987 through 1991. Use a signed number to represent the change from 1989 to 1990.

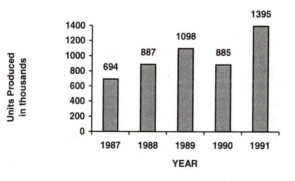

     29. _____

30.   One country's wheat prices and production figures
       for a recent ten-year period are given below.

Year	Price ($ Per Bushel)	Production (Millions Of Bushels)
1982	2.70	2200
1983	2.30	2000
1984	2.95	1750
1985	3.80	2200
1986	3.90	2400
1987	3.60	2800
1988	3.55	2800
1989	3.50	2450
1990	3.35	2600
1991	3.20	2750

Determine the change in production
from 1984 to 1985 by subtraction.   30. _____

__For Exercises 31-32, refer to the following pie chart.__

Sales Income Allocation

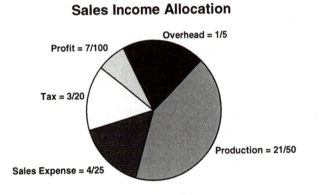

31.   What fractional part of total sales income
       went to areas other than production?   31. _____

32.   If 7.8 million dollars was the total sales
       income, how much of it went for taxes?   32. _____

33.   Give an example which shows that
       subtraction is not a commutative operation.   33. _____

CHAPTER ONE, FORM C   BEGINNING ALGEBRA

NAME_____   SECTION_____

For Exercises 1-3, write your answer in lowest terms.

1. Write $\dfrac{416}{1408}$ in lowest terms.   1. _____

2. Add: $\dfrac{3}{8} + \dfrac{2}{3} + \dfrac{7}{12}$.   2. _____

3. Divide: $\dfrac{42}{55} \div \dfrac{30}{35}$.   3. _____

4. Decide whether $(-5)^2 + 12^2 = 13^2$
    is true or false.   4. _____

5. Graph the group of numbers
    $$|3|, |-2|, -1\frac{1}{8}, -|-3|$$
    on the given number line.   5.

6. Select the smaller number: $-0.998, -0.997$ 6. _____

7. Write the following in symbols;
    then simplify the expression.

    The product of 12 and –5, divided
    by the difference between 3 and –2.   7. _____

8. In the expression $\dfrac{a \div b}{a + b}$, if the values of
    of both a and b are negative,
    is the value of the expression
    positive or negative?   8. _____

For Exercises 9-15, perform the indicated operations whenever possible. Leave all fractional answers in lowest terms.

9. $-9 + (-8 + 3) + (5 - 10)$ 9. _____

10. $4\frac{4}{5} - 5\frac{5}{6}$ 10. _____

11. $-8 - \left[-7 + (12 - 7)\right] - (-4)$ 11. _____

12. $9 + (2)^3 + \left(4^2 - 10\right)$ 12. _____

13. $(-7)(-9) + 4(-3) + (-5)^2$ 13. _____

14. $\dfrac{13(-4-2)}{8(-9)+(-7-4)(2-5)}$ 14. _____

15. $\dfrac{-8\left[4 + (-2 + 7)\right]}{-3\left[4 - (-9)\right] - 3(-9)}$ 15. _____

For Exercises 16-17, find the solution for each equation from the set $\{-5, -4, -3, -2, -1, 0, 1, 2, 3, 4, 5\}$ by guessing or by trial and error.

16. $\dfrac{14}{t} = -7$ 16. _____

17. $9b - 3 = -30$ 17. _____

For Exercises 18-19, evaluate the expressions, given $m = -4$ and $p = 6$.

18. $2m^2 - 4p$ 18. _____

19. $\dfrac{-3m + 5p}{2(p - m)}$ 19. _____

20.   On April 2, the temperature in St. Paul, Minnesota, was $-37°$ F. On June 10, the temperature in Grand Rapids, Michigan was $113°$ higher. What was the temperature in Grand Rapids, Michigan, on June 10?   20. _____

For Exercises 21-25, match the property in Column I with its example in Column II.

COLUMN I   COLUMN II

21.   Associative   A. $16 + 6p = 2(8 + 3p)$ 21. _____

22.   Commutative   B. $(-3)\left(-\frac{1}{3}\right) = 1$ 22. _____

23.   Distributive   C. $d \cdot 1 = d$ 23. _____

24.   Identity   D. $5a \cdot 2 = 2 \cdot 5a$ 24. _____

25.   Inverse   E. $2x + (3y + z) = (2x + 3y) + z$ 25. _____

In Exercises 26-27, simplify by combining like terms.

26. $2(6a - 1) - (2a - 3) + 3(7a - 5)$ 26. _____

27. $-5t - 3t + 16t - t + 9t$ 27. _____

28.   Use the distributive property to rewrite
      the expression. Simplify if possible.

      $5(4r) + 5(-3s)$ 28. _____

29.   The following bar graph shows the revenue, in millions
      of dollars, received by Andrew Computer Corporation
      during the years 1987 through 1991. Use a signed number
      to represent the change from 1989 to 1990.

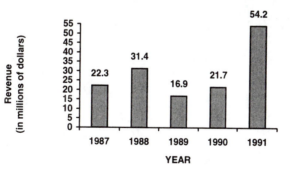

29. _____

30. One country's wheat prices and production figures for a recent ten-year period are given below.

Year	Price ($ Per Bushel)	Production (Millions Of Bushels)
1982	2.70	2200
1983	2.30	2000
1984	2.95	1750
1985	3.80	2200
1986	3.90	2400
1987	3.60	2800
1988	3.55	2800
1989	3.50	2450
1990	3.35	2600
1991	3.20	2750

Determine the change in production
from 1988 to 1989 by subtraction.   30. _____

For Exercises 31-33, refer to the following pie chart.

Favorite All-Time Situation Comedies

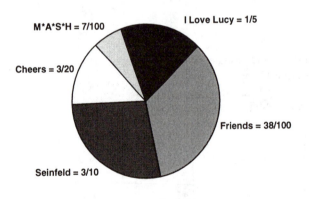

31. What fractional portion of people interviewed
liked a show other than M*A*S*H?   31. _____

32. If 50,000 people were interviewed,
how many of them liked Cheers best?   32. _____

33. Give an example of a number that is
a whole number but not a natural number.   33. _____

CHAPTER ONE, FORM D

BEGINNING ALGEBRA

NAME_____

SECTION_____

For Exercises 1-3, write your answer in lowest terms.

1. Write $\dfrac{594}{828}$ in lowest terms.

    1. _____

2. Add: $\dfrac{1}{8} + \dfrac{2}{3} + \dfrac{11}{12}$.

    2. _____

3. Divide: $\dfrac{45}{52} \div \dfrac{63}{91}$.

    3. _____

4. Decide whether $(-6)^2 + (-8)^2 = 10^2$ is true or false.

    4. _____

5. Graph the group of numbers

    $-|3|, |-1|, -4\dfrac{1}{8}, -|-1|$

    on the given number line.

    5. _____

6. Select the smaller number: $-0.676, \ -0.667$

    6. _____

7. Write the following in symbols; then simplify the expression.

    The product of 5 and –24, divided by the difference between –4 and 8.

    7. _____

8. In the expression $\dfrac{(a-b)(b-a)}{a \cdot b}$, if the value of a is negative and the value of b is positive, is the value of the expression positive or negative?

    8. _____

For Exercises 9-15, perform the indicated operations whenever possible. Leave all fractional answers in lowest terms.

9. $-9+(-5-9)-(5-2)$ 9. _____

10. $3\dfrac{3}{4}-5\dfrac{5}{8}$ 10. _____

11. $4-\left[(-3-2)-(4-1)\right]-(-4)$ 11. _____

12. $-9+(2)^3+\left(10^2-4\right)$ 12. _____

13. $(-6)(-3)+4(-7)+(-8)^2$ 13. _____

14. $\dfrac{8(-6-6)}{7(-8)-(6-16)(2-6)}$ 14. _____

15. $\dfrac{-8\left[2-(-2+9)\right]}{-7[1-(-7)]-3(-22)}$ 15. _____

For Exercises 16-17, find the solution for each equation from the set {-5, -4, -3, -2, -1, 0, 1, 2, 3, 4, 5} by guessing or by trial and error.

16. $-\dfrac{18}{p}=-6$ 16. _____

17. $-5k+3=28$ 17. _____

For Exercises 18-19, evaluate the expressions, given n = -2 and r = 4.

18. $3n^3-5r$ 18. _____

19. $\dfrac{-3n+5r}{2(r-n)+1}$ 19. _____

20. On August 15, the temperature in Houston, Texas, was $99°$ F. On December 25, the temperature in Anchorage, Alaska was $118°$ lower. What was the temperature in Anchorage, Alaska, on December 25?   20. _____

For Exercises 21-25, match the property in Column I with its example in Column II.

COLUMN I   COLUMN II

21. Associative   A. $5z \cdot 6 = 6 \cdot 5z$ 21. _____

22. Commutative   B. $26 + 13p = 13(2 + p)$ 22. _____

23. Distributive   C. $(4)\left(\dfrac{1}{4}\right) = 1$ 23. _____

24. Identity   D. $3x + (4y + 5z) = (3x + 4y) + 5z$ 24. _____

25. Inverse   E. $f \cdot 1 = f$ 25. _____

In Exercises 26-27, simplify by combining like terms.

26. $3(7m - 1) - (m - 4) + 2(3m - 4)$ 26. _____

27. $-5c - 3c + 13c - 9c + c$ 27. _____

28.   Use the distributive property to rewrite
      the expression. Simplify if possible.

      $4(5r) + 4(-3t)$ 28. _____

29.   The following bar graph shows the revenue, in millions
      of dollars, received by Andrew Computer Corporation
      during the years 1987 through 1991. Use a signed number
      to represent the change from 1990 to 1991.

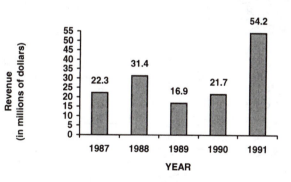

                                                             29. _____

30.   One country's wheat prices and production figures
        for a recent ten-year period are given below.

Year	Price ($ Per Bushel)	Production (Millions Of Bushels)
1982	2.70	1950
1983	2.30	2200
1984	2.95	1750
1985	3.80	2200
1986	3.90	2400
1987	3.60	2800
1988	3.55	2800
1989	3.50	2450
1990	3.35	2600
1991	3.20	2750

Determine the change in production
from 1982 to 1983 by subtraction.   30. _____

For Exercises 31-33, refer to the following pie chart.

Favorite All-Time Situation Comedies

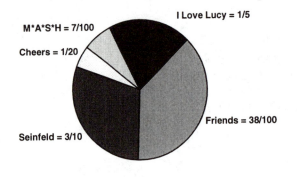

31.   What fractional portion of people interviewed
        liked a show other than Cheers?   31. _____

32.   If 23,000 people were interviewed,
        how many of them liked Seinfeld best?   32. _____

33.   Give an example of a number that does not
        have a multiplicative inverse.   33. _____

CHAPTER ONE, FORM E   BEGINNING ALGEBRA

NAME_____   SECTION_____

For Exercises 1-3, choose the best answer.

1. Write $\dfrac{840}{5280}$ in lowest terms.

   (a) $\dfrac{5}{11}$ (b) $\dfrac{7}{11}$ (c) $\dfrac{7}{44}$ (d) $\dfrac{5}{44}$ 1._____

2. Add: $\dfrac{1}{6} + \dfrac{3}{8} + \dfrac{5}{12}$.

   (a) $\dfrac{9}{26}$ (b) $\dfrac{23}{24}$ (c) $\dfrac{3}{8}$ (d) $\dfrac{1}{4}$ 2._____

3. Divide: $\dfrac{8}{42} \div \dfrac{24}{45}$.

   (a) $\dfrac{5}{14}$ (b) $\dfrac{32}{315}$ (c) $\dfrac{1}{3}$ (d) $\dfrac{42}{15}$ 3._____

Decide whether each statement is true or false.

4. $3[-10 - 6(-7)] \leq 96$ (a) True   (b) False   4._____

5. $\dfrac{17 + 3(1+2)}{4(2-1) - (-9)} < 2$ (a) True   (b) False   5._____

Select the smaller number from each list.

6. $-|8|, -(-12), -|-12|, -(-8)$

   (a) $-|8|$ (b) $-(-12)$ (c) $-|-12|$ (d) $-(-8)$ 6._____

7. $-.061, -6.1, -.601, -.61$

   (a) $-.061$ (b) -6.1 (c) $-.601$ (d) $-.61$ 7._____

For Exercises 8-9, write a numerical expression for the phrase and simplify.

8.   The product of –8 and –9, divided by the sum of –15 and 3

      (a) 4   (b) –36   (c) 6   (d) –6   8. _____

9.   12 less than the difference between 7 and –3

      (a) –15   (b) –22   (c) 2   (d) –2   9. _____

For Exercises 10-16, perform the indicated operations whenever possible.

10. $-8-(3-10)+(-6)$

      (a) –7   (b) –21   (c) –5   (d) 5   10. _____

11. $-3\frac{1}{6}+2\frac{3}{4}$

      (a) $5\frac{11}{12}$ (b) $-\frac{5}{12}$ (c) $-1\frac{1}{2}$ (d) $-\frac{2}{3}$ 11. _____

12. $-8-[-4+(6-8)]$

      (a) –6   (b) 10   (c) 2   (d) –2   12. _____

13. $4^2-(-5)-(3^3-4)$

      (a) 16   (b) 8   (c) –2   (d) 1   13. _____

14. $(-3)(-14)+8(-2)+(-3)^2$

      (a) 35   (b) –67   (c) 49   (d) 67   14. _____

15. $\dfrac{-3-(-2)}{4(-5)+(3-5)(2-12)}$

      (a) $\frac{1}{40}$ (b) $-\frac{1}{40}$ (c) $\frac{1}{8}$ (d) Undefined   15. _____

16. $\dfrac{-8[2-(-1+8)]}{-7[3-(-5)]-6(-11)}$

    (a) 5   (b) 4   (c) –5   (d) Undefined   16. _____

For Exercises 16-17, find the solution for each equation from the set {–5, –4, –3, –2, –1, 0, 1, 2, 3, 4, 5} by guessing or by trial and error.

17. $2y+3=-7$

    (a) –2   (b) 3   (c) –3   (d) –5   17. _____

18. $-3x-5=7$

    (a) –5   (b) 4   (c) –4   (d) –3   18. _____

For Exercises 19-20, evaluate the expressions, given x = –4 and y = –3.

19. $3x-4y^2$

    (a) –48   (b) –73   (c) 24   (d) 0   19. _____

20. $\dfrac{3x-4y}{y-3}$

    (a) 4   (b) 0   (c) –4   (d) Undefined   20. _____

21. At 12 noon and 1:00 p.m., Doris read a gauge in the lab where she works. The first reading was -1.58 and the second reading was -5.06. By how much did the reading increase or decrease?

    (a) Increased 3.48   (b) Increased 6.64
    (c) Decreased 3.48   (d) Decreased 6.64   21. _____

22. Use the distributive property to rewrite the expression. Simplify if possible.

$$-\frac{5}{6}(12a+18b-6c)$$

    (a) $10a+15b-5c$ (b) $-10a-15b+5c$
    (c) $-60a-90b+30c$ (d) $-20abc$ 22. _____

For Exercises 23-27, select the letter of the property that matches each example.

(a) Associative   (b) Commutative   (c) Inverse   (d) Identity   (e) Distributive

23. $-7+7=0$ 23. _____

24. $3(4) + 3(5) = 3\,(4 + 5)$ 24. _____

25. $(-8)(1)=-8$ 25. _____

26. $3+(-5)=(-5)+3$ 26. _____

27. $6(5\cdot 8) = (6\cdot 5)(8)$ 27. _____

In exercise 28, simplify by combining like terms.

28. $5(2x-1)-(x-12)+2(3x-5)$

        (a) $15x-3$ (b) $15x-29$ (c) $15x+7$ (d) $15x-17$ 28. _____

29.   If a is positive and b is negative, what can you determine about
        the value of the following expression?

$$\frac{(a - b)(b - a)}{a\cdot b}$$

        (a) Positive   (b) Negative
        (c) Zero   (d) Cannot be determined   29. _____

30.   The following bar graph shows the approximate
        viewing audience, in millions, for the NCAA
        finals during the years 1986 through 1990. Use
        a signed number to represent the change in viewers
        from 1988 to 1989.

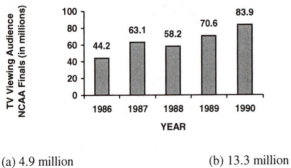

        (a) 4.9 million   (b) 13.3 million
        (c) 39.7 million   (d) 12.4 million   30. _____

31.   One country's wheat prices and production figures
for a recent eight-year period are given below.

Year	Price ($ Per Bushel)	Production (Millions Of Bushels)
1982	2.70	2200
1983	2.30	2000
1984	2.95	1750
1985	3.80	2200
1986	3.90	2400
1987	3.60	2800
1988	3.55	2800
1989	3.50	2450

Determine the change in price per bushel from 1982 to 1983 by subtraction.

(a) $0.65   (b) $0.40   (c) −$0.40   (d) −$0.65   31. _____

__For Exercises 32-33, refer to the following pie chart.__

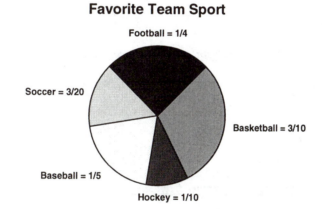

Favorite Team Sport

32.   What fractional part of all people surveyed chose
football or hockey as their favorite team sport?

(a) $\frac{1}{5}$ (b) $\frac{1}{4}$ (c) $\frac{7}{20}$ (d) $\frac{9}{25}$ 32. _____

33.   If 1050 people were surveyed, how many said that
baseball was their favorite sport?

(a) 315   (b) 210   (c) 262   (d) 200   33. _____

CHAPTER ONE, FORM F   BEGINNING ALGEBRA

NAME_____   SECTION_____

For Exercises 1-3, choose the best answer.

1. Write $\dfrac{520}{7150}$ in lowest terms.

    (a) $\dfrac{4}{11}$ (b) $\dfrac{6}{11}$ (c) $\dfrac{6}{55}$ (d) $\dfrac{4}{55}$ 1. _____

2. Add: $\dfrac{1}{3} + \dfrac{5}{8} + \dfrac{1}{12}$.

    (a) $\dfrac{25}{24}$ (b) $\dfrac{7}{23}$ (c) $\dfrac{17}{24}$ (d) $\dfrac{13}{12}$ 2. _____

3. Divide: $\dfrac{12}{64} \div \dfrac{24}{56}$.

    (a) $\dfrac{4}{7}$ (b) $\dfrac{7}{16}$ (c) $\dfrac{1}{2}$ (d) $\dfrac{4}{7}$ 3. _____

For Exercises 4-5, decide whether each statement is true or false.

4. $4[-10 - 3(-7)] \le -96$ (a) True   (b) False   4. _____

5. $\dfrac{27 + 3(1 + 2)}{4(5 - 3) - (-1)} > 2$ (a) True   (b) False   5. _____

For Exercises 6-7, select the smaller number from each list.

6. $-|-5|, \; -(-12), \; -|12|, \; -(-5)$

    (a) $-|-5|$ (b) $-(-12)$ (c) $-|12|$ (d) $-(-5)$ 6. _____

7. $-6.4, \; -.64, \; -.604, \; -.064$

    (a) -6.4 (b) $-.64$ (c) $-.604$ (d) $-.064$ 7. _____

For Exercises 8-9, write a numerical expression for the phrase and simplify.

8.   The product of –5 and –12, divided by the sum of –3 and –1

      (a) 15   (b) –15   (c) 30   (d) –30   8. _____

9.   –7 less than the difference between 12 and –4

      (a) 9   (b) 23   (c) 15   (d) –1   9. _____

For Exercises 10-16, perform the indicated operations whenever possible.

10. $-7 - (4 - 10) + (-5)$

      (a) 6   (b) –18   (c) –6   (d) –4   10. _____

11. $-2\frac{3}{8} + 3\frac{1}{4}$

      (a) $1\frac{7}{8}$ (b) $-1\frac{7}{8}$ (c) $5\frac{5}{8}$ (d) $\frac{7}{8}$ 11. _____

12. $-7 - [-6 + (4 - 8)]$

      (a) –17   (b) –5   (c) –3   (d) 3   12. _____

13. $5^2 - (-4) - (2^3 - 6)$

      (a) 12   (b) 19   (c) 21   (d) 27   13. _____

14. $(-2)(-11) + 7(-3) + (-4)^2$

      (a) 17   (b) –15   (c) 15   (d) –59   14. _____

15. $\dfrac{-5 - (-7)}{3(-4) + (5 - 6)(2 - 10)}$

      (a) $\frac{1}{2}$ (b) $-\frac{1}{2}$ (c) $\frac{3}{5}$ (d) 3   15. _____

16. $\dfrac{-7[3+(-4+6)]-1}{-9[3-(-5)]-8(-9)}$

(a) $\dfrac{1}{3}$ (b) –1   (c) 1   (d) Undefined   16. _____

For Exercises 17-18, find the solution for each equation from the set {–5, –4, –3, –2, –1, 0, 1, 2, 3, 4, 5} by guessing or by trial and error.

17. $5y+7=-8$

(a) 1   (b) –1   (c) 3   (d) –3   17. _____

18. $-2x+5=-5$

(a) 5   (b) –5   (c) 0   (d) 1   18. _____

For Exercises 19-20, evaluate the expressions, given x = -4 and y = -3.

19. $3x^2-4y$

(a) –36   (b) –60   (c) 60   (d) 12   19. _____

20. $\dfrac{2x-3y}{4-x}$

(a) $\dfrac{1}{8}$ (b) $-\dfrac{17}{8}$ (c) $\dfrac{17}{8}$ (d) Undefined   20. _____

21.   When a 25 mile per hour wind gust blows through Chicago on a day when the standing temperature is actually $5°$ F, the temperature of the air on a person's skin is $46°$ F colder due to the wind chill factor. What is the wind chill temperature?

(a) $-46°$ F   (b) $-21°$ F   (c) $-51°$ F   (d) $-41°$ F   21. _____

22.   Use the distributive property to rewrite the expression. Simplify if possible.
$$-(-2a+18b-6c)$$

(a) $-2a+18b-6c$ (b) $2a-18b+6c$
(c) $2a-18b-6c$ (d) $-10abc$ 22. _____

For Exercises 23-27, select the letter of the property that matches each example.

(a) Associative   (b) Commutative   (c) Inverse   (d) Identity   (e) Distributive

23. $6(a + 8) = 6a + 48$ 23. _____

24. $-\frac{1}{3}(-3) = 1$ 24. _____

25. $9 + [(-2) + 5] = [9 + (-2)] + 5$ 25. _____

26. $0 + (-6) = -6$ 26. _____

27. $7(-8) = (-8)7$ 27. _____

In exercise 28, simplify by combining like terms.

28. $9(2x - 3) - (12x - 1) + 2(3x - 2)$

       (a) $36x - 32$ (b) $36x - 30$ (c) $12x - 32$ (d) $12x - 30$ 28. _____

29.   If a is positive and b is negative, what can you determine about
the value of the following expression?

$$\frac{(a \div b)(b \cdot a)}{(a - b) \cdot b}$$

(a) Positive   (b) Negative
(c) Zero   (d) Cannot be determined   29. _____

30.   The following bar graph shows the approximate viewing audience,
in millions, for the NCAA finals during the years 1986 through 1990.
Use a signed number to represent the change in viewers from 1987 to 1988.

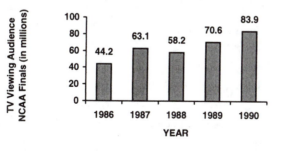

(a) –12.4 million   (b) –26.4 million
(c) –4.9 million   (d) –7.5 million   30. _____

31.   One country's wheat prices and production figures
       for a recent eight-year period are given below.

Year	Price ($ Per Bushel)	Production (Millions Of Bushels)
1982	2.70	2200
1983	2.30	2000
1984	2.95	1750
1985	3.80	2200
1986	3.90	2400
1987	3.60	2800
1988	3.55	2800
1989	3.50	2450

Determine the change in production from 1984 to 1985 by subtraction.
    (a) 450 million bushels   (b) −450 million bushels
    (c) 900 million bushels   (d) −900 million bushels   31. _____

For Exercises 32-33, refer to the following pie chart.

Mr. Quigley's Budget

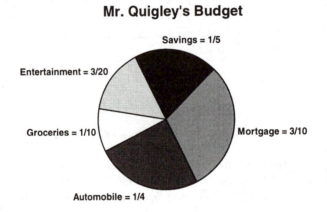

32.   What fractional part of Mr. Quigley's Budget is not
       used for Groceries or Entertainment?

    (a) $\frac{1}{4}$ (b) $\frac{3}{4}$ (c) $\frac{3}{10}$ (d) $\frac{7}{10}$ 32. _____

33.   If Mr. Quigley earns $2100, how much does he put into savings?

    (a) $420   (b) $525   (c) $210   (d) $630   33. _____

CHAPTER TWO, FORM A

BEGINNING ALGEBRA

NAME_____   SECTION_____

For Exercises 1-7, solve the equation.

1. $6x - 9 = 4x - 7$

1._____

2. $5(d - 2) - (3d - 1) = 3d - 6$

2._____

3. $7.4x - 11 = 3.4x + 17.8$

3._____

4. $3j - (j - 4) + 3 = 7 + 2j$

4._____

5. $8y - 7 = 6(2y + 5) - 3y$

5._____

6. $-\dfrac{7}{4}p = 28$

6._____

7. $0.3(w + 70) = 0.7(w + 40) + 25$

7._____

For Exercises 8-9, solve the problem.

8.   If 7 times a number is added to 18, the result is -17.
Find the number.

8._____

9.   A log 78 feet long is to be cut into three pieces of
varying lengths. The longest piece is to be four times
the length of the shortest piece and the middle-sized
piece is to be 6 feet less than twice the length of the
shortest piece. How long is the shortest piece?

9._____

10.   The formula for the area of a trapezoid is

$$A = \frac{1}{2}(B + b)h \,.$$

(a) Solve for b.

10. (a) _____

(b) If $A = 80$, $B = 5$, and $h = 8$,
    find the value of b.

10. (b) _____

11.   Suppose that in the equation

$$\frac{a}{b} = \frac{c}{d},$$

a and b are positive numbers and d is a negative
number. Must c be positive or negative?   11. _____

For Exercises 12-13, find the measure of each marked angle.

12.

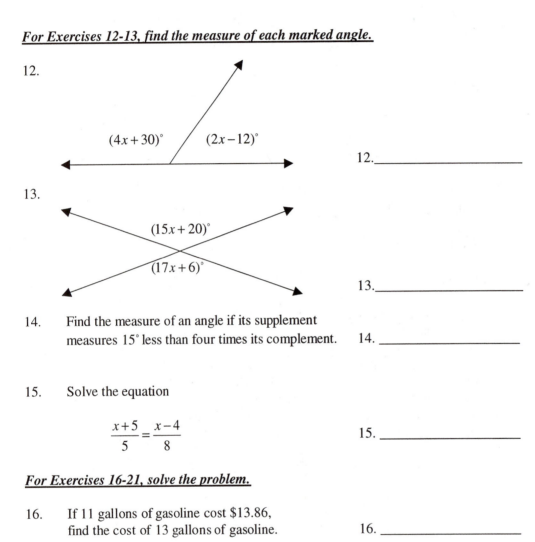

12. _____

13.

13. _____

14.   Find the measure of an angle if its supplement
measures 15° less than four times its complement.   14. _____

15.   Solve the equation

$$\frac{x+5}{5} = \frac{x-4}{8}$$

15. _____

For Exercises 16-21, solve the problem.

16.   If 11 gallons of gasoline cost $13.86,
find the cost of 13 gallons of gasoline.   16. _____

17.   Which is the better buy: two 100-sheet rolls
of paper towel for $0.79 or one 150-sheet roll
for $0.69?   17. _____

18. How many gallons of pure antifreeze must be mixed with 20 gallons of 25% antifreeze to get a mixture that is 40% antifreeze?

18. _____

19. The distance between Buffalo, New York and Chicago, Illinois is 600 miles. On a U.S. road atlas, this distance is represented by 10 inches. On the same atlas, the distance between Chicago and Denver, Colorado is 18 inches. What is the distance between Chicago and Denver?

19. _____

20. Kate has $780 in five-dollar bills and twenty-dollar bills. If she has four more twenties than fives, how many five-dollar bills does she have?

20. _____

21. Two investments produce a total annual interest income of $201. If $1200 more was invested at 7% than at 6%, how much was invested at 6%?

21. _____

For Exercises 22-24, solve the inequality and graph the solution.

22. $6(5-w)-3 \geq -3(w-7)-w$

22. _____

23. $-3 \leq 6y-3 \leq 15$

23. _____

24. $-3 < \dfrac{3}{2}x+6 < 9$

24. _____

25. Dave Lambie has grades of 84 and 90 on his first two Biology tests. What possible scores could he make on his third test so that his average is at least 88?

25. _____

CHAPTER TWO, FORM B   BEGINNING ALGEBRA

NAME_____   SECTION_____

For Exercises 1-7, solve the equation.

1. $8x + 14 = 3x - 1$ 1. _____

2. $6(4 - 5d) = 2 - 2(3 + d)$ 2. _____

3. $2.4a - 6.8 = 1.4a + 5.9$ 3. _____

4. $6 - (y + 2) = 8 - y$ 4. _____

5. $\dfrac{4x}{5} = -20$ 5. _____

6. $-5(3x + 6) = -3(5x + 10)$ 6. _____

7. $0.04(k + 10) + 0.06(k - 30) = 7.6$ 7. _____

For Exercises 8-9, solve the problem.

8.   If 8 is subtracted from five times a number,
      the result is 2 more than six times the number.
      Find the number.   8. _____

9.   A ribbon 106 inches long is cut into three. The
      middle-sized piece is 4 inches longer than twice
      the shortest piece. The longest piece is 3 times
      as long as the shortest piece. How long is the
      shortest piece?   9. _____

10.   A formula for an equation of a straight line is

$$y = mx + b.$$

      (a) Solve for m.   10. (a) _____
      (b) If $y = 3$, $b = 15$, and $x = -3$,
          find the value of m.   10. (b) _____

11. Suppose that in the equation

$$\frac{a}{b} = \frac{c}{d},$$

a and d are negative numbers and c is a positive
number. Must b be positive or negative?

11. _____

For Exercises 12-13, find the measure of each marked angle.

12.

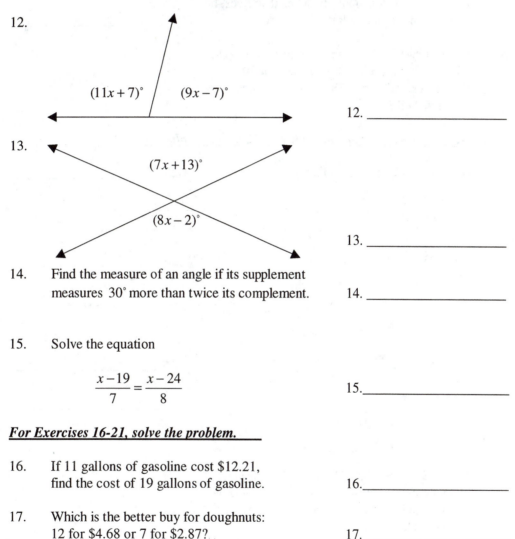

$(11x+7)°$ $(9x-7)°$

12. _____

13.

$(7x+13)°$

$(8x-2)°$

13. _____

14. Find the measure of an angle if its supplement
measures 30° more than twice its complement.

14. _____

15. Solve the equation

$$\frac{x-19}{7} = \frac{x-24}{8}$$

15. _____

For Exercises 16-21, solve the problem.

16. If 11 gallons of gasoline cost $12.21,
find the cost of 19 gallons of gasoline.

16. _____

17. Which is the better buy for doughnuts:
12 for $4.68 or 7 for $2.87?

17. _____

18. How many gallons of a 25% chemical solution must be added to 60 liters of a 50% solution to get a 30% solution?

18. _____

19. Two trains leave Houston at the same time. One travels west at 70 miles per hour and the other travels east at 65 miles per hour. In how many hours will they be 1215 miles apart?

19. _____

20. Jamie has $305 in five-dollar bills and twenty-dollar bills. If she has four more twenties than fives, how many twenty-dollar bills does she have?

20. _____

21. Rachael invested a certain amount of money at 9% and $800 more at 11% The annual interest that she earned was $188. How much did she invest at 11%?

21. _____

For Exercises 22-24, solve the inequality and graph the solution.

22. $4 + 3(p + 2) < 4(p - 2) - 7p$

22. _____

23. $-6 \leq 7y + 1 < 15$

23. _____

24. $5 < \frac{2}{5}t + 3 < 7$

24. _____

25. Candy Myers worked 7 hours on Wednesday and 9 hours on Thursday. What are the possible numbers of hours she can work on Friday to average at least 7 hours per day for those three days?

25. _____

CHAPTER TWO, FORM C   **BEGINNING ALGEBRA**

NAME_____   SECTION_____

For Exercises 1-7, solve the equation.

1. $2x - 3 = 7x - 8$ 1. _____

2. $3(a + 5) = -2a + 5 + 2(a - 10)$ 2. _____

3. $8.6y + 13.8 = 6.6y - 14.6$ 3. _____

4. $4r - 3(3r - 1) = 3 - 5r$ 4. _____

5. $\dfrac{4}{5}n = -10$ 5. _____

6. $3(5x + 10) = 5(3x - 6)$ 6. _____

7. $0.5(x - 10) + 0.7(x + 20) = 57$ 7. _____

For Exercises 8-9, solve the problem.

8.   If 3 is subtracted from four times a number, the
     result is 10 less than five times the number.
     Find the number.   8. _____

9.   During the 1987-88 NBA season, Michael
     Jordan scored 92 points fewer than twice the
     number of points Brad Daugherty scored. Larry
     Bird scored 795 more points than Daugherty.
     Altogether, the three players scored 6623 points.
     How many points did Daugherty score? 9. _____

10. The formula for the area of a triangle is

$$A = \frac{1}{2}bh .$$

(a) Solve for h.   10. (a) _____
(b) If $A = 140$ and $b = 40$,
    find the value of h.   10. (b) _____

11. Write an applied problem that can be solved by
     using the formula for the perimeter of a
     rectangle. It is not necessary to solve your problem.
     Use a separate sheet of paper, if necessary.   11. _____

For Exercises 12-13, find the measure of each marked angle.

12.

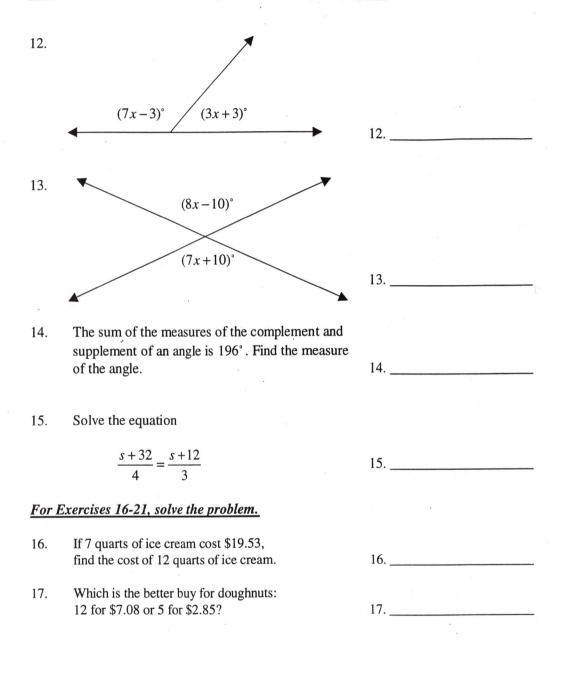

12. _____

13.

13. _____

14. The sum of the measures of the complement and
     supplement of an angle is $196°$. Find the measure
     of the angle.   14. _____

15. Solve the equation

$$\frac{s+32}{4} = \frac{s+12}{3}$$ 15. _____

For Exercises 16-21, solve the problem.

16. If 7 quarts of ice cream cost $19.53,
     find the cost of 12 quarts of ice cream.   16. _____

17. Which is the better buy for doughnuts:
     12 for $7.08 or 5 for $2.85?   17. _____

18. A certain metal alloy is 30% silver. How many ounces of this alloy must be mixed with 50 ounces of an alloy that is 80% silver to get a metal alloy that is 50% silver?

18. _____

19. Two planes leave Chicago's O'Hare airport at the same time. One flies east to Washington, D.C. at 550 miles per hour. The other flies west to Las Vegas, Nevada at 475 miles per hour. In how many hours will they be 2050 miles apart?

19. _____

20. Brad has $700 in ten-dollar bills and twenty-dollar bills. If he has two more twenties than tens, how many twenty-dollar bills does he have?

20. _____

21. Whitney receives $147 annual interest income on two investments. If the amount she invested at 6% was $350 more than the amount she invested at 8%, how much did she invest at 6%?

21. _____

For Exercises 22-24, solve the inequality and graph the solution.

22. $-9q + 6(q - 5) > -4q - 5(2q - 5)$

22. _____

23. $-3 \le 4r - 7 < 9$

23. _____

24. $1 < \dfrac{3}{4}t - 8 < 4$

24. _____

25. Gail Brown earned $331 and $296 the first two weeks that she worked as a caterer. What are the possible amounts she can earn the third week in order to average at least $325 per week?

25. _____

CHAPTER TWO, FORM D   BEGINNING ALGEBRA

NAME_____   SECTION_____

For Exercises 1-7, solve the equation.

1. $5a + 11 = 8a - 19$ 1. _____

2. $3 - 5(z + 1) = 2 + (z - 4)$ 2. _____

3. $-(d + 7) + 4d - 7 = 3d - 7$ 3. _____

4. $5.3m + 7.9 = 4.3m + 10.2$ 4. _____

5. $-\dfrac{5}{6}n = 2$ 5. _____

6. $3(8x - 12) = 4(6x - 9)$ 6. _____

7. $0.2(x - 20) + 0.6(x + 20) = 32$ 7. _____

For Exercises 8-9, solve the problem.

8.   If a number is subtracted from 94, the result
      is 19 more than –5. Find the number.   8. _____

9.   Diane ordered 474 costumes for her church Easter
      pageant. She ordered 56 more costumes from Castle
      Company than from Ajax Costumes. She ordered 42
      fewer costumes from Light Fantastic than 3 times the
      number from Ajax. How many costumes did she order
      from Ajax?   9. _____

10.   The formula for the area of a trapezoid is

$$A = \frac{1}{2}(B + b)h.$$

      (a) Solve for b.   10. (a) _____
      (b) If $A = 240$, $B = 34$, and $h = 12$,
          find the value of b.   10. (b) _____

11. Write an applied problem that can be solved by
     using the formula for the area of a triangle. It is
     not necessary to solve your problem. Use a
     separate sheet of paper, if necessary.   11. _____

For Exercises 12-13, find the measure of each marked angle.

12.

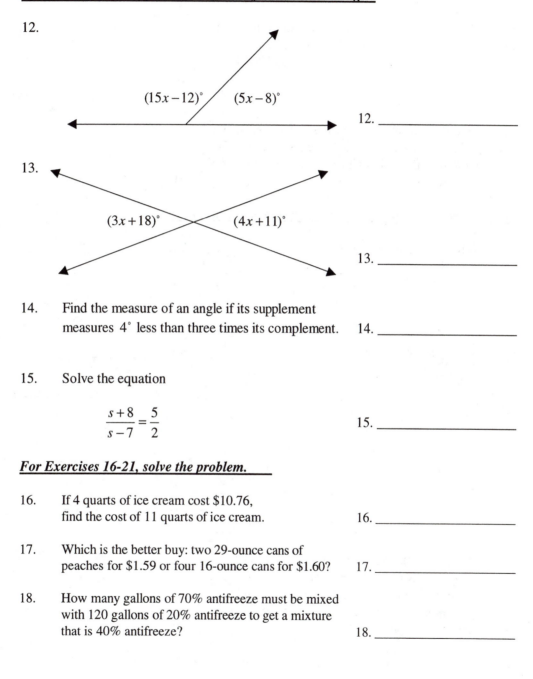

12. _____

13.

13. _____

14. Find the measure of an angle if its supplement
     measures $4°$ less than three times its complement.   14. _____

15. Solve the equation

$$\frac{s+8}{s-7} = \frac{5}{2}$$ 15. _____

For Exercises 16-21, solve the problem.

16. If 4 quarts of ice cream cost $10.76,
     find the cost of 11 quarts of ice cream.   16. _____

17. Which is the better buy: two 29-ounce cans of
     peaches for $1.59 or four 16-ounce cans for $1.60? 17. _____

18. How many gallons of 70% antifreeze must be mixed
     with 120 gallons of 20% antifreeze to get a mixture
     that is 40% antifreeze?   18. _____

19. Two planes leave New York's Kennedy airport at the same time. One flies east to Paris, France at 550 miles per hour. The other flies west to Tokyo, Japan at 475 miles per hour. In how many hours will they be 7175 miles apart?

19._____

20. The Crown Theater took in $800 in twenty-dollar bills and ten-dollar bills. There were 32 more tens than twenties. How many ten-dollar bills were there?

20. _____

21. Two investments produce annual interest income of $270. If the amount invested at 9% was $800 more than the amount invested at 7.5%, how much was invested at 9%?

21. _____

For Exercises 22-24, solve the inequality and graph the solution.

22. $-10 - 2(5b - 4) \geq 4b - 8(b + 1)$

22. _____

23. $-13 \leq 9c - 4 < 14$

23. _____

24. $-7 < \frac{1}{3}t - 5 < 1$

24. _____

25. Steve Buchelt sold his first two professional photographs for $190 and $215. What are the possible amounts he can charge for his third photograph in order to average at least $200 per photograph?

25. _____

CHAPTER TWO, FORM E   BEGINNING ALGEBRA

NAME_____   SECTION_____

For Exercises 1-7, solve the equation.

1. $6x + 12 = 8x + 2$

   (a) -5 (b) 5 (c) -1 (d) 1 1._____

2. $3y + 4(2y + 1) - (5y - 1 - 5) = 9$

   (a) $-\dfrac{1}{6}$ (b) $\dfrac{19}{6}$ (c) 19 (d) 6 2._____

3. $\dfrac{5m}{6} = -60$

   (a) -72 (b) 72 (c) -50 (d) 50 3._____

4. $5.7g + 16.2 = 7.9 + 4.7g$

   (a) 24.1 (b) 11.7 (c) -23.2 (d) -8.3 4._____

5. $0.3(x + 20) + 0.9(x - 10) = 21$

   (a) 20 (b) -10
   (c) All real numbers   (d) No solution   5._____

6. $-2 + (7 - k) = -(-5 + k)$

   (a) 5 (b) 0
   (c) All real numbers   (d) No solution   6._____

7. $3(z - 1) - z - 7 = 2z + 6$

   (a) 4 (b) 0
   (c) All real numbers   (d) No solution   7._____

For Exercises 8-9, solve the problem.

8.   The difference between 2 times a number and -8 is -12.
      Find the number.

   (a) 2 (b) -2 (c) 10 (d) -10 8._____

9.   A 36-foot pipe is divided into three pieces. One piece is twice as long as the shortest piece, and the third piece is 1 foot longer than four times the shortest piece. Find the length of the shortest piece.

(a) 10 feet   (b) 21 feet

(c) 5 feet   (d) 4 feet   9. _____

10.   The formula for the perimeter of a rectangle is

$$P = 2L + 2W .$$

If $P = 110$ and $L = 20$, find the value of W.

(a) 35   (b) 45   (c) 70   (d) 90   10. _____

11.   Solve the formula $A = \dfrac{1}{2}bh$ for b.

(a) $b = \dfrac{2A}{h}$ (b) $b = 2Ah$

(c) $b = 2A - h$ (d) $b = \dfrac{2h}{A}$ 11. _____

__For Exercises 12-13, find the measure of each marked angle.__

12.

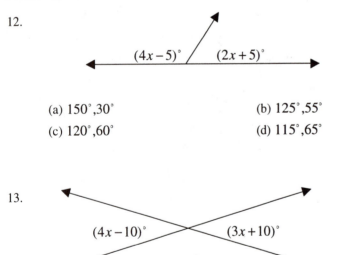

(a) $150°, 30°$ (b) $125°, 55°$

(c) $120°, 60°$ (d) $115°, 65°$ 12. _____

13.

(a) $20°, 20°$ (b) $70°, 70°$

(c) $40°, 40°$ (d) $30°, 30°$ 13. _____

14. The measure of the supplement of an angle is 4 times the measure of its complement. Find the measure of the angle.

(a) $40°$

(b) $45°$

(c) $50°$

(d) $60°$

14. _____

15. Solve the equation

$$\frac{x+8}{x-9} = \frac{2}{3}$$

(a) -9 (b) -42 (c) 43 (d) 6

15. _____

For Exercises 16-21, solve the problem.

16. If 8 boxes of tissue cost $10.00, how much will 15 boxes of tissue cost?

(a) $15.00

(b) $18.75

(c) $19.75

(d) $18.40

16. _____

17. Andrew invested some of his money at 7.5% and $350 less at 8%. His total annual interest income is $437. How much did Andrew invest at 7.5%?

(a) $2988.71

(b) $3000

(c) $2650

(d) $8215

17. _____

18. How many liters of pure acid must be mixed with 20 liters of a 15% acid solution to get a 20% mixture?

(a) 10 liters

(b) 12.5 liters

(c) 100 liters

(d) 1.25 liters

18. _____

19. Which of the following is the best buy for canned pears?

(a) A 28-ounce can for $0.89

(b) A 16-ounce can for $0.49

(c) A 19-ounce can for $0.57

(c) An 8-ounce can for $0.26

19. _____

20. Kelly has a purse full of quarters and dimes. There are 58 more quarters than dimes and the total value of the change is $19.05. How many dimes are there?

(a) 71   (b) 13   (c) 129   (d) 20

20. _____

21. If 9 gallons of gasoline cost $10.17, find the cost of 19 gallons of gasoline.

(a) $16.33   (b) $21.47   (c) $20.34   (d) $4.82

21. _____

For Exercises 22-24, determine the graph of the solution of the inequality.

22. $5(3x-4)-10x \le 4(2x-1)-16$ 22._____

(a)

(b)

(c)

(d)

23. $-4 \le 2b-6 < 8$ 23._____

(a)

(b)

(c)

(d)

24. $-5 < \dfrac{1}{2}x-3 < -1$ 24._____

(a)

(b)

(c)

(d)

25.   If $3y < 2y$, then

(a) y may be positive or negative
(b) y must be zero
(c) y must be positive
(d) y must be negative   25._____

CHAPTER TWO, FORM F

BEGINNING ALGEBRA

NAME_____   SECTION_____

___

For Exercises 1-7, solve the equation.

1. $7x + 11 = 9x + 15$

     (a) 2   (b) –2   (c) $-\dfrac{1}{4}$ (d) $\dfrac{13}{8}$ 1._____

2. $2a + 2(3a + 2) - (3a - 9 - 3) = 9$

     (a) $-\dfrac{7}{11}$ (b) $-\dfrac{7}{5}$ (c) $\dfrac{17}{5}$ (d) -7 2._____

3. $\dfrac{2m}{5} = -20$

     (a) –50   (b) 50   (c) –8   (d) 8   3._____

4. $7.8t - 13.1 = 8.8t + 6.9$

     (a) 20.0   (b) -20.0 (c) 6.3   (d) -6.3 4._____

5. $-(x - 3) = 13 - (x + 10)$

     (a) 0   (b) 6
     (c) All real numbers   (d) No solution   5._____

6. $0.06(x - 30) + 0.27(x + 10) = 10.8$

     (a) 30   (b) 3
     (c) All real numbers   (d) No solution   6._____

7. $2(4w + 2) = 4(2w - 1)$

     (a) $-\dfrac{1}{2}$ (b) 0
     (c) All real numbers   (d) No solution   7._____

For Exercises 8-9, solve the problem.

8.   If 5 is subtracted from three times a number, the result
     is 7 more than twice the number. What is the number?

     (a) 2   (b) -2   (c) 12   (d) -12   8._____

9.   A 60-foot pipe is divided into three pieces. One piece is twice as long as the shortest piece, and the third piece is 4 feet less than five times the shortest piece. Find the length of the shortest piece.

(a) 8 feet   (b) 7 feet
(c) 6 feet   (d) 4 feet   9. _____

10.   A formula for an equation of a straight line is

$$y = mx + b.$$

If $y = -31$, $m = 4$, and $b = -3$, find the value of x.

(a) –7   (b) 7   (c) –9   (d) 9   10. _____

11.   Solve the formula $V = \dfrac{1}{3}\pi r^2 h$ for h.

(a) $h = \dfrac{\pi r^2}{3V}$ (b) $h = V - \dfrac{1}{3}\pi r^2$

(c) $h = \dfrac{3V}{\pi r^2}$ (d) $h = \dfrac{3\pi}{Vr^2}$ 11. _____

For Exercises 12-13, find the measure of each marked angle.

12.

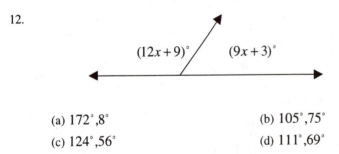

$(12x + 9)^\circ$ $(9x + 3)^\circ$

(a) $172^\circ, 8^\circ$ (b) $105^\circ, 75^\circ$
(c) $124^\circ, 56^\circ$ (d) $111^\circ, 69^\circ$ 12. _____

13.

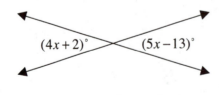

$(4x + 2)^\circ$ $(5x - 13)^\circ$

(a) $62^\circ, 62^\circ$ (b) $19^\circ, 19^\circ$
(c) $28^\circ, 28^\circ$ (d) $118^\circ, 118^\circ$ 13. _____

14. Find the measure of an angle if its supplement measures $10°$ less than three times its complement.

   (a) $40°$ (b) $140°$

   (c) $50°$ (d) $60°$ 14. _____

15. Solve the equation

$$\frac{w+4}{8} = \frac{w+10}{5}$$

   (a) -2 (b) 2 (c) -20 (d) $-\dfrac{22}{3}$ 15. _____

For Exercises 16-21, solve the problem.

16. If 9 boxes of tissue cost $11.25, how much will 16 boxes of tissue cost?

   (a) $21.25   (b) $20.00

   (c) $19.50   (d) $20.25   16. _____

17. Matthew invested some of his money at 6.5% and $450 less at 8%. His total annual interest income is $167. How much did Matthew invest at 8%?

   (a) $1400   (b) $1850

   (c) $950   (d) $500   17. _____

18. How many gallons of pure antifreeze must be mixed with 30 gallons of 15% antifreeze to get a mixture that is 40% antifreeze?

   (a) 1.25 gallons   (b) 10 gallons

   (c) 12.5 gallons   (d) 15 gallons   18. _____

19. Which of the following is the best buy for trash bags?

   (a) A 10-count box for $1.25   (b) A 25-count box for $2.49

   (c) A 20-count box for $1.99   (c) A 15-count box for $1.60   19. _____

20. Jason has a pocket full of quarters and dimes. There are 58 more quarters than dimes and the total value of the change is $20.45. How many dimes are there?

   (a) 75   (b) 17   (c) 58   (d) 34   20. _____

21.   Two cars leave from the same point, traveling in opposite
        directions. If one travels at a constant speed of 52 miles
        per hour and the other travels at a constant speed of 58 miles
        per hour, in how many hours will they be 385 miles apart?

        (a) 3 hours   (b) 4 hours   (c) $3\frac{1}{2}$ hours   (d) $4\frac{1}{2}$ hours   21. _____

For Exercises 22-24, determine the graph of the solution of the inequality.

22. $7(3-e)+8 \geq 4(3e-1)-8e$ 22. _____

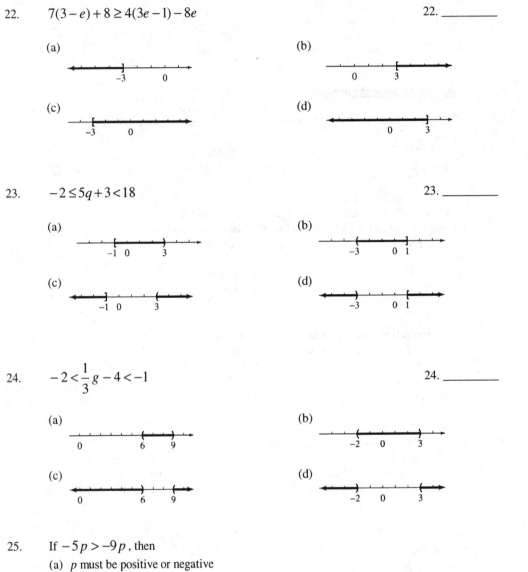

        (a)

        (b)

        (c)

        (d)

23. $-2 \leq 5q+3 < 18$ 23. _____

        (a)

        (b)

        (c)

        (d)

24. $-2 < \frac{1}{3}g - 4 < -1$ 24. _____

        (a)

        (b)

        (c)

        (d)

25.   If $-5p > -9p$, then
        (a) p must be positive or negative
        (b) p must be zero
        (c) p must be positive
        (d) p must be negative   25. _____

CHAPTER THREE, FORM A   BEGINNING ALGEBRA

NAME_____   SECTION_____

__The graph below shows the total amount of sales for Andrew Computer Corporation__
__from 1992-1998. Use the graph to respond to Exercises 1 and 2.__

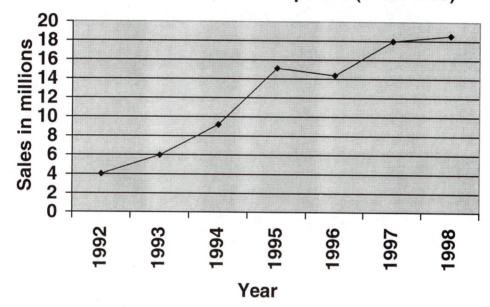

Sales for Andrew Computers (1992-1998)

1.   Which one of the following is the best estimate for total sales in 1995?

      (a) $15,000,000   (b) $14,100,000   (c) $16,000,000   1. _____

2.   Estimate the percent increase in sales from 1992 to 1993.   2. _____

__For Exercises 3-4, complete the ordered pairs for the given equation.__

3. $y = 4x - 1$ $(0, \ ), (-3, \ )(\ , 15)$ 3. _____

4. $6x - 2y = 3$ $(0, \ ), (\ , 3)(\ , 0)$ 4. _____

5.   Is $(3, -2)$ a solution of $3x - 2y = 5$?   5. _____

6.   At the x-intercept for a given linear equation, the
      x-value is -5. What is the y-value at this point?   6. _____

For Exercises 7-8, graph the linear equation. Give the x- and y-intercepts.

7. $2x - y = 3$ 7. x-intercept: _____
                                                           y-intercept: _____

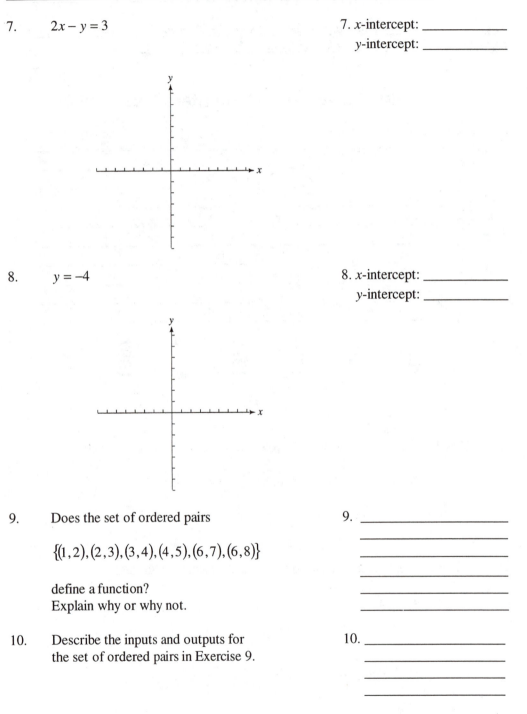

8. $y = -4$ 8. x-intercept: _____
                                                           y-intercept: _____

9.   Does the set of ordered pairs   9. _____
                                                           _____
        $\{(1,2),(2,3),(3,4),(4,5),(6,7),(6,8)\}$ _____
                                                           _____
        define a function?   _____
        Explain why or why not.   _____

10.   Describe the inputs and outputs for   10. _____
        the set of ordered pairs in Exercise 9.   _____
                                                           _____
                                                           _____

In Exercises 11-16, find the slope of each line.

11.   Through $(-2, 7)$ and $(3, -1)$.   11. _____

12. $6x = 3y - 5$ 12. _____

13.

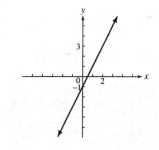

                                                            13. _____

14.   (These are two different views of the same line.)

(a)   (b)

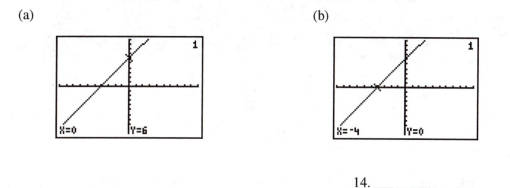

                                                            14. _____

15.   A line parallel to the graph of $y = -3x - 6$.   15. _____

16.   A line perpendicular to the graph of $y = -3x - 6$. 16. _____

CHAPTER THREE, FORM B   **BEGINNING ALGEBRA**

NAME_____   SECTION_____

The graph below shows the total amount of sales for Andrew Computer Corporation from 1992-1998. Use the graph to respond to Exercises 1 and 2.

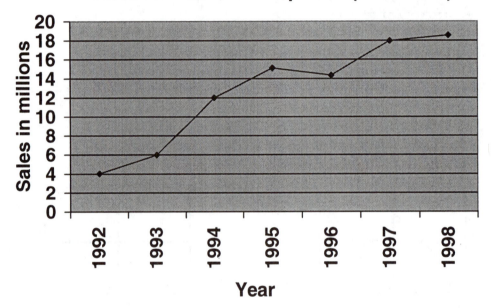

1.   Which one of the following is the best estimate for total sales in 1992?

      (a) $4,000,000   (b) $6,000,000   (c) $12,000,000   1._____

2.   Estimate the percent increase in sales from 1993 to 1994.   2._____

For Exercises 3-4, complete the ordered pairs for the given equation.

3. $y = -x - 5$ $(0, \ ), (-3, \ )(\ , 15)$ 3._____

4. $3x - 2y = 6$ $(0, \ ), (\ , 3)(\ , 0)$ 4._____

5.   Is $(3, -2)$ a solution of $3x + 2y = -5$?   5. _____

6.   At the y-intercept for a given linear equation, the
       y-value is 2. What is the x-value at this point?   6. _____

For Exercises 7-8, graph the linear equation. Give the x- and y-intercepts.

7. $x - 2y = -8$ 7. x-intercept: _____

                                                   y-intercept: _____

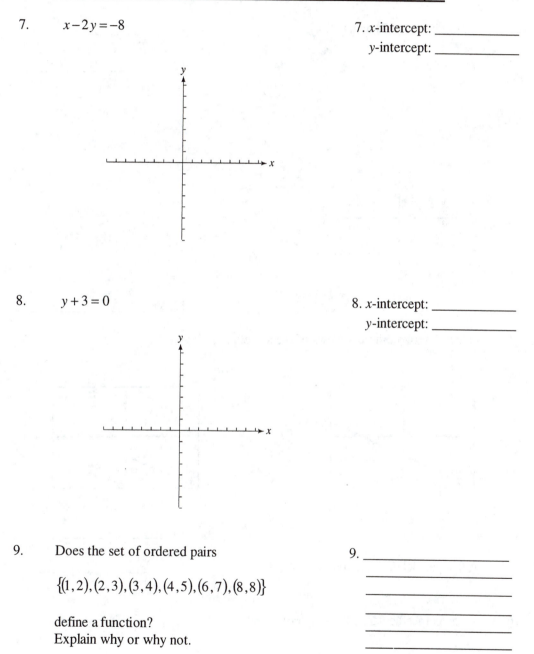

8. $y + 3 = 0$ 8. x-intercept: _____

                                                     y-intercept: _____

9.   Does the set of ordered pairs   9. _____

        $\{(1,2), (2,3), (3,4), (4,5), (6,7), (8,8)\}$ _____
                                                        _____

        define a function?   _____
        Explain why or why not.   _____

10. Describe the inputs and outputs for
    the set of ordered pairs in Exercise 9.

10. _____
    _____
    _____
    _____

In Exercises 11-16, find the slope of each line.

11. Through $(-2,-7)$ and $(-3,1)$.

11. _____

12. $7x = -3y + 1$

12. _____

13.

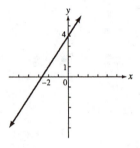

13. _____

14. (These are two different views of the same line.)

(a)   (b)

14. _____

15. A line parallel to the graph of $2x = -3y - 6$.

15. _____

16. A line perpendicular to the graph of $2x = -3y - 6$.   16. _____

CHAPTER THREE, FORM C   **BEGINNING ALGEBRA**

NAME_____   SECTION_____

The graph below shows the total amount of sales for Andrew Computer Corporation from 1992-1998. Use the graph to respond to Exercises 1 and 2.

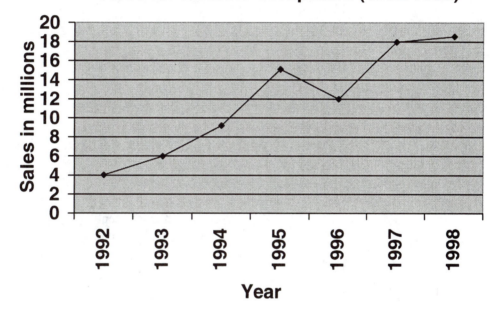

Sales for Andrew Computers (1992-1998)

1.   Which one of the following is the best estimate for total sales in 1994?

(a) $8,000,000   (b) $9,000,000   (c) $10,000,000   1._____

2.   Estimate the percent decrease in sales from 1995 to 1996.   2._____

For Exercises 3-4, complete the ordered pairs for the given equation.

3. $2y = -x - 5$ $(0, \ ), (-3, \ )(\ , 15)$ 3._____

4. $3x + 2y = -9$ $(0, \ ), (\ , 3)(\ , 0)$ 4._____

5.   Is $(2 , -3)$ a solution of $2x - 3y = -5$?   5. _____

6.   To find the y-intercept for the graph $2x - 4y = -5$
        we let _____ = 0.   6. _____

For Exercises 7-8, graph the linear equation. Give the x- and y-intercepts.

7. $5x - 2y = 4$ 7. *x*-intercept: _____
                                                            y-intercept: _____

8. $x - 5 = 0$ 8. *x*-intercept: _____
                                                            y-intercept: _____

9.   Does the set of ordered pairs   9. _____
                                                            _____
        $\{(1,2),(2,3),(3,4),(4,5),(4,6),(6,7)\}$ _____
                                                            _____
        define a function?   _____
        Explain why or why not.   _____

10. Describe the inputs and outputs for
the set of ordered pairs in Exercise 9.

10. _____
_____
_____
_____

In Exercises 11-16, find the slope of each line.

11. Through $(-5,-6)$ and $(-6,5)$.

11. _____

12. $-2x = -3y + 1$

12. _____

13.

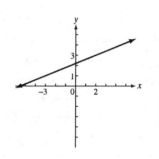

13. _____

14. (These are two different views of the same line.)

(a)

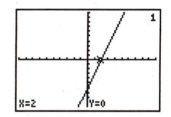

(b)

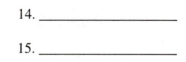

14. _____

15. A line parallel to the graph of $11x = 9y - 6$.

15. _____

16. A line perpendicular to the graph of $11x = 9y - 6$.

16. _____

CHAPTER THREE, FORM D   **BEGINNING ALGEBRA**

NAME_____   SECTION_____

The graph below shows the total amount of sales for Andrew Computer Corporation from 1992-1998. Use the graph to respond to Exercises 1 and 2.

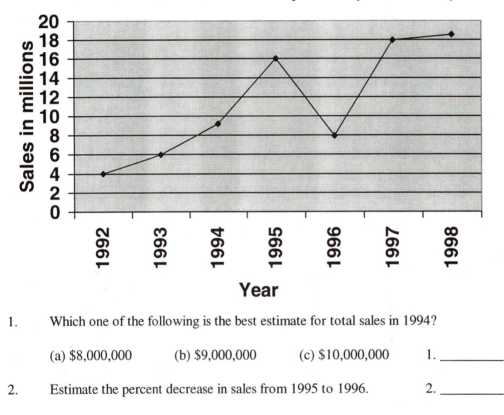

1.   Which one of the following is the best estimate for total sales in 1994?

       (a) $8,000,000   (b) $9,000,000   (c) $10,000,000   1. _____

2.   Estimate the percent decrease in sales from 1995 to 1996.   2. _____

For Exercises 3-4, complete the ordered pairs for the given equation.

3. $y = -5x - 2$ $(0, \ ), (-3, \ )(\ , 15)$ 3. _____

4. $3x + 4y = -12$ $(0, \ ), (\ , 3)(\ , 0)$ 4. _____

5.   Is $(2, -3)$ a solution of $2x + 3y = -5$?

5. _____

6.   To find the x-intercept for the graph $x - 3y = -9$
we let' _____ = 0.

6. _____

For Exercises 7-8, graph the linear equation. Give the x- and y-intercepts.

7. $2x - 3y = 12$

7. x-intercept: _____

   y-intercept: _____

8. $x + 2 = 0$

8. x-intercept: _____

   y-intercept: _____

9.   Does the set of ordered pairs

   $\{(1,2),(2,3),(3,4),(4,5),(5,6),(6,11)\}$

   define a function?
   Explain why or why not.

9. _____

   _____

   _____

   _____

   _____

10.   Describe the inputs and outputs for
        the set of ordered pairs in Exercise 9.

10. _____
    _____
    _____
    _____

In Exercises 11-16, find the slope of each line.

11.   Through $(-5,-6)$ and $(-5,6)$.

11. _____

12. $2x = -3y + 4$

12. _____

13.

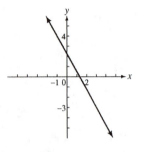

13. _____

14.   (These are two different views of the same line.)

(a)   (b)

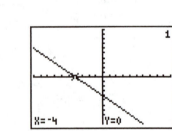   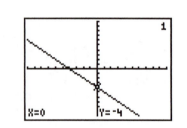

14. _____

15.   A line parallel to the graph of $-x = 9y - 6$.

15. _____

16.   A line perpendicular to the graph of $-x = 9y - 6$.

16. _____

CHAPTER THREE, FORM E   **BEGINNING ALGEBRA**

NAME_____   SECTION_____

*The graph below shows the total amount of sales for Andrew Computer Corporation from
1992-1998. Use the graph to respond to Exercises 1 and 2.*

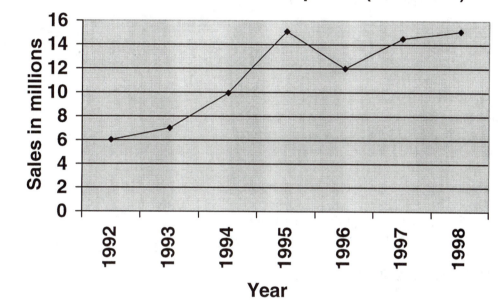

Sales for Andrew Computers (1992-1998)

1.   Between which two years did the sales amount
      show the greatest change?

      (a) 1992-1993   (b) 1993-1994
      (c) 1994-1995   (d) 1996-1997   1. _____

2.   Estimate the percent decrease in sales from
      1995 to 1996.

      (a) 3%   (b) 12%
      (c) 15%   (d) 20%   2. _____

For Exercises 3-5, find the two ordered pairs that are both solutions for the given equation.

3. $y = -5x - 2$

(a) $(0, -2)$ and $(1, -3)$ (b) $(0, -2)$ and $(-1, -3)$
(c) $(-2, -12)$ and $(-1, 3)$ (d) $(-2, 8)$ and $(-1, 3)$ 3. _____

4. $3x + 4y = -12$

(a) $(0, -3)$ and $(4, 0)$ (b) $(-4, 0)$ and $(0, -3)$
(c) $(-8, 9)$ and $(0, 3)$ (d) $(8, -9)$ and $(4, 0)$ 4. _____

5. $x - 4 = 1$

(a) $(5, -5)$ and $(-4, 1)$ (b) $(-5, 5)$ and $(-4, 1)$
(c) $(5, -4)$ and $(-5, 1)$ (d) $(5, -4)$ and $(5, 1)$ 5. _____

6.   Which of the following statements describes the intercepts
of the graph of a line whose equation has the form
$x = c$, $(c \neq 0)$?

(a) The line has an x-intercept but no y-intercept.
(b) The line has a y-intercept but no x-intercept.
(c) The line has an x-intercept and a y-intercept.
(d) The line has no x-intercept or y-intercept.   6. _____

For Exercises 7-9, graph the linear equation.

7. $x - 4 = 0$ 7. _____

(a)   (b)   (c)   (d)

8. $2x - y = 4$ 8. _____

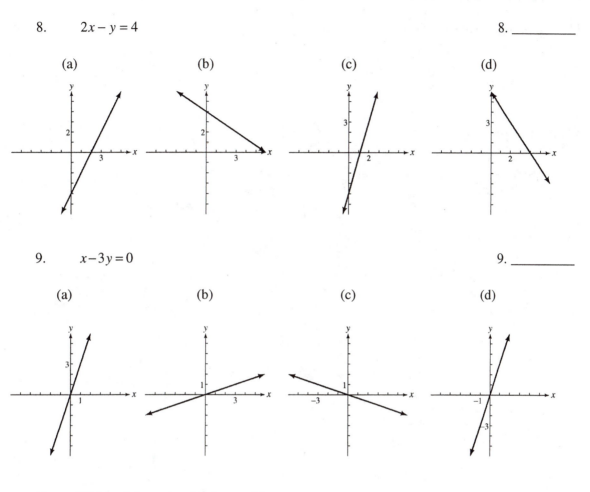

(a)   (b)   (c)   (d)

9. $x - 3y = 0$ 9. _____

(a)   (b)   (c)   (d)

10.   Which of the ordered pairs would prevent

$\{(1,2),(3,5),(4,6),(5,7),(7,8)\}$

from being a function?

(a) $(2,3)$ (b) $(3,4)$ (c) $(6,7)$ (d) $(8,9)$ 10. _____

In Exercises 11-16, find the slope of each line.

11. Through $(-5,-6)$ and $(5,6)$.

    (a) $\dfrac{6}{5}$ (b) 0   (c) $\dfrac{5}{6}$ (d) Undefined   11. _____

12. $2x = -3y + 4$

    (a) $-\dfrac{3}{2}$ (b) $-\dfrac{2}{3}$ (c) $\dfrac{4}{3}$ (d) 2   12. _____

13. $y + 6 = 0$

    (a) -6 (b) 6   (c) 0   (d) Undefined   13. _____

14.

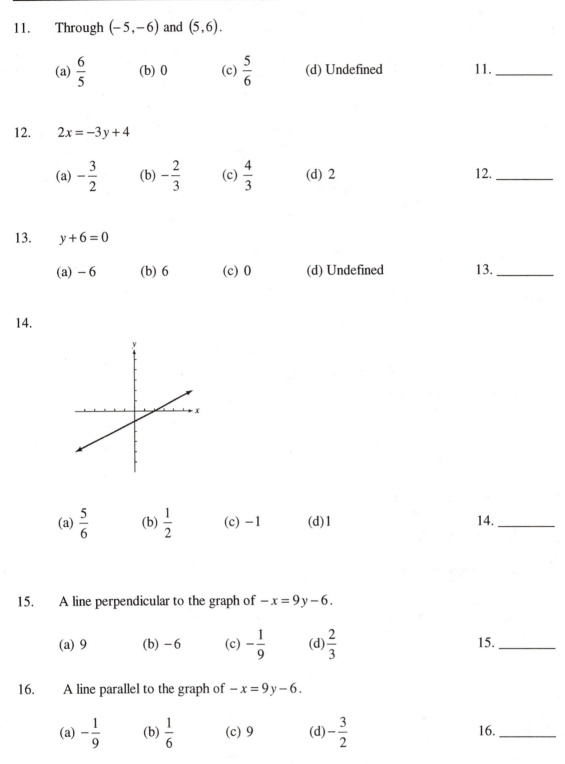

    (a) $\dfrac{5}{6}$ (b) $\dfrac{1}{2}$ (c) -1 (d) 1   14. _____

15. A line perpendicular to the graph of $-x = 9y - 6$.

    (a) 9   (b) -6 (c) $-\dfrac{1}{9}$ (d) $\dfrac{2}{3}$ 15. _____

16. A line parallel to the graph of $-x = 9y - 6$.

    (a) $-\dfrac{1}{9}$ (b) $\dfrac{1}{6}$ (c) 9   (d) $-\dfrac{3}{2}$ 16. _____

CHAPTER THREE, FORM F   **BEGINNING ALGEBRA**

NAME_____   SECTION_____

The graph below shows the total amount of sales for Andrew Computer Corporation from 1992-1998. Use the graph to respond to Exercises 1 and 2.

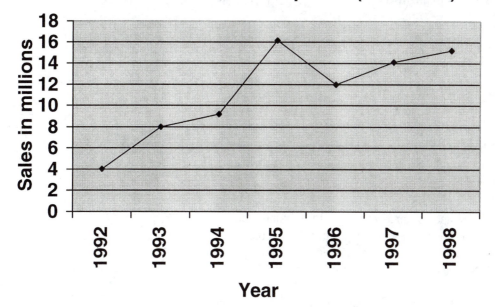

1.   Between which two years did the sales amount
      show the greatest change?

      (a) 1992-1993   (b) 1993-1994
      (c) 1995-1996   (d) 1996-1997   1. _____

2.   Estimate the percent increase in sales from
      1992 to 1993.

      (a) 4%   (b) 25%
      (c) 50%   (d) 100%   2. _____

For Exercises 3-5, find the two ordered pairs that are both solutions for the given equation.

3. $y = -3x - 5$

   (a) $(0,-5)$ and $(1,-8)$ (b) $(0,-5)$ and $(-1,-8)$
   (c) $(-2,-11)$ and $(-1,8)$ (d) $(-2,1)$ and $(-1,-8)$ 3. _____

4. $4x - 3y = -12$

   (a) $(0,4)$ and $(3,0)$ (b) $(-3,0)$ and $(0,-4)$
   (c) $(-9,8)$ and $(0,4)$ (d) $(9,16)$ and $(-3,0)$ 4. _____

5. $x + 4 = -1$

   (a) $(5,-5)$ and $(4,-1)$ (b) $(-5,5)$ and $(4,-1)$
   (c) $(-5,-4)$ and $(-5,1)$ (d) $(5,4)$ and $(5,-1)$ 5. _____

6. Which of the following statements describes the intercepts of the graph of a line whose equation has the form $y = c$, $(c \neq 0)$?

   (a) The line has an x-intercept but no y-intercept.
   (b) The line has a y-intercept but no x-intercept.
   (c) The line has an x-intercept and a y-intercept.
   (d) The line has no x-intercept or y-intercept.   6. _____

For Exercises 7-9, graph the linear equation.

7. $y - 4 = 0$ 7. _____

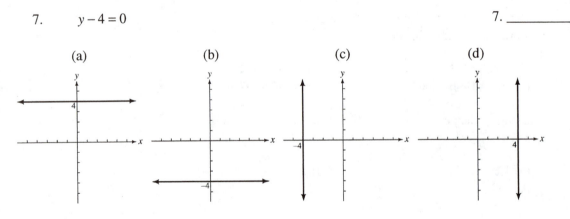

(a)   (b)   (c)   (d)

8. $x - 2y = 4$ 8. _____

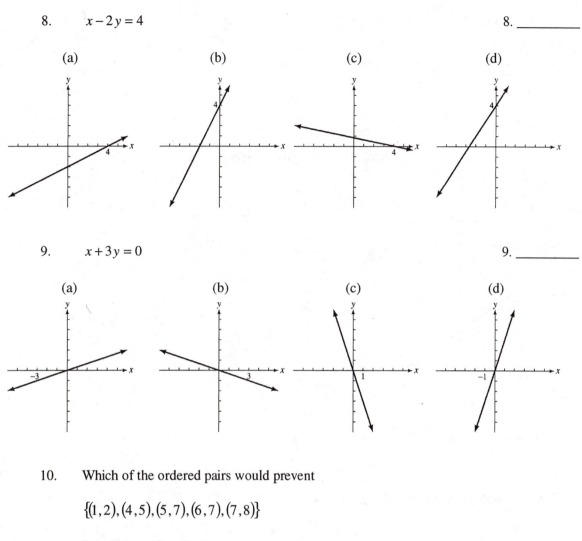

(a)   (b)   (c)   (d)

9. $x + 3y = 0$ 9. _____

(a)   (b)   (c)   (d)

10.   Which of the ordered pairs would prevent

$\{(1,2),(4,5),(5,7),(6,7),(7,8)\}$

from being a function?

(a) $(2,3)$ (b) $(3,4)$ (c) $(5,6)$ (d) $(8,9)$ 10. _____

In Exercises 11-16, find the slope of each line.

11.   Through $(-5,-6)$ and $(-5,6)$.

(a) -1 (b) 0 (c) -5 (d) Undefined   11. _____

12. $3x = -4y + 2$

(a) $-\dfrac{3}{4}$ (b) $\dfrac{1}{2}$ (c) $\dfrac{2}{3}$ (d) $-\dfrac{4}{3}$ 12. _____

13. $x - 4 = 0$

(a) -4 (b) 4 (c) 0 (d) Undefined   13. _____

14.

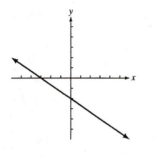

(a) $-\dfrac{3}{2}$ (b) $-\dfrac{2}{3}$ (c) $\dfrac{2}{9}$ (d) $\dfrac{3}{2}$ 14. _____

15.   A line parallel to the graph of $-2x = 5y - 6$.

(a) $-\dfrac{2}{5}$ (b) $-\dfrac{5}{6}$ (c) $\dfrac{6}{5}$ (d) $\dfrac{5}{2}$ 15. _____

16.   A line perpendicular to the graph of $-2x = 5y - 6$.

(a) $-\dfrac{2}{5}$ (b) $-\dfrac{5}{6}$ (c) $\dfrac{6}{5}$ (d) $\dfrac{5}{2}$ 16. _____

CHAPTER FOUR, FORM A   **BEGINNING ALGEBRA**

NAME_____   SECTION_____

_____

For Exercises 1-2, combine like terms whenever possible and write the polynomial in descending order. Give the degree of the simplified polynomial, and determine whether the simplified polynomial is a monomial, binomial, trinomial, or none of these.

1. $14y^2 + 2y - 9y^2 - 1$ 1. _____

2. $3a^5 + 6a^5 - 2a^4 - 7a^5$ 2. _____

3.   Create a set of ordered pairs that lie
      on the graph of $y = 3x^2 + 2$. Then   3.
      graph the equation.

For Exercises 4-12, perform the indicated operation.

4. $(8j^2 + 5j - 7) - (4j^2 - 3) - (2j^2 - 5j + 2)$ 4. _____

5. $(-6a^3b^2 + 12ab^4 - a^2b^2) - (8ab^4 + 2a^3b^2 + ab)$ 5. _____

6.   Subtract:
$$11x^5 - 6x^3 + 8x$$
$$\underline{3x^5 - 5x^3 -\ \ x}$$ 6. _____

7. $-2a^2(3a^3 + 4a^2 - 7a + 11)$ 7. _____

8. $(b-7)(b-3)$ 8. _____

9. $(7r-9)(5r+3)$ 9. _____

10. $(6y-5z)^2$ 10. _____

11. $(2a-3b)(2a+3b)$ 11. _____

12. $(2t+1)(t^2+t-12)$ 12. _____

13.   What polynomial expression represents
      the area of this square?

 $3x+4$

                                         13. _____

For Exercises 14-16, evaluate the expression.

14. 2^{-6} 14. _____

15. $3^{-1}+6^{-1}$ 15. _____

16. -3^0 16. _____

For Exercises 17-19, simplify and write the answer using only positive exponents. Assume that all variables represent nonzero numbers.

17. $\dfrac{4^8 \cdot 4^{-3}}{4^{-5}}$ 17. _____

18. $\dfrac{(2a^3b)^3(ab^3)^2}{(ab)^4}$ 18. _____

19. $\dfrac{(3k^2)^{-1}(2k^{-3})^4}{3k^4}$ 19. _____

20. Do you agree or disagree with the following statement?

$$-3^4 = (-3)^4$$

Justify your answer.

20. _____

_____

_____

_____

_____

For Exercises 21-23, perform the division.

21. $\dfrac{4r^3 - 12r^2 + 8r + 16}{-4r}$

21. _____

22. $\dfrac{3a^2 - 11a - 20}{a - 5}$

22. _____

23. $(2g^4 + 4g^2 - 12) \div (g - 1)$

23. _____

24. (a) Write $26{,}000{,}000$ using scientific notation.

24. (a) _____

(b) Write 4.5×10^{-6} without using exponents.

24. (b) _____

(c) Write the quotient and write your answer without using scientific notation:

$$\dfrac{7.5 \times 10^{-4}}{3 \times 10^{-6}}$$

24. (c) _____

25. Use scientific notation to solve the following problem. Write your answer without exponents.

The weight of a fruit fly is about 1.3×10^{-4} pounds. How much would $100{,}000$ fruit flies weigh?

25. _____

CHAPTER FOUR, FORM B   BEGINNING ALGEBRA

NAME_____   SECTION_____

For Exercises 1-2, combine like terms whenever possible and write the polynomial in descending order. Give the degree of the simplified polynomial, and determine whether the simplified polynomial is a monomial, binomial, trinomial, or none of these.

1. $8y^2 + 2y - 7y^2 + 4 - y^2$ 1. _____

2. $13a^6 - a^6 - 4a^3$ 2. _____

3.   Create a set of ordered pairs that lie
       on the graph of $y = 2x^2 - 1$. Then   3.
       graph the equation.

For Exercises 4-12, perform the indicated operation.

4. $(p^2 - 2p + 8) - (2p^2 + 2p - 6) - (3p^2 - 4p - 1)$ 4. _____

5. $(7a^3b^2 - 8ab^4 + 6ab) - (-8ab^4 - 2ab + a^3b^2)$ 5. _____

6.   Subtract:
$$8x^5 - 3x^2 + 5$$
$$-8x^5 + 10x^2 - 9$$
                                                          6. _____

7. $-5c^2(4 - 5c + 3c^2)$ 7. _____

8. $(z+3)(z+9)$

8. _____

9. $(5m+4)(2m-3)$

9. _____

10. $(7x-y)(7x+y)$

10. _____

11. $(4+8d)^2$

11. _____

12. $(3z-2)(z^2+5z+7)$

12. _____

13.   What polynomial expression represents
the area of this rectangle?

$x+4$

$4x-5$

13. _____

For Exercises 14-16, evaluate the expression.

14. 4^{-3}

14. _____

15. $\left(-\dfrac{2}{3}\right)^0$

15. _____

16. $8^{-1}+2^{-1}$

16. _____

For Exercises 17-19, simplify and write the answer using only positive exponents. Assume that all variables represent nonzero numbers.

17. $\dfrac{9^4 \cdot 9^{-2}}{9^{-3}}$

17. _____

18. $\dfrac{(2a^3b^2)^4(3a^2b^2)^2}{(-2ab)^4}$

18. _____

19. $\dfrac{(3^{-1}x^{-1}y^{-1}z)^{-3}}{(z^2x^{-1})^{-2}}$

19. _____

20.   Do you agree or disagree with the   20. _____
      following statement?   _____
                                                        _____
                    $(-3)^0 = -1$ _____
      Justify your answer.

For Exercises 21-23, perform the division.

21. $\dfrac{8x^2y^3 - 4xy^2 + 12xy}{4xy}$ 21. _____

22. $\dfrac{9v^2 + 6v + 1}{3v + 1}$ 22. _____

23. $(6h^3 + 5h^2 - 34h + 13) \div (3h - 5)$ 23. _____

24.   (a) Write $5{,}800{,}000$ using scientific notation.   24. (a) _____

      (b) Write 9.1×10^{-3} without using exponents.   24. (b) _____

      (c) Write the quotient and write your answer
          without using scientific notation:

                    $\dfrac{71.3 \times 10^4}{2.3 \times 10^7}$ 24. (c) _____

25.   Use scientific notation to solve the following
      problem. Write your answer without exponents.

      If the federal government loses 2.3×10^5
      dollars every day, how much money will
      it lose in 700 days?   25. _____

CHAPTER FOUR, FORM C   **BEGINNING ALGEBRA**

NAME_____   SECTION_____

For Exercises 1-2, combine like terms whenever possible and write the polynomial in descending order. Give the degree of the simplified polynomial, and determine whether the simplified polynomial is a monomial, binomial, trinomial, or none of these.

1. $2a^3 - 3a^2 + 5a^3 - 7a^2$ 1. _____

2. $3 - 2x + x^2 - 3x^4 + 4x^3$ 2. _____

3.   Create a set of ordered pairs
        that lie on the graph of $y = -x^2 + 1$.
        Then graph the equation.   3.

For Exercises 4-12, perform the indicated operation.

4. $(6p^2 - 2p - 1) + (3p^2 + 5p - 1) - (p^2 - 6p + 1)$ 4. _____

5. $(3a^2b - 8ab^2 + 6a^2b^2) - (-8ab^4 - 2ab + a^3b^2)$ 5. _____

6.   Subtract:
$$2t^3 - 3t^2 + t + 1$$
$$\underline{5t^3 + 6t^2 - 8t + 1}$$ 6. _____

7. $-6p^2(p^4 + 3p^3 - p - 11)$ 7. _____

8. $(y-4)(y-5)$ 8. _____

9. $(8x-1)(3x+2)$ 9. _____

10. $(m-11)^2$ 10. _____

11. $(2x-7y)(2x+7y)$ 11. _____

12. $(w+3)(2w^2-w-4)$ 12. _____

13.   What polynomial expression represents
      the perimeter of this triangle?

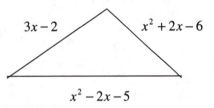

3x − 2 x^2+2x-6

x^2-2x-5 13. _____

For Exercises 14-16, evaluate the expression.

14. $(10)^{-3}$ 14. _____

15. $(-7)^0+(8)^0$ 15. _____

16. $3^{-3}+3^3$ 16. _____

For Exercises 17-19, simplify and write the answer using only positive exponents. Assume that all variables represent nonzero numbers.

17. $\dfrac{5^2 \cdot 5^{-7}}{5^3}$ 17. _____

18. $\dfrac{(4a^3b)^3(2ab^2)^2}{(8ab)^3}$ 18. _____

19. $\dfrac{c^9(c^3)^4}{(c^3)^5(c^2)^{-6}}$ 19. _____

20. Do you agree or disagree with the following statement?

$$(3a + 4b)^2 = (3a)^2 + (4b)^2 = 9a^2 + 16b^2$$

Justify your answer.

20. _____
_____
_____
_____

For Exercises 21-23, perform the division.

21. $\dfrac{5x^4 - 10xy^3 - 15x + 25}{5x^2}$

21. _____

22. $\dfrac{2v^2 - 11v + 16}{2v + 3}$

22. _____

23. $(2h^4 + 5h^2 + 3) \div (h^2 + 1)$

23. _____

24. (a) Write $58,000,000$ using scientific notation.

24. (a) _____

(b) Write 5.2×10^{-5} without using exponents.

24. (b) _____

(c) Write the quotient and write your answer without using scientific notation:

$$\frac{2.56 \times 10^{-4}}{0.8 \times 10^{-7}}$$

24. (c) _____

25. Use scientific notation to solve the following problem. Write your answer without exponents.

Each new car that comes off the assembly line is fueled with 1.27 gallons of gasoline. If 100,000 cars are produced in a month, how much gasoline will the auto manufacturer need to order?

25. _____

CHAPTER FOUR, FORM D   BEGINNING ALGEBRA

NAME_____   SECTION_____

__For Exercises 1-2, combine like terms whenever possible and write the polynomial in descending order. Give the degree of the simplified polynomial, and determine whether the simplified polynomial is a monomial, binomial, trinomial, or none of these.__

1. $4a^3 - a^4 + a^4 - 3a^2 - a$ 1._____

2. $2 + 6x^3 + x^2 - 1 - x^2 + 4x^3$ 2._____

3.   Create a set of ordered pairs that
      lie on the graph of $y = -x^2 - 1$.
      Then graph the equation.   3.

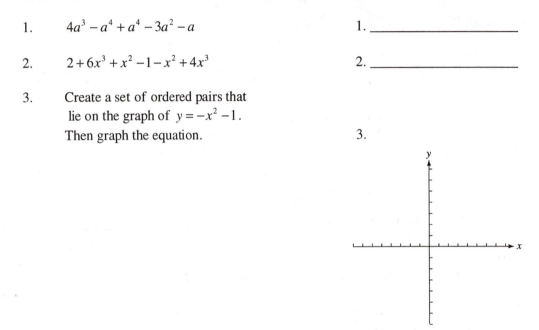

__For Exercises 4-12, perform the indicated operation.__

4. $(6p^2 - 3p + 1) + (-4p^2 + 3p + 4) - (2p^2 - 6p - 6)$ 4._____

5. $(5a^3b^2 - 8ab + 6ab^4) - (-2ab^4 + ab + 5a^3b^2)$ 5._____

6.   Subtract:
$$5t^3 + 6t^2 - 8t + 1$$
$$2t^3 - 3t^2 + t + 1$$ 6._____

7. $-2p^2(3p^2 - 4p - 5)$ 7._____

8. $(b-4)(b+5)$ 8._____

9. $(8x-3)(3x+8)$ 9._____

10. $(7m-6n)^2$ 10._____

11. $(9p+2q)(9p-2q)$ 11._____

12. $(3x-7)(2x^2-x+5)$ 12._____

13.   What polynomial expression represents
      the perimeter of this rectangle?

$y-3$

$2y^2+3y-1$ 13._____

For Exercises 14-16, evaluate the expression.

14. $\left(\dfrac{3}{5}\right)^{-2}$ 14._____

15. $-(4)^0-(4)^0$ 15._____

16. $4^{-1}+2^{-1}$ 16._____

For Exercises 17-19, simplify and write the answer using only positive exponents. Assume that all variables represent nonzero numbers.

17. $\dfrac{4^{-3}\cdot 4^5}{4^{-7}}$ 17._____

18. $\dfrac{(3a^3b^2)^2(2ab^2)^3}{(ab)^4}$ 18._____

19. $\dfrac{(a^{-2}b^{-2})^{-3}(ab^2)^4}{(a^3b)^{-2}}$ 19._____

20. Do you agree or disagree with the
    following statement?

    $$(-5)^3 = -5^3$$

    Justify your answer.

20. _____
_____
_____
_____

For Exercises 21-23, perform the division.

21. $\dfrac{7x^2y^3 + 14x^2y - 28xy}{7xy}$

21. _____

22. $\dfrac{12v^3 - 11v^2 + 9v + 18}{4v + 3}$

22. _____

23. $(3h^3 - h^2 + 7h - 4) \div (h + 1)$

23. _____

24. (a) Write $760,000,000$ using scientific notation.

24. (a) _____

   (b) Write 8.1×10^{-5} without using exponents.

24. (b) _____

   (c) Write the quotient and write your answer
       without using scientific notation:

   $$\dfrac{14.4 \times 10^{-4}}{1.2 \times 10^{-7}}$$

24. (c) _____

25. Use scientific notation to solve the following
    problem. Write your answer without exponents.

    A biology experiment created 3.2×10^9 bacteria
    If these bacteria were evenly dispersed between
    the 10^5 different samples, how many bacteria
    would be in one sample?

25. _____

CHAPTER FOUR, FORM E   BEGINNING ALGEBRA

NAME_____   SECTION_____

__For Exercises 1-2, consider the following polynomial:__ $8y^5 - 2y^3 + 2y + 4$

1.   Give the degree of the polynomial.

    (a) 3   (b) 5   (c) 8   (d) 4   1. _____

2.   Choose the word that best describes the polynomial.

    (a) Monomial   (b) Binomial
    (c) Trinomial   (d) None of these   2. _____

3.   Graph the solution of the equation $y = x^2 - 3$.   3. _____

(a)

(b)

(c)

(d)

For Exercises 4-12, perform the indicated operations.

4. $(3x^2 + 2x - 5) - (-7x^2 - 2x + 5) + (x^2 + x + 6)$

(a) $-3x^2 + 4x + 6$ (b) $11x^2 + 5x - 4$

(c) $-3x^2 + x + 6$ (d) $11x^2 + x + 6$ 4. _____

5. $(-7p^3r^2 + 8p^2r^3 - p^4r) - (3p^4r - 9p^3r^2 + pr)$

(a) $4p^3r^2 + 8p^2r^2 + 2p^4r^2 - pr$ (b) $2p^3r^2 + 8p^3r^4 - 4p^4r$

(c) $2p^3r^2 + 8p^2r^3 - 4p^4r - pr$ (d) $-10p^3r^2 - p^2r^3 - 4p^4r$ 5. _____

6.   Subtract: $2t^2 - 5t + 21$
                    $\underline{t^2 - 5t + 7}$

(a) $3t^2 + 14$ (b) $3t^2 - 10t + 28$

(c) $t^2 - 10t + 28$ (d) $t^2 + 14$ 6. _____

7. $-8a^3(4a^2 - 5a + 7)$

(a) $32a^5 - 40a^4 + 56a^3$ (b) $-32a^5 - 40a^4 - 56a^3$

(c) $-32a^5 + 40a^4 + 56a^3$ (d) $-32a^5 + 40a^4 - 56a^3$ 7. _____

8. $(3x - 2y)(3x + 2y)$

(a) $9x^2 - 4y^2$ (b) $9x^2 + 12xy - 16y^2$

(c) $9x^2 - 12xy - 16y^2$ (d) $9x^2 - xy - 16y^2$ 8. _____

9. $(x - 4)(x + 9)$

(a) $x^2 + 5x - 5$ (b) $x^2 - 36x + 5$

(c) $x^2 + 5x - 36$ (d) $x^2 + 13x - 36$ 9. _____

10. $(x-1)(4x^2-2x-1)$

      (a) $4x^3-2x^2+x+1$ (b) $4x^3-6x^2+x+1$

      (c) $4x^3-6x^2-x+1$ (d) $4x^3+2x^2+x+1$ 10. _____

11. $(9-y)^2$

      (a) $81-18y-y^2$ (b) $81-18y+y^2$

      (c) $81+y^2$ (d) $18-18y+y^2$ 11. _____

12. $(7y-3)(7y+3)$

      (a) $49y^2+42y-9$ (b) $49y^2-42y-9$

      (c) $49y^2-9$ (d) $49y^2+4y-9$ 12. _____

13.   Find the polynomial expression that represents
      the area of this square.

$3x-4$

      (a) $9x^2+16$ (b) $9x^2+24x-16$

      (c) $12x-16$ (d) $9x^2-24x+16$ 13. _____

For Exercises 14-16, evaluate the expression.

14. $(-2)^{-3}$

      (a) 8 (b) -8 (c) $\dfrac{1}{8}$ (d) $-\dfrac{1}{8}$ 14. _____

15. -33^0

      (a) 1 (b) -1 (c) 0 (d) -33 15. _____

16. $5^{-1} + 4^{-1}$

    (a) –9   (b) $\dfrac{1}{9}$ (c) 9   (d) $\dfrac{9}{20}$ 16. _____

For Exercises 17-19, simplify and write the answer using only positive exponents. Assume that all variables represent nonzero numbers.

17. $\dfrac{5^7 \cdot 5^{-5}}{5^{-3}}$

    (a) 5^5 (b) $\dfrac{1}{5}$ (c) –5   (d) 5^9 17. _____

18. $\dfrac{(a^3)^{-2}(z^2)^4}{a^0(z^{-3})^{-2}}$

    (a) $\dfrac{z^2}{a^7}$ (b) $\dfrac{z^{14}}{a^6}$ (c) $\dfrac{z^2}{a^6}$ (d) $\dfrac{a^6}{z^2}$ 18. _____

19. $\dfrac{(3x^3 y^2)^3 (2x^2 y^3)^2}{(xy)^5}$

    (a) $108x^8 y^7$ (b) $108x^5 y^5$

    (c) $361x^{12} y^{10}$ (c) $36x^8 y^7$ 19. _____

20.   Without calculating, what do you know about the value of 451^{-5} ?

    (a) It is a very large positive number
    (b) It is a very small positive fraction
    (c) It is a negative integer
    (d) It is a negative fraction   20. _____

For Exercises 21-23, perform the division.

21. $(-12x^3 y^3 - 16x^5 y^4 + 20xy) \div (-4xy)$

    (a) $3x^2 y^2 + 4x^4 y^3 - 5$ (b) $3x^2 y^2 + 4x^4 y^3 + 20x^2 y^2$

    (c) $-3x^2 y^2 - 4x^4 y^3 + 5$ (d) $3x^4 y^4 + 4x^6 y^5 - 5x^2 y^2$ 21. _____

22. $(3x^2 - 6x - 3) \div (x + 1)$

(a) $3x^2 - 3x - 9 + \dfrac{6}{x+1}$ (b) $3x^2 - 3x - 9$

(c) $3x - 9 + \dfrac{6}{x+1}$ (d) $3x^2 - 3x - 3$ 22. _____

23. $\dfrac{2z^3 - 3z^2 + 8z + 6}{2z + 1}$

(a) $z^2 - 2z + 5 + \dfrac{1}{2z+1}$ (b) $z^2 - 2z + 6$

(c) $z^2 - z + 5 + \dfrac{1}{2z+1}$ (d) $z^2 - z + 6$ 23. _____

24.   Find the following quotient and write your answer
without using scientific notation.

$$\frac{(6.3 \times 10^3)(4 \times 10^{-5})}{(2.1 \times 10^{-4})}$$

(a) 0.12   (b) 1.2   (c) 1200   (d) 120   24. _____

25.   Use scientific notation to solve the following problem.
Write your answer without using exponents.

The Daily Sun uses 2.3×10^4 pounds of newsprint each day.
How many pounds of newsprint will be used in 1000 days?

(a) 2,300,000   (b) 23,000,000
(c) 46,000,000   (d) 4,600,000   25. _____

CHAPTER FOUR, FORM F   **BEGINNING ALGEBRA**

NAME_____   SECTION_____

For Exercises 1-2, consider the following polynomial: $3x + 8x^5 - 4$

1.   Give the degree of the polynomial.

     (a) 3   (b) 5   (c) 8   (d) -4 1. _____

2.   Choose the word that best describes the polynomial.

     (a) Monomial   (b) Binomial
     (c) Trinomial   (d) None of these   2. _____

3.   Graph the solution of the equation $y = 1 - x^2$.   3. _____

    (a)   (b)

    (c)   (d)

For Exercises 4-12, perform the indicated operations.

4. $(-3y^2z^2 + 7yz^4 - y^3z) - (-y^3z + 8yz^4 + 2y^2z^2)$

(a) $4y^2z^2 + 15yz^4 + y^2z^2$ (b) $-5y^2z^2 - yz^4$

(c) $-5y^2z^2 - yz^4 + 2y^3z$ (d) $-y^2z^2 - yz^4$ 4. _____

5. $(6p^2 - 3p + 1) + (2p^2 - 6p - 1) - (-4p^2 + 3p - 1)$

(a) $12p^2 - 15p - 1$ (b) $4p^2 - 3p - 1$

(c) $12p^2 - 12p + 1$ (d) $12p^2 - 3p - 1$ 5. _____

6.   Subtract: $5t^2 + 13t - 4$
                         $16t^2 + 4t - 3$
                         _____

(a) $-11t^2 + 9t - 1$ (b) $21t^2 + 9t - 7$

(c) $-11t^2 + 17t - 7$ (d) $21t^2 + 17t - 7$ 6. _____

7. $9x^2(3x^5 - 2x + 1)$

(a) $27x^7 - 18x^3 + 9x^2$ (b) $27x^7 - 18x^3 + 1$

(c) $27x^7 - 18x^2 + 9x$ (d) $27x^{10} - 18x^3 + 1$ 7. _____

8. $(x - 7)(x + 6)$

(a) $x^2 - x + 42$ (b) $x^2 - x - 42$

(c) $x^2 - 13x - 42$ (d) $x^2 - 13x + 42$ 8. _____

9. $(4m + 3n)(3m - 7n)$

(a) $12m^2 - 19mn - 21n^2$ (b) $12m^2 - 19mn + 21n^2$

(c) $12m^2 - 37mn + 21n^2$ (d) $12m^2 - 37mn - 21n^2$ 9. _____

10. $(a-1)(4a^2-2a-1)$

   (a) $4a^3-2a^2+a+1$ (b) $4a^3-6a^2+a+1$
   (c) $4a^3-6a^2-a+1$ (d) $4a^3+2a^2+a+1$ 10. _____

11. $(7-y)^2$

   (a) $49-14y-y^2$ (b) $49-14y+y^2$
   (c) $49+y^2$ (d) $14-14y+y^2$ 11. _____

12. $(3y-7)(3y+7)$

   (a) $9y^2+42y-49$ (b) $9y^2-42y-49$
   (c) $9y^2-49$ (d) $9y^2+4y-49$ 12. _____

13.   Find the polynomial expression that represents
      the perimeter of this square.

$3x-4$

   (a) $9x^2+16$ (b) $9x^2+24x-16$
   (c) $12x-16$ (d) $9x^2-24x+16$ 13. _____

For Exercises 14-16, evaluate the expression.

14. $(-3)^{-2}$

   (a) 9   (b) -9 (c) $\dfrac{1}{9}$ (d) $-\dfrac{1}{9}$ 14. _____

15. $(-23)^0$

   (a) 1   (b) -1 (c) 0   (d) -23 15. _____

16. $3^{-1} - 6^{-1}$

(a) -3 (b) $-\dfrac{1}{3}$ (c) $\dfrac{1}{3}$ (d) $\dfrac{1}{6}$ 16. _____

For Exercises 17-19, simplify and write the answer using only positive exponents. Assume that all variables represent nonzero numbers.

17. $(4^{-3})^{-6}$

(a) 4^{9} (b) $\dfrac{1}{4^{9}}$ (c) 4^{18} (d) $\dfrac{1}{4^{18}}$ 17. _____

18. $\dfrac{d^{3}e^{0}f^{-4}}{d^{5}e^{3}f^{-7}}$

(a) $\dfrac{f^{3}}{d^{2}e^{3}}$ (b) $\dfrac{d^{2}e^{3}}{f^{3}}$

(c) $\dfrac{f^{3}}{d^{2}e^{2}}$ (d) $\dfrac{f^{11}}{d^{2}e^{3}}$ 18. _____

19. $\dfrac{(3a^{3}b^{2})^{3}(2a^{2}b^{3})^{2}}{(ab)^{5}}$

(a) $108a^{8}b^{7}$ (b) $108a^{5}b^{5}$
(c) $361a^{12}b^{10}$ (c) $36a^{8}b^{7}$ 19. _____

20.   Without calculating, what do you know about the value of -451^{-4} ?

(a) It is a very large positive number
(b) It is a very small positive fraction
(c) It is a negative integer
(d) It is a negative fraction   20. _____

For Exercises 21-23, perform the division.

21. $(64g^{4}h^{3} - 8g^{2}h^{4} + 12gh^{5}) \div (4gh)$

(a) $16g^{5}h^{4} - 2g^{3}h^{5} + 3g^{2}h^{6}$ (b) $16g^{3}h^{2} - 2gh^{3} + 3gh^{4}$
(c) $16g^{3}h^{2} - 2gh^{3} + 3h^{4}$ (d) $16g^{3}h^{2} + 2g^{3}h^{5} + 3gh^{4}$ 21. _____

22. $(2t^3 + t^2 + 4) \div (t - 1)$

(a) $2t^2 - t + \dfrac{4}{t-1}$ (b) $2t^2 + 3t + 3 + \dfrac{1}{t-1}$

(c) $2t^2 + 3t + 3 + \dfrac{7}{t-1}$ (d) $2t^2 + 3t + \dfrac{4}{t-1}$ 22. _____

23. $\dfrac{3a^3 - 11a^2 + 25a - 25}{3a - 5}$

(a) $a^2 - 4a + 5$ (b) $a^2 - 2a - 5$

(c) $a^2 - 2a + 5$ (d) $a^2 - 6a + 5$ 23. _____

24.   Find the following quotient and write your answer without using scientific notation.

$$\frac{(7.2 \times 10^{-4})}{(0.4 \times 10^4)(9 \times 10^{-7})}$$

(a) 20   (b) 0.2
(c) 0.0000002   (d) 0.00002   24. _____

25.   Use scientific notation to solve the following problem. Write your answer without using exponents.

The total mileage traveled by all Speedy Buses was 1.75×10^7 miles. If each cab travels 35,000 miles, how many cabs are in the fleet?

(a) 1750   (b) 175   (c) 5,000   (d) 500   25. _____

CHAPTER FIVE, FORM A

BEGINNING ALGEBRA

NAME_____

SECTION_____

1.   Which one of the following is the correct completely factored form of $3w^2 + 6w - 24$?

(a) $(3w - 6)(w + 4)$ (b) $3(w - 2)(w + 4)$

(c) $(3w + 6)(w - 4)$ (d) $3(w^2 + 2w - 8)$ 1. _____

For Exercises 2-14, factor the polynomial completely. If it cannot be factored, write "prime".

2. $21a^2 + 58a + 21$ 2. _____

3. $d^2 - 2d + 6$ 3. _____

4. $c^2 - 9c + 20$ 4. _____

5. $8w^4 - 28w^3 - 16w^2$ 5. _____

6. $12x^2 - x - 6$ 6. _____

7. $t^3 - 27$ 7. _____

8. $16h^2 + 81j^2$ 8. _____

9. $15x^2 + 31x - 24$ 9. _____

10. $x^4 - 81$ 10. _____

11. $8w^3 + 125x^3$ 11. _____

12. $wx + 3xz - 2wy - 6yz$ 12. _____

13. $2a^3 + 20a^2 + 50a$ 13. _____

14. $9x^2 - 60xy + 100y^2$ 14. _____

15.   Since the square of x is x^2 and the square of $\dfrac{2}{3}$ 15. _____

        is $\dfrac{4}{9}$, why isn't the square of $\left(x + \dfrac{2}{3} \right)$ equal to   _____

        $x^2 + \dfrac{4}{9}$?   _____

For Exercises 16-19, solve the equation.

16. $g^2 - 3g - 18 = 0$ 16. _____

17. $t^2 = 4(2t - 3)$ 17. _____

18. $81d^2 - 4 = 0$ 18. _____

19. $64x^3 = 16x$ 19. _____

For Exercises 20-22, solve each problem.

20.   If a baseball is thrown upward from ground
        level at an initial velocity of 64 feet per second ,
        after t seconds, its height h is given by the formula
        $h = -16t^2 + 64t$. After how many seconds will its
        height be 48 feet?   20. _____

21.   To reach a damsel-in-distress, a knight places
        a 13-foot ladder against the castle wall. The
        base of the ladder is 7 feet less than the distance
        from the top of the ladder to the ground. How
        high on the castle will the ladder reach?

        (a) Let x represent the height on the castle that
            the ladder reaches. Write an expression for
            the distance of the ladder to the wall.   21. (a) _____

        (b) How high on the castle will the ladder reach?
            (Hint: use the Pythagorean formula.)   21. (b) _____

22.   The automotive industry uses the equation
        $y = -0.04x^2 + 0.93x + 21$ to calculate trends in fuel
        economy, where y is the average miles per gallon
        and x represents the number of years after 1978.
        What was the average miles per gallon in 1991?
        (Round your answer to the nearest whole number.)   22. _____

For Exercises 23-24, solve each inequality and graph the solutions.

23. $(3x+5)(2x-1) > 0$ 23. _____

24. $y^2 - 7y + 12 \leq 0$ 24. _____

25.   Why isn't " $y = \dfrac{3}{4}$ " the correct response to

        "Solve the equation $y^2 = \dfrac{9}{16}$ "?   25. _____

CHAPTER FIVE, FORM B

BEGINNING ALGEBRA

NAME_____   SECTION_____

1.   Which one of the following is the correct
     completely factored form of $-10w^2 - 28w + 6$?

   (a) $-2(5w-1)(w-3)$ (b) $(2-10w)(w+3)$

   (c) $-2(5w-1)(w+3)$ (d) $(10w-2)(w-3)$ 1._____

For Exercises 2-14, factor the polynomial completely. If it cannot be factored, write "prime".

2. $x^2 - 11x + 30$ 2._____

3. $18y^4 - 42y^3$ 3._____

4. $c^2 - 8c + 20$ 4._____

5. $4y^2 + 25y + 25$ 5._____

6. $20x^2 - 7x - 6$ 6._____

7. $54t^3 + 16v^3$ 7._____

8. $ab - a - b + 1$ 8._____

9. $-8z^2 + 48z - 72$ 9._____

10. $125 - 8x^3$ 10._____

11. $64c^2 - 16c + 1$

11._____

12. $18x^3 - 39x^2 - 15x$

12._____

13. $b^2c^2 - 2b^2c + b^2$

13._____

14. $81x^4 - 16$

14._____

15.   Since the square of y is y^2 and the square of 11 is 121, why isn't the square of $(y+11)$ equal to $y^2 + 121$?

15._____

_____

_____

For Exercises 16-19, solve the equation.

16. $h^2 + 2h - 35 = 0$

16._____

17. $t^2 = 9(t-2)$

17._____

18. $9d^2 - 49 = 0$

18._____

19. $64r^3 = r$

19._____

For Exercises 20-22, solve each problem.

20.   If a golf ball is propelled upward from ground level at an initial velocity of 144 feet per second , after t seconds, its height h is given by the formula $h = -16t^2 + 144t$. After how many seconds will its height be 320 feet?

20._____

21.   Rachael is flying a kite that is 30 feet farther
       above her hand than its horizontal distance
       from her. The string from her hand to the kite
       is 150 feet long.

       (a) Write an expression for the horizontal
           distance of the kite from her hand.   21. (a) _____

       (b) How high in the air is the kite?
           (Hint: use the Pythagorean formula.)   21. (b) _____

22.   The automotive industry uses the equation
       $y = -0.04x^2 + 0.93x + 21$ to calculate trends in fuel
       economy, where y is the average miles per gallon
       and x represents the number of years after 1978.
       What was the average miles per gallon in 1987?
       (Round your answer to the nearest whole number.)   22. _____

For Exercises 23-24, solve each inequality and graph the solutions.

23. $(2y - 5)(3y + 2) > 0$ 23. _____

24. $c^2 + 11c + 30 \leq 0$ 24. _____

25.   Why isn't "$c = \dfrac{1}{4}$" the correct response to

       "Solve the equation $c^2 = \dfrac{1}{16}$"?   25. _____

CHAPTER FIVE, FORM C   **BEGINNING ALGEBRA**

NAME_____   SECTION_____

1.   Which one of the following is the correct
      completely factored form of $4w^2 - 64$?

    (a) $4(w-4)(w+4)$ (b) $2(w+4)(2w-8)$

    (c) $(2w-8)(2w+8)$ (d) $4(w-4)^2$ 1._____

___

For Exercises 2-14, factor the polynomial completely. If it cannot be factored, write "prime".

2. $x^2 - 8x + 15$ 2._____

3. $a^2 - 3a + 4$ 3._____

4. $45b^3c^2 - 30b^4c^3 + 5b^2c^2$ 4._____

5. $15y^2 - 7y - 2$ 5._____

6. $2y^3 + 2yz^2 - 5y^2z - 5z^3$ 6._____

7. $16s^2 + 24st + 9t^2$ 7._____

8. $1 + 125g^3$ 8._____

9. $36z^2 + 64$ 9._____

10. $3r^2 + r - 4$ 10._____

11. $8a^3 - b^3$ 11. _____

12. $6d^2 - 54$ 12. _____

13. $10d^2 + 7d + 1$ 13. _____

14. $81x^4 - 256$ 14. _____

15. Since the square of y is y^2 and the square of   15. _____
     -6 is 36, why isn't the square of $(y-6)$ _____
     equal to $y^2 + 36$?   _____

For Exercises 16-19, solve the equation.

16. $h^2 - 3h - 40 = 0$ 16. _____

17. $4x(x+1) = -1$ 17. _____

18. $25f^2 = 64$ 18. _____

19. $y^2 - 25y = 0$ 19. _____

For Exercises 20-22, solve each problem.

20. If a model rocket is propelled upward from ground
     level at an initial velocity of 208 feet per second ,
     after t seconds, its height h is given by the formula
     $h = -16t^2 + 208t$. After how many seconds will its
     height be 640 feet?   20. _____

21.   An airplane is flying at an altitude that is 7000 feet less than its horizontal distance from the edge of the runway. The slanted distance from the airplane to the edge of the runway is exactly 13,000 feet.

(a) Let x represent the horizontal distance of the airplane. Write an expression for the altitude of the airplane.

21. (a) _____

(b) What is the altitude of the airplane? (Hint: use the Pythagorean formula.)

21. (b) _____

22.   The automotive industry uses the equation $y = -0.04x^2 + 0.93x + 21$ to calculate trends in fuel economy, where y is the average miles per gallon and x represents the number of years after 1978. What was the average miles per gallon in 1993? (Round your answer to the nearest whole number.)

22. _____

For Exercises 23-24, solve each inequality and graph the solutions.

23. $(2s - 5)(3s - 4) > 0$

23. _____

24. $k^2 + 5k + 4 \le 0$

24. _____

25.   Why isn't "$d = \dfrac{1}{3}$" the correct response to "Solve the equation $d^2 = \dfrac{1}{9}$"?

25. _____

CHAPTER FIVE, FORM D   BEGINNING ALGEBRA

NAME_____   SECTION_____

Which one of the following is the correct
completely factored form of $9w^2 - 81$?

(a) $(3w-9)(3w+9)$ (b) $3(w+3)(3w-9)$

(c) $9(w-3)(w+3)$ (d) $3(w-3)(w+3)$ 1. _____

For Exercises 2-14, factor the polynomial completely. If it cannot be factored, write
"prime".

2. $c^2 - 8c + 7$ 2. _____

3. $e^2 - 3e + 4$ 3. _____

4. $45y^3x^2 - 30y^4x^3 + 5y^2x^2$ 4. _____

5. $6a^2 - 17a + 10$ 5. _____

6. $2b^3 + 2bd^2 - 5b^2d - 5d^3$ 6. _____

7. $36s^2 + 84st + 49t^2$ 7. _____

8. $27 + 125p^3$ 8. _____

9. $25q^2 + 100$ 9. _____

10. $3r^2 - 8r + 4$ 10. _____

11. $27a^3 - 8b^3$ 11. _____

12. $3h^2 - 48$ 12. _____

13. $8d^2 - 2d - 1$ 13. _____

14. $32x^4 - 2$ 14. _____

15.   Since the square of a is a^2 and the square of   15. _____
       -7 is 49, why isn't the square of $(a-7)$ _____
       equal to $a^2 + 49$?   _____

For Exercises 16-19, solve the equation.

16. $6(x^2 - 1) = 5x$ 16. _____

17. $h^2 - 5h - 36 = 0$ 17. _____

18. $y^2 - 81y = 0$ 18. _____

19. $49p^2 = 64$ 19. _____

For Exercises 20-22, solve each problem.

20.   If a model rocket is propelled upward from ground
       level at an initial velocity of 112 feet per second,
       after t seconds, its height h is given by the formula
       $h = -16t^2 + 112t$. After how many seconds will its
       height be 192 feet?   20. _____

21. Two cars leave Cedar Lake, Indiana, at the same time, one heading due north, and the other one heading due east. After a few hours, they are 170 miles apart. The car traveling north had traveled 70 miles farther than the car traveling east.

    (a) Let x represent the distance traveled by the car going east. Write an expression for the car traveling north.

21. (a) _____

    (b) How far has each car traveled? (Hint: use the Pythagorean formula.)

21. (b) _____

22. The automotive industry uses the equation $y = -0.04x^2 + 0.93x + 21$ to calculate trends in fuel economy, where y is the average miles per gallon and x represents the number of years after 1978. What will the average miles per gallon be in 2002? (Round your answer to the nearest whole number.)

22. _____

For Exercises 23-24, solve each inequality and graph the solutions.

23. $(4x - 5)(3x - 1) < 0$

23. _____

24. $d^2 + 2d - 15 \geq 0$

24. _____

25. Why isn't "$s = \dfrac{5}{4}$" the correct response to "Solve the equation $s^2 = \dfrac{25}{16}$"?

25. _____

CHAPTER FIVE, FORM E

BEGINNING ALGEBRA

NAME_____

SECTION_____

For Exercises 1-14, factor the polynomial completely. If it cannot be factored, choose "prime."

1. $x^2 + 10x - 24$

    (a) $(x-6)(x-2)$ (b) $(x+12)(x-2)$

    (c) $(x+8)(x-3)$ (d) $(x+6)(x+4)$ 1. _____

2. $8z^2 - 6z - 1$

    (a) $(4z-1)(2z-1)$ (b) $(2z+1)(4z-1))$

    (c) $(8z+1)(z+1)$ (d) Prime   2. _____

3. $3y^2 + 21y + 18$

    (a) $3(y+6)(y+1)$ (b) $3(y^2+7y+6)$

    (c) $3(y+5)(y+2)$ (d) $3(y+3)(y+2)$ 3. _____

4. $8a^3 - 27$

    (a) $(2a-3)(4a^2+6a+9)$ (b) $(2a+3)(4a^2+6a-9)$

    (c) $(2a+3)(4a^2+6a-9)$ (d) $(2a-3)(4a^2-6a+9)$ 4. _____

5. $w^{10} - 4w^9 x - 21w^8 x^2$

    (a) $w^5(w-7x)(w+3x)$ (b) $w^8(w+3x)(w-7x)$

    (c) $w^8(w+4x)(w-8x)$ (d) $w^9(w+3x)(w-7x)$ 5. _____

6. $-4r^2 - 12r + 40$

    (a) $(8-4r)(r+5)$ (b) $-4(r+5)(r-2)$

    (c) $(-4r-20)(r-2)$ (d) Prime   6. _____

7. $x^2 + 16x + 64$

    (a) $(x+8)(x-8)$ (b) $(x+8)^2$

    (c) $(x-8)^2$ (d) Prime   7. _____

8. $9k^2 + 30kp + 25p^2$

    (a) $(3k+5p)(3k-5p)$ (b) $(3k+5p)(3k+25p)$

    (c) $(3k+5p)(k+p)$ (d) $(3k+5p)^2$ 8. _____

9. $w^2 + 400$

    (a) $(w+20)(w-20)$ (b) $(w+20)^2$

    (c) $(w+40)(w+10)$ (d) Prime   9. _____

10. $x^2 - 19x - 42$

    (a) $(x+21)(x-2)$ (b) $(x-21)(x+2)$

    (c) $(x+7)(x-6)$ (d) Prime   10._____

11. $16w^4 - 81$

    (a) $(4w^2-9)(2w+3)^2$ (b) $(4w^2-9)^2$

    (c) $(4w^2+9)(2w+3)(2w-3)$ (d) $(2w-3)^4$ 11._____

12. $2x^3 - 3x^2y + 2xy^2 - 3y^3$

    (a) $(x^2-y^2)(2x+3y)$ (b) $(x^2+y^2)(2x-3y)$

    (c) $x^2(2x-3y)+y^2(2x-3y)$ (d) Prime   12._____

13. $6q^2 - 7qr - 5r^2$

    (a) $(3q+5r)(2q-r)$ (b) $(3q-5r)(2q+r)$

    (c) $(3q+r)(2q-5r)$ (d) $(3q-r)(2q+5r)$ 13._____

14. $3q^2 - 5q - 8$

    (a) $(3q-4)(q-2)$ (b) $(3q+2)(q-4)$

    (c) $(3q-8)(q+1)$ (d) Prime   14._____

15.   Which of the following is not a factor of $5t^2 - 40t + 75$?

    (a) $(t-5)$ (b) $(t-3)$

    (c) 5   (d) $(t-15)$ 15._____

For Exercises 16-19, solve the equation.

16. $y^2 - 3y - 10 = 0$

    (a) $1, -10$ (b) $-1, 10$

    (c) $5, -2$ (d) $-5, 2$ 16. _____

17. $9x^2 - 64 = 0$

    (a) $\dfrac{8}{3}, -\dfrac{8}{3}$ (b) $\dfrac{3}{8}, -\dfrac{3}{8}$

    (c) $\dfrac{8}{9}, -\dfrac{8}{9}$ (d) $\dfrac{8}{3}$ 17. _____

18. $z^2 = 7z$

    (a) $0, 7$ (b) $-7, 7$

    (c) 0 (d) -7 18. _____

19. $w(w+2) = 80$

    (a) $0, -2$ (b) $-8, 10$

    (c) $8, -10$ (d) $3, 16$ 19. _____

20.   Which of the following is *not* a solution
     to the equation $9h^3 = h$?

    (a) 0 (b) 3 (c) $\dfrac{1}{3}$ (d) $-\dfrac{1}{3}$ 20. _____

For Exercises 21-23, solve the problem.

21.   If a ball is dropped from a 144-foot building, its position (in feet above the ground) is given by $s = -16t^2 + 112t + 144$, where t is time in seconds after it was dropped. How long will it take for the ball to hit the ground?

(a) 8.1 seconds   (b) 5.3 seconds
(c) 1.7 seconds   (d) 12.0 seconds   21. _____

22.   A 26-foot long support wire for a radio tower stretches from its mooring point on the tower to a point on the ground x feet from the base of the tower. The height at which the wire attaches to the tower is 4 feet less than twice the distance from the tower to its anchor on the ground. At what height does the wire attach to the tower?

(a) 15 feet   (b) 13.2 feet
(c) 6.6 feet   (d) 22.4 feet   22. _____

23.   The automotive industry uses the equation $y = -0.04x^2 + 0.93x + 21$ to calculate trends in fuel economy, where y is the average miles per gallon and x represents the number of years after 1978. What will the average miles per gallon be in 2003?

(a) 19.25 mpg   (b) 43.25 mpg
(c) 25.0 mpg   (d) 25.95 mpg   23. _____

24.   Solve the inequality

$$12t^2 + 7t - 12 \geq 0$$

(a) $x \leq -\dfrac{3}{4}$ or $x \geq \dfrac{4}{3}$ (b) $x \leq -\dfrac{4}{3}$ or $x \geq \dfrac{3}{4}$

(c) $-\dfrac{3}{4} \leq x \leq \dfrac{4}{3}$ (d) $-\dfrac{4}{3} \leq x \leq \dfrac{3}{4}$ 24. _____

25.   Graph the solutions of the inequality   25. _____

$$3c^2 + c - 14 \leq 0$$

(a)   (b)

   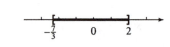

(c)   (d)

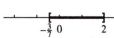   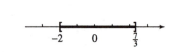

CHAPTER FIVE, FORM F

BEGINNING ALGEBRA

NAME_____   SECTION_____

_____

For Exercises 1-14, factor the polynomial completely. If it cannot be factored, choose "prime."

1. $y^2 + 9y - 36$

  (a) $(y-6)(y+6)$ (b) $(y+12)(y-3)$
  (c) $(y+18)(y-2)$ (d) $(y-12)(y+3)$ 1. _____

2. $8s^2 - 6s + 1$

  (a) $(4s-1)(2s-1)$ (b) $(2s+1)(4s-1))$
  (c) $(8s+1)(s+1)$ (d) Prime   2. _____

3. $4r^2 - 28r + 40$

  (a) $4(r-5)(r-2)$ (b) $4(r^2 - 7r + 10)$
  (c) $4(r+5)(r+2)$ (d) $4(r-10)(r-1)$ 3. _____

4. $27x^3 + 64$

  (a) $(3x+4)(9x^2 - 12x + 16)$ (b) $(3x-4)(9x^2 + 12x - 16)$
  (c) $(3x+4)(9x^2 + 12x + 16)$ (d) $(3x-4)(9x^2 - 12x - 16)$ 4. _____

5. $a^{10} - 4a^9 b - 21a^8 b^2$

  (a) $a^9(a+3b)(a-7b)$ (b) $a^8(a+4b)(a-8b)$
  (c) $a^8(a+3b)(a-7b)$ (d) $a^5(a-7b)(a+3b)$ 5. _____

6. $-6r^2 - 18r - 24$

  (a) $-6(r+4)(r-1)$ (b) $-6(r-4)(r+1)$
  (c) $-6(r^2 + 3r + 4)$ (d) Prime   6. _____

7. $j^2 - 14j + 49$

(a) $(j+7)(j-7)$ (b) $(j+7)^2$
(c) $(j-7)^2$ (d) Prime   7._____

8. $49k^2 + 56kp + 16p^2$

(a) $(7k+2p)(7k+8p)$ (b) $(7k+4p)^2$
(c) $(7k+4p)(7k-4p)$ (d) $(7k+4p)(k+p)$ 8._____

9. $4w^2 + 400$

(a) $4(w+10)(w-10)$ (b) $4(w+10)^2$
(c) $4(w^2+100)$ (d) Prime   9._____

10. $x^2 - 19x - 42$

(a) $(x+21)(x-2)$ (b) $(x-21)(x+2)$
(c) $(x+7)(x-6)$ (d) Prime   10._____

11. $w^4 - 81$

(a) $(w^2+9)(w+3)(w-3)$ (b) $(w-3)^4$
(c) $(w^2-9)(w+3)^2$ (d) $(w^2-9)^2$ 11._____

12. $2b^3 - 3b^2d + 2bd^2 - 3d^3$

(a) $b^2(2b-3d) + y^2(2b-3d)$ (b) $(b^2+d^2)(2b-3d)$
(c) $(b^2-d^2)(2b+3d)$ (d) Prime   12._____

13. $8q^2 - 2qr - 5r^2$

(a) $(4q-5r)(2q+r)$ (b) $(4q-r)(2q+5r)$
(c) $(4q+5r)(2q-r)$ (d) Prime   13._____

14. $2x^2 - 5x - 7$

(a) $(2x+7)(x-1)$ (b) $(2x-1)(x+7)$
(c) $(2x-7)(x+1)$ (d) $(2x+1)(x-7)$ 14._____

15.   Which of the following is not a factor of $6t^2 - 60t + 144$?

    (a) $(t-4)$ (b) $(t-6)$

    (c) 6   (d) $(t-2)$ 15._____

For Exercises 16-19, solve the equation.

16. $y^2 - 15y + 56 = 0$

    (a) $-7, -8$ (b) $-7, 8$

    (c) $7, -8$ (d) $7, 8$ 16._____

17. $49x^2 - 4 = 0$

    (a) $\dfrac{2}{7}, -\dfrac{2}{7}$ (b) $\dfrac{2}{7}$

    (c) $\dfrac{7}{2}, -\dfrac{7}{2}$ (d) $\dfrac{7}{2}$ 17._____

18. $p^2 = 9p$

    (a) $0, 9$ (b) $-9, 9$

    (c) 0   (d) 9   18._____

19. $t(t+2) = 80$

    (a) $8, -10$ (b) $3, 16$

    (c) $0, -2$ (d) $-8, 10$ 19._____

20.   Which of the following is **not** a solution
       to the equation $4x^3 = x$?

    (a) 0   (b) 2   (c) $-\dfrac{1}{2}$ (d) $\dfrac{1}{2}$ 20._____

For Exercises 21-23, solve the problem.

21. If a model rocket is propelled upward from ground level at an initial velocity of 96 feet per second , after t seconds, its height h is given by the formula $h = -16t^2 + 96t$. After how many seconds will its height be 144 feet?

(a) 24 seconds   (b) 9 seconds

(c) 3 seconds   (d) 1.5 seconds   21. _____

22. The length of the hypotenuse of a right triangle is 3 meters more than twice the length of the shorter leg. The longer leg is 7 meters longer than the shorter leg. Find the length of the hypotenuse.

(a) 11 meters   (b) 13 meters

(c) 17 meters   (d) 23 meters   22. _____

23. The automotive industry uses the equation $y = -0.04x^2 + 0.93x + 21$ to calculate trends in fuel economy, where y is the average miles per gallon and x represents the number of years after 1978. What will the average miles per gallon be in 2000?

(a) 26.4 mpg   (b) 40.58 mpg

(c) 22.0 mpg   (d) 22.1 mpg   23. _____

24. Solve the inequality

$$6t^2 + 13t - 5 < 0$$

(a) $t < -\dfrac{5}{2}$ or $t > \dfrac{1}{3}$ (b) $t < -1$ or $t > -\dfrac{5}{6}$

(c) $-\dfrac{5}{2} < t < \dfrac{1}{3}$ (d) $-1 < t < -\dfrac{5}{6}$ 24. _____

25. Graph the solutions of the inequality   25. _____

$$2c^2 - c - 10 \geq 0$$

(a)   (b)

(c)   (d)

CHAPTER SIX, FORM A   BEGINNING ALGEBRA

NAME_____   SECTION_____

---

1. Find any values for which
    $\dfrac{2y+3}{y^2-6y+5}$ is undefined.   1. _____

2. Find the numerical value of $\dfrac{5c-2}{2c^2-7c-15}$

    when (a) $c=-2$ and (b) $c=5$.   2. (a) _____

                                                    2. (b) _____

3. Write four rational expressions equivalent to:

    $$-\frac{4-3x}{5x+6}$$

                                                    3. (a) _____

                                                    3. (b) _____

                                                    3. (c) _____

                                                    3. (d) _____

For Exercises 4-5, write the rational expression in lowest terms.

4. $\dfrac{24a^3b^5}{-6a^2b}$ 4. _____

5. $\dfrac{3x^2-7x-6}{3x^2+11x+6}$ 5. _____

For Exercises 6-8, multiply or divide. Write the answer in lowest terms.

6. $\dfrac{9(x+y)}{5} \div \dfrac{10(x+y)}{3}$ 6. _____

7. $\dfrac{g^2+7g+10}{g^2+2g-15} \cdot \dfrac{4g-4}{3g+6}$ 7. _____

8. $\dfrac{2a^2 - 5a - 12}{a^2 - 9a + 18} \div \dfrac{4a^2 - 9}{a^2 - 10a + 24}$

8. _____

For Exercises 9-10, find the least common denominator for each list of fractions.

9. $\dfrac{5}{6b}, \dfrac{-7}{15b^3}, \dfrac{8}{3b^2}$

9. _____

10. $\dfrac{x-3}{4x^2 + 15x - 4}, \dfrac{2x+7}{8x^2 - 6x + 1}$

10. _____

For Exercises 11-12, rewrite each rational expression with the given denominator.

11. $\dfrac{9}{7x^3} = \dfrac{}{63x^5}$

11. _____

12. $\dfrac{8}{6m - 42} = \dfrac{}{12m - 84}$

12. _____

For Exercises 13-16, add or subtract. Write each answer in lowest terms.

13. $\dfrac{3x+1}{x-4} - \dfrac{x+9}{x-4}$

13. _____

14. $\dfrac{3}{a-3} + \dfrac{5}{5a - 15}$

14. _____

15. $\dfrac{w+6}{2w-5} - \dfrac{w-4}{5-2w}$

15. _____

16. $\dfrac{r-1}{r^2 + 2r - 8} - \dfrac{2-3r}{r^2 + 3r - 4}$

16. _____

For Exercises 17-18, simplify the complex fraction.

17. $\dfrac{\dfrac{10y^3}{3z^2}}{\dfrac{15y}{8z^3}}$

17. _____

18. $\dfrac{\dfrac{1}{c+1}-1}{\dfrac{1}{c+1}+1}$

18. _____

19. What are the two numbers that would have to be rejected as possible solutions for the equation

$$\frac{x-3}{x-2}-\frac{x+3}{x+2}=\frac{x-4}{x^2-4}$$

19. _____

For Exercises 20-21, solve the equation and check your answer.

20. $\dfrac{9}{x}=5-\dfrac{1}{x}$

20. _____

21. $\dfrac{1}{t+5}+\dfrac{3}{t-5}=\dfrac{10}{25-t^2}$

21. _____

22. Solve the formula $H=\dfrac{b}{D+d}$ for D.

22. _____

For Exercises 23-25, solve each problem.

23. A boat goes 6 miles per hour in still water. It takes as long to go 40 miles upstream as 80 miles downstream. Find the speed of the current.

23. _____

24. It takes Brad 2 hours to mow his lawn. It takes Gail 3 hours to mow the same lawn. How long will it take them to mow the lawn if they do the job together?

24. _____

25. If the temperature is constant, the pressure of a gas in a container varies inversely as the volume of the container. If the pressure is 16 pounds per square foot in a container of 300 cubic feet, what is the pressure of the gas if the container is compressed to 25 cubic feet?

25. _____

CHAPTER SIX, FORM B   BEGINNING ALGEBRA

NAME_____   SECTION_____

---

1.   Find any values for which
     $\dfrac{x+3}{x^2+5x+4}$ is undefined.   1. _____

2.   Find the numerical value of $\dfrac{5c-2}{2c^2-7c-15}$

     when (a) $c=2$ and (b) $c=-5$.   2. (a) _____

                                             2. (b) _____

3.   Write four rational expressions equivalent to:

     $$-\frac{7+8p}{5p-1}$$

                                             3. (a) _____

                                             3. (b) _____

                                             3. (c) _____

                                             3. (d) _____

For Exercises 4-5, write the rational expression in lowest terms.

4. $\dfrac{60a^3b^6}{-4a^2b}$ 4. _____

5. $\dfrac{12x^2+13x+3}{4x^2-13x-12}$ 5. _____

For Exercises 6-8, multiply or divide. Write the answer in lowest terms.

6. $\dfrac{9(a-b)}{5} \div \dfrac{25(a-b)}{6}$ 6. _____

7. $\dfrac{64g^2-9}{3g^2+15g} \div \dfrac{16g^2-2g-3}{2g^2+9g-5}$ 7. _____

8. $\dfrac{36x^2-1}{5x^2-45x}\cdot\dfrac{2x^2-17x-9}{12x^2+4x-1}$

8. _____

For Exercises 9-10, find the least common denominator for each list of fractions.

9. $\dfrac{25}{3x},\dfrac{7}{15x^2},\dfrac{18}{25x^4}$

9. _____

10. $\dfrac{x-3}{15x^2+2x-1},\dfrac{2x+7}{10x^2+13x-3}$

10. _____

For Exercises 11-12, rewrite each rational expression with the given denominator.

11. $\dfrac{5}{9c^3}=\dfrac{}{72c^9}$

11. _____

12. $\dfrac{8v}{7v+56}=\dfrac{}{21v+168}$

12. _____

For Exercises 13-16, add or subtract. Write each answer in lowest terms.

13. $\dfrac{4t+1}{t-8}-\dfrac{t+25}{t-8}$

13. _____

14. $\dfrac{3}{a-5}+\dfrac{22}{5a-25}$

14. _____

15. $\dfrac{w-2}{w-3}-\dfrac{w-7}{3-w}$

15. _____

16. $\dfrac{1}{r^2-25}-\dfrac{r+2}{r^2+4r-5}$

16. _____

For Exercises 17-18, simplify the complex fraction.

17. $\dfrac{4-\dfrac{1}{z^2}}{2-\dfrac{1}{z}}$

17. _____

18. $\dfrac{\dfrac{4b-3}{a^2 b^2}}{\dfrac{b+2}{ab^2}}$

18. _____

19. What are the two numbers that would have to be rejected as possible solutions for the equation

$$\frac{r-2}{r-3} - \frac{r+2}{r+3} = \frac{r-2}{r^2-9}$$

19. _____

For Exercises 20-21, solve the equation and check your answer.

20. $\dfrac{9}{x} - \dfrac{1}{4} = \dfrac{1}{5}$

20. _____

21. $\dfrac{6}{t} - \dfrac{6}{t-1} = -1$

21. _____

22. Solve the formula $H = \dfrac{2+b}{D+b}$ for b.

22. _____

For Exercises 23-25, solve each problem.

23. Rod's car travels 20 mph faster than Bobbi's van. In the same time that the car travels 280 miles, the van travels 200 miles. Find the speed of each vehicle.

23. _____

24. It takes Steve 4 hours to paint a room. It takes Diane 5 hours to do the same job. How long will it take them, working together, to paint the entire room?

24. _____

25. The current in a simple electrical outlet varies inversely as the resistance. If the current is 72 amps when the resistance is 8 ohms, find the resistance when the current is 48 amps.

25. _____

CHAPTER SIX, FORM C   **BEGINNING ALGEBRA**

NAME_____   SECTION_____

1. Find any values for which
    $\dfrac{b-7}{b^2+3b+2}$ is undefined.   1. _____

2. Find the numerical value of $\dfrac{5c-2}{4c^2+c-3}$

    when (a) $c=-1$ and (b) $c=-5$.   2. (a) _____

                                                       2. (b) _____

3. Write four rational expressions equivalent to:

    $$-\frac{7-6n}{5n+1}$$

                                                       3. (a) _____

                                                       3. (b) _____

                                                       3. (c) _____

                                                       3. (d) _____

For Exercises 4-5, write the rational expression in lowest terms.

4. $\dfrac{42f^3g^5}{-6f^2g^3}$ 4. _____

5. $\dfrac{4p^2+8p-5}{10p^2+21p-10}$ 5. _____

For Exercises 6-8, multiply or divide. Write the answer in lowest terms.

6. $\dfrac{7(g-h)}{4}\div\dfrac{28(g-h)}{14}$ 6. _____

7. $\dfrac{9g^2-25}{g^2-2g-8}\div\dfrac{3g^2+g-10}{6g^2-24g}$ 7. _____

8. $\dfrac{16x^2 - 9}{3x^2 - 21x} \cdot \dfrac{12x^2 - 90x + 42}{8x^2 - 10x + 3}$

8. _____

For Exercises 9-10, find the least common denominator for each list of fractions.

9. $\dfrac{11}{13t^5},\ \dfrac{9}{14t^2},\ \dfrac{37}{91t^4}$

9. _____

10. $\dfrac{4x - 3}{49x^2 + 7x - 2},\ \dfrac{x + 7}{14x^2 - 45x - 14}$

10. _____

For Exercises 11-12, rewrite each rational expression with the given denominator.

11. $\dfrac{3}{-7d^3} = \dfrac{}{84d^5}$

11. _____

12. $\dfrac{7v}{4v - 12} = \dfrac{}{8v - 24}$

12. _____

For Exercises 13-16, add or subtract. Write each answer in lowest terms.

13. $\dfrac{4h + 1}{h - 7} - \dfrac{h + 22}{h - 7}$

13. _____

14. $\dfrac{3}{4a - 20} + \dfrac{21}{2a - 10}$

14. _____

15. $\dfrac{q - 1}{q - 4} - \dfrac{q - 6}{4 - q}$

15. _____

16. $\dfrac{1}{j^2 - 9} - \dfrac{j + 2}{j^2 - 7j + 12}$

16. _____

For Exercises 17-18, simplify the complex fraction.

17. $\dfrac{16 - \dfrac{1}{t^2}}{4 + \dfrac{1}{t}}$

17. _____

18. $\dfrac{\dfrac{4b+8}{c^2 d^2}}{\dfrac{b+2}{c^4 d^2}}$

18. _____

19. What are the two numbers that would have to be rejected as possible solutions for the equation

$$\frac{k-7}{k-6} - \frac{k+7}{k+6} = \frac{k-7}{k^2-36}$$

19. _____

For Exercises 20-21, solve the equation and check your answer.

20. $\dfrac{4}{p} + \dfrac{1}{8} = \dfrac{1}{6}$

20. _____

21. $\dfrac{24}{t} - \dfrac{24}{t-2} = -2$

21. _____

22. Solve the formula $A = P(1 + rt)$ for t.

22. _____

For Exercises 23-25, solve each problem.

23. The speed of a passenger train is 16 mph faster than the speed of a freight train. The freight train travels 340 miles in the same time that it takes the passenger train to travel 420 miles. Find the speed of each train.

23. _____

24. Doug can type a report in 4 hours. Jennifer can type the same report in 2 hours. How long would it take the two of them, working together, to type the report?

24. _____

25. The current in a simple electrical outlet varies inversely as the resistance. If the current is 75 amps when the resistance is 8 ohms, find the current when the resistance is 10 ohms.

25. _____

CHAPTER SIX, FORM D   BEGINNING ALGEBRA

NAME_____   SECTION_____

1.   Find any values for which
$\dfrac{d+4}{d^2-9d+20}$ is undefined.   1. _____

2.   Find the numerical value of $\dfrac{c+8}{3c^2+4c-4}$

when (a) $c=-2$ and (b) $c=5$.   2. (a) _____

2. (b) _____

3.   Write four rational expressions equivalent to:

$$-\frac{1-6s}{5s+4}$$

3. (a) _____

3. (b) _____

3. (c) _____

3. (d) _____

For Exercises 4-5, write the rational expression in lowest terms.

4. $\dfrac{40g^8h^5}{-8g^6h^3}$ 4. _____

5. $\dfrac{3p^2+2p-16}{3p^2-4p-4}$ 5. _____

For Exercises 6-8, multiply or divide. Write the answer in lowest terms.

6. $\dfrac{9(g-j)}{5}\div\dfrac{15(g-j)}{27}$ 6. _____

7. $\dfrac{3g^2-12}{g^2-g-6}\div\dfrac{2g-4}{g^2-6g+9}$ 7. _____

8. $\dfrac{2z^2+11z-21}{2z^2+7z-49}\cdot\dfrac{6z^2-13z-5}{6z^2-7z-3}$

8. _____

For Exercises 9-10, find the least common denominator for each list of fractions.

9. $\dfrac{17}{4t^5},\dfrac{11}{36t^2},\dfrac{37}{45t^4}$

9. _____

10. $\dfrac{4x-3}{9x^2-16},\dfrac{x+7}{6x^2+x-12}$

10. _____

For Exercises 11-12, rewrite each rational expression with the given denominator.

11. $\dfrac{3b}{13d^3}=\dfrac{}{52b^2d^5}$

11. _____

12. $\dfrac{7x}{5x-10}=\dfrac{}{20x-40}$

12. _____

For Exercises 13-16, add or subtract. Write each answer in lowest terms.

13. $\dfrac{4h+1}{h-3}-\dfrac{h+10}{h-3}$

13. _____

14. $\dfrac{5}{3a+15}+\dfrac{27}{6a+30}$

14. _____

15. $-\dfrac{3}{6-7q}-\dfrac{8q+2}{7q-6}$

15. _____

16. $\dfrac{2j}{j^2+5j+6}+\dfrac{j+1}{j^2+2j-3}$

16. _____

For Exercises 17-18, simplify the complex fraction.

17. $\dfrac{81-\dfrac{1}{y^2}}{9+\dfrac{1}{y}}$

17. _____

18. $\dfrac{\dfrac{5a+15}{s^2t^2}}{\dfrac{a+3}{s^4t^6}}$

18. _____

19.   What are the two numbers that would have to
be rejected as possible solutions for the equation

$$\frac{k}{k-5} - \frac{k}{k+5} = \frac{k-7}{k^2-25}$$

19. _____

For Exercises 20-21, solve the equation and check your answer.

20. $\dfrac{7}{3p-1} = \dfrac{6}{2p+3}$

20. _____

21. $\dfrac{v}{v-3} = \dfrac{1}{v-3} + 2$

21. _____

22.   Solve the formula $L = \dfrac{76}{4G+H}$ for G.

22. _____

For Exercises 23-25, solve each problem.

23.   Steve can paddle his canoe 8 miles upstream
against the current in the same time it would
take him to paddle 12 miles downstream. The
speed of the current is 3 miles per hour. How
fast can Steve paddle in still water?

23. _____

24.   Mike can tune up a car in 4 hours. His friend
Rick can do the job in 6 hours. How long would
it take them if they worked together?

24. _____

25.   The current in a simple electrical outlet varies
inversely as the resistance. If the current is 45
amps when the resistance is 12 ohms, find the
current when the resistance is 75 ohms.

25. _____

CHAPTER SIX, FORM E   **BEGINNING ALGEBRA**

NAME_____   SECTION_____

---

1.   Find any values for which $\dfrac{x+5}{x^3 - 7x^2 + 12x}$ is undefined.

   (a) 0, 3, 4   (b) 0, $-3, -4$

   (c) 0, 3, 4, -5 (d) 0, -5 1._____

2.   Find the numerical value of $\dfrac{c-2}{2c^2 - 7c - 15}$ when $c = -2$.

   (a) 0   (b) $\dfrac{4}{9}$ (c) $-\dfrac{4}{7}$ (d) Undefined   2._____

3.   Pick the rational expression equivalent to the following:

$$-\frac{4-3x}{5x+6}$$

   (a) $\dfrac{-4-3x}{5x+6}$ (b) $\dfrac{-4+3x}{-5x-6}$

   (c) $\dfrac{3x-4}{-5x-6}$ (d) $\dfrac{3x-4}{5x+6}$ 3._____

For Exercises 4-5, write the rational expression in lowest terms.

4. $\dfrac{36c^5 d^4}{96cd^8}$

   (a) $\dfrac{c^4}{3d^4}$ (b) $\dfrac{3c^4 d^4}{8}$ (c) $\dfrac{c^4 d^4}{3}$ (d) $\dfrac{3c^4}{8d^4}$ 4._____

5. $\dfrac{1-w}{w^2 - 1}$

   (a) $-(w+1)$ (b) $w+1$ (c) $-\dfrac{1}{w+1}$ (d) $\dfrac{1}{w+1}$ 5._____

For Exercises 6-8, multiply or divide. Write the answer in lowest terms.

6. $\dfrac{9(t+d)}{5} \div \dfrac{10(t+d)}{6}$

    (a) $3(t+d)^2$ (b) $\dfrac{25}{27}$

    (c) $\dfrac{27}{25}$ (d) $\dfrac{27}{25(t+d)}$ 6. _____

7. $\dfrac{c^2+7c+10}{c^2+2c-15} \cdot \dfrac{4c+12}{3c+15}$

    (a) $\dfrac{4(c+2)(c+3)}{3(c-3)(c+5)}$ (b) $\dfrac{4(c+2)(c+5)}{3}$

    (c) $\dfrac{4(c+2)(c+3)}{3(c-3)}$ (d) $\dfrac{4(c+2)}{3}$ 7. _____

8. $\dfrac{a^2-b^2}{2a-2b} \div \dfrac{8}{a+b}$

    (a) $\dfrac{1}{16}$ (b) $\dfrac{(a+b)^2}{16}$

    (c) $\dfrac{(a-b)(a+b)}{16}$ (d) $\dfrac{a+b}{8}$ 8. _____

For Exercises 9-10, find the least common denominator for each list of fractions.

9. $\dfrac{5}{6ab^3},\ \dfrac{-7}{2a^2},\ \dfrac{8}{4b^2}$

    (a) $2a$ (b) $12a^2b^3$

    (c) $48a^3b^5$ (d) $24a^3b^3$ 9. _____

10. $\dfrac{x-3}{2x^2-9x+4},\ \dfrac{2x+7}{2x^2+9x-5}$

    (a) $(2x-1)^2(x-4)(x+5)$ (b) $(x-4)(x+5)$

    (c) $(2x-1)(x-4)(x+5)$ (d) $2x-1$ 10. _____

For Exercises 11-12, rewrite each rational expression with the given denominator.

11. $\dfrac{9}{7z^3} = \dfrac{}{63z^{12}}$

   (a) $\dfrac{18z^9}{63z^{12}}$ (b) $\dfrac{81z^9}{63z^{12}}$ (c) $\dfrac{81z^4}{63z^{12}}$ (d) $\dfrac{18z^4}{63z^{12}}$ 11. _____

12. $\dfrac{8}{4p+16} = \dfrac{}{4p^2-64}$

   (a) $\dfrac{p+4}{4p^2-64}$ (b) $\dfrac{8p+32}{4p^2-64}$

   (c) $\dfrac{p-4}{4p^2-64}$ (d) $\dfrac{8p-32}{4p^2-64}$ 12. _____

For Exercises 13-16, add or subtract. Write each answer in lowest terms.

13. $\dfrac{3r+1}{r-4} - \dfrac{r+9}{r-4}$

   (a) $\dfrac{2r+10}{r-4}$ (b) $\dfrac{4r-8}{r-4}$

   (c) $\dfrac{4r+10}{r-4}$ (d) 2   13. _____

14. $\dfrac{t+2}{3t-9} + \dfrac{3-2t}{5t-15}$

   (a) $\dfrac{19-t}{15(t-3)}$ (b) $\dfrac{5-t}{15(t-3)}$

   (c) $\dfrac{3t+8}{15(t-3)}$ (d) $\dfrac{5-t}{8(t-3)}$ 14. _____

15. $\dfrac{3w+2}{4w-5} - \dfrac{w-1}{5-4w}$

   (a) $\dfrac{w+1}{w-5}$ (b) $\dfrac{2w+3}{4w-5}$

   (c) $\dfrac{4w+1}{4w-5}$ (d) $\dfrac{2w+1}{4w-5}$ 15. _____

16. $\dfrac{r-1}{r^2-r-2}+\dfrac{r-2}{r^2-4}$

(a) $\dfrac{2r^2-4}{(r-2)(r+2)(r+1)}$ (b) $r-1$

(c) $\dfrac{2}{(r+2)(r+1)}$ (d) $\dfrac{2}{r+1}$ 16. _____

For Exercises 17-18, simplify the complex fraction.

17. $\dfrac{\dfrac{3-t}{6}}{\dfrac{9-t^2}{15}}$

(a) $\dfrac{5}{2}(3+t)$ (b) $\dfrac{5}{2(3+t)}$ (c) $\dfrac{5}{2(3-t)^3}$ (d) $\dfrac{5}{2(3-t)}$ 17. _____

18. $\dfrac{\dfrac{x}{3}-y}{y-\dfrac{x}{3}}$

(a) $\left(\dfrac{x}{3}-y\right)^2$ (b) $\dfrac{x-y}{x+y}$ (c) -1 (d) $\dfrac{x-3y}{x+3y}$ 18. _____

19. What are the numbers that would have to be rejected as possible solutions for the equation

$$\dfrac{a-3}{a-2}-\dfrac{a+3}{a+2}=\dfrac{a-4}{a^2-4}$$

(a) $3,-3,4$ (b) $2,-2,4$ (c) $2,-2$ (d) $0,2,-2$ 19. _____

For Exercises 20-21, solve the equation.

20. $\dfrac{6}{x}=\dfrac{5}{x+1}$

(a) -1 (b) -6 (c) $\dfrac{1}{11}$ (d) No Solution   20. _____

21. $\dfrac{2}{t+2} - \dfrac{7}{t-3} = \dfrac{5}{t^2 - t - 6}$

   (a) -5 (b) 5 (c) $\dfrac{3}{5}$ (d) $-\dfrac{3}{5}$

   21. _____

22. Solve the formula $H = \dfrac{b}{D+d}$ for D.

   (a) $D = \dfrac{b-d}{H}$ (b) $D = \dfrac{b}{H} - d$

   (c) $D = b - Hd - H$ (d) $D = Hb - d$

   22. _____

For Exercises 23-25, solve each problem.

23. Steve can paddle his canoe 2 miles upstream against
    the current in the same time it would take him to
    paddle 6 miles downstream. Steve can paddle two
    mph in still water. What is the speed of the current?

   (a) 4 mph   (b) 3 mph   (c) -1 mph   (d) 1 mph

   23. _____

24. It takes Brad 2 hours to mow his lawn. It takes
    Gail 3 hours to mow the same lawn. How long
    will it take them to mow the lawn if they do the
    job together?

   (a) $2\dfrac{1}{2}$ hours   (b) $1\dfrac{1}{5}$ hours   (c) $1\dfrac{1}{6}$ hours   (d) $\dfrac{5}{6}$ hours

   24. _____

25. The current in a simple electrical circuit varies inversely
    as the resistance. If the current is 72 amps when the
    resistance is 8 ohms, find the resistance when the current
    is 48 amps.

   (a) 12 ohms   (b) $\dfrac{1}{12}$ ohms   (c) 432 ohms   (d) $5\dfrac{1}{3}$ ohms

   25. _____

CHAPTER SIX, FORM F   BEGINNING ALGEBRA

NAME_____   SECTION_____

1.   Find any values for which $\dfrac{p+5}{p^2-3p+2}$ is undefined.

   (a) 1, 2, -5 (b) 1, 2
   (c) -5 (d) $-1,\ -2$ 1._____

2.   Find the numerical value of $\dfrac{d+16}{2d^2-7d-15}$ when $d=-2$.

   (a) 2   (b) $-\dfrac{14}{9}$ (c) $-\dfrac{14}{37}$ (d) $\dfrac{18}{7}$ 2._____

3.   Pick the rational expression equivalent to the following:

$$-\frac{3-5t}{9t+7}.$$

   (a) $\dfrac{-3-5t}{9t+7}$ (b) $\dfrac{3-5t}{-9t-7}$

   (c) $\dfrac{-3+5t}{-9t-7}$ (d) $\dfrac{-3-5t}{9t+7}$ 3._____

For Exercises 4-5, write the rational expression in lowest terms.

4. $\dfrac{35b^3c^2}{10a^3b^9}$

   (a) $\dfrac{7a^3c^2}{2b^6}$ (b) $\dfrac{7c^2}{2a^3b^6}$ (c) $\dfrac{7}{2a^3b^6c^2}$ (d) $\dfrac{7b^6c^2}{2a^3}$ 4._____

5. $\dfrac{w^2-1}{w-1}$

   (a) $(w+1)$ (b) $w-1$ (c) $\dfrac{1}{w+1}$ (d) $\dfrac{1}{w-1}$ 5._____

For Exercises 6-8, multiply or divide. Write the answer in lowest terms.

6. $\dfrac{3t}{t+4} \div \dfrac{4t}{t+4}$

(a) $\dfrac{3}{4}$ (b) $\dfrac{4}{3}$

(c) $\dfrac{3t}{t+4}$ (d) $\dfrac{12t^2}{(t+4)^2}$ 6. _____

7. $\dfrac{c^2+9c+20}{5c-10} \cdot \dfrac{-10c+20}{c^2-25}$

(a) $\dfrac{-2(c+4)(c+5)}{(c-5)^2}$ (b) $\dfrac{-2(c-4)}{c+5}$

(c) $\dfrac{-2(c+4)(c+5)}{(c-5)(c-2)}$ (d) $\dfrac{-2(c+4)}{c-5}$ 7. _____

8. $\dfrac{2a+1}{16-a^2} \div \dfrac{4a+2}{a-4}$

(a) $-\dfrac{1}{8+2a}$ (b) $\dfrac{2(a-4)}{16-a^2}$

(c) $\dfrac{a-4}{2(16-a^2)}$ (d) $\dfrac{1}{4a+2}$ 8. _____

For Exercises 9-10, find the least common denominator for each list of fractions.

9. $\dfrac{-5}{4z^3},\ \dfrac{7}{36z^5},\ \dfrac{-13}{45z^2}$

(a) $180z^5$ (b) $72z^5$

(c) z^2 (d) $90z^5$ 9. _____

10. $\dfrac{q-3}{9q^2-16},\ \dfrac{2q+7}{6q^2+q-12}$

(a) $(3q+4)(3q-4)(2q-3)$ (b) $(3q-4)^2(2q+3)$

(c) $(3q+4)(3q-4)(2q+3)$ (d) $(3q+4)^2(2q+3)$ 10. _____

For Exercises 11-12, rewrite each rational expression with the given denominator.

11. $\dfrac{6b}{13a^2} = \dfrac{}{52a^4b^3}$

(a) $\dfrac{4a^2b^3}{52a^4b^3}$ (b) $\dfrac{24a^3b^4}{52a^4b^3}$ (c) $\dfrac{24a^2b^4}{52a^4b^3}$ (d) $\dfrac{24a^2b}{52a^4b^3}$ 11. _____

12. $\dfrac{8d}{5d-10} = \dfrac{}{20d-40}$

(a) $\dfrac{4}{20d-40}$ (b) $\dfrac{20}{20d-40}$

(c) $\dfrac{4d}{20d-40}$ (d) $\dfrac{32d}{20d-40}$ 12. _____

For Exercises 13-16, add or subtract. Write each answer in lowest terms.

13. $\dfrac{8-3s}{s^3} + \dfrac{19+s}{s^3}$

(a) $\dfrac{27-2s}{s^3}$ (b) $\dfrac{11-4s}{s^3}$

(c) $\dfrac{27-4s}{s^3}$ (d) $\dfrac{27-2s}{2s^3}$ 13. _____

14. $\dfrac{x-5}{x-1} - \dfrac{x-3}{1-x}$

(a) $\dfrac{2(x+4)}{1-x}$ (b) $\dfrac{-2(x-4)}{1-x}$

(c) $\dfrac{-2}{x-1}$ (d) $\dfrac{-8}{x-1}$ 14. _____

15. $\dfrac{3w}{w+2} - \dfrac{5}{w-3}$

(a) $\dfrac{3w-5}{(w+2)(w-3)}$ (b) $\dfrac{3w^2-8w-10}{(w+2)(w-3)}$

(c) $\dfrac{3w^2-14w+10}{(w+2)(w-3)}$ (d) $\dfrac{3w^2-14w-10}{(w+2)(w-3)}$ 15. _____

16. $\dfrac{2y}{y^2 - 3y - 10} + \dfrac{3}{y^2 - 2y - 8}$

(a) $\dfrac{3y^2 + 7y + 6}{(y-5)(y-4)(y+2)}$ (b) $\dfrac{(2y-3)(y+3)}{(y-5)(y-4)(y+2)}$

(c) $\dfrac{2y^2 - 5y - 15}{(y-5)(y-4)(y+2)}$ (d) $\dfrac{2y+3}{(2y-9)(y+2)}$ 16. _____

For Exercises 17-18, simplify the complex fraction.

17. $\dfrac{\dfrac{4-d}{8}}{\dfrac{16-d^2}{24}}$

(a) $\dfrac{3}{4+d}$ (b) $\dfrac{3}{4-d}$ (c) $\dfrac{64-d^3}{3}$ (d) $\dfrac{4+d}{3}$ 17. _____

18. $\dfrac{\dfrac{x+3}{x}}{\dfrac{1}{x} + \dfrac{1}{5}}$

(a) $\dfrac{(x+3)(x+5)}{5x^2}$ (b) $\dfrac{(x+3)(x+5)}{5}$

(c) $\dfrac{5(x+3)}{x+5}$ (d) $\dfrac{x+3}{5(x+5)}$ 18. _____

19.   What are the numbers that would have to be rejected as possible solutions for the equation

$$\dfrac{a-3}{a+3} - \dfrac{a+3}{a-1} = \dfrac{a-4}{a^2 + 2a - 3}$$

(a) $-1, 0$ (b) $-3, 0$ (c) $1, -3$ (d) $0, 1, -3$ 19. _____

For Exercises 20-21, solve the equation.

20. $\dfrac{19}{j} - \dfrac{5}{8} = \dfrac{1}{6}$

(a) 14   (b) 17   (c) 21   (d) 24   20. _____

21. $\dfrac{6}{m} = \dfrac{10}{m-3} + 13$

(a) $2, \dfrac{4}{13}$ (b) $-2, \dfrac{13}{9}$ (c) $2, \dfrac{9}{13}$ (d) $-2, 3$ 21. _____

22. Solve the formula $A = \dfrac{1}{2}h(b_1 + b_2)$ for h.

(a) $h = \dfrac{b_1 + b_2}{A}$ (b) $h = \dfrac{2A}{b_1 + b_2}$

(c) $h = \dfrac{2A}{b_1} + \dfrac{2A}{b_2}$ (d) $h = \dfrac{b_1 + b_2}{2A}$ 22. _____

For Exercises 23-25, solve each problem.

23. A boat can go 20 miles against the wind in the same time
     that it can go 60 miles with the wind. The wind is blowing
     4 mph. Find the speed of the boat with no wind.

(a) 4 mph   (b) 8 mph   (c) 40 mph   (d) 80 mph   23. _____

24. Connie can type a report in 2 hours. Jerry can type the
     same report in 4 hours. How long will it take them
     to type the report if they do the job together?

(a) $\dfrac{8}{3}$ hours   (b) $\dfrac{4}{3}$ hours   (c) 3 hours   (d) 6 hours   24. _____

25. The current in a simple electrical circuit varies inversely
     as the resistance. If the current is 72 amps when the
     resistance is 8 ohms, find the resistance when the current
     is 36 amps.

(a) 16 ohms   (b) $\dfrac{1}{16}$ ohms   (c) 320 ohms (d) 4 ohms   25. _____

CHAPTER SEVEN, FORM A   **BEGINNING ALGEBRA**

NAME_____   SECTION_____

For Exercises 1-3, write an equation in slope-intercept form.

1.   Through $(-3,-1)$ with slope -2.   1._____

2.

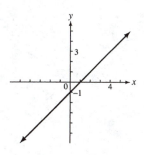

2._____

3.

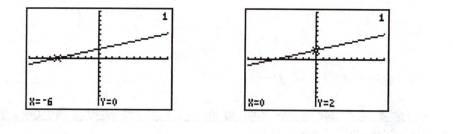

3._____

4.   Write the equation $y = -\dfrac{3}{4}x - 7$ in standard form.   4._____

In Exercises 5-6, graph the linear inequality.

5. $3x - y \geq 6$

5.

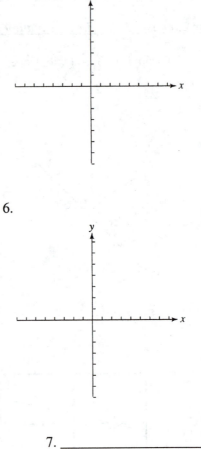

6. $x + 2y \leq -3$

6.

7.   If $f(x) = -5x + 7$, find $f(-3)$.

7. _____

In Exercises 8-11, decide whether each of the following represents a function. If it does, give the domain and the range.

8.

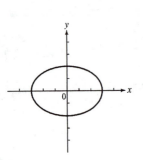

8. _____

9. $x = y^2 + 2$ 9. _____

10. $y = -3x - 4$ 10. _____

11. $\{(1,2),(1,3),\ (2,3),(3,4),(4,5),(6,7),(7,8)\}$ 11. _____

12.   Is the following statement true or false?   12. _____
                                                   _____
      All lines are graphs of functions.   _____
                                                   _____
      Justify your answer.   _____

For Exercises 13-16, the table gives data on the number of automobiles leased nationally between 1986 and 1992.

1986	11654
1987	14516
1988	17638
1989	19493
1990	17782
1991	20063
1992	24589

13.   Does this set define a function?
      If so what is $f(1991)$?   13. _____

14.   If the set of data defines a function f
      and $f(x) = 14516$, what is x?   14. _____

15.   The data points are closely approximated
      by the line $y = 1848.2x - 3658338.8$.
      What is the slope of this line?   15. _____

16.   What does the slope of the line tell you
      about the annual change in the number
      of automobiles leased each year?   16. _____

CHAPTER SEVEN, FORM B   BEGINNING ALGEBRA

NAME_____   SECTION_____

For Exercises 1-3, write an equation in slope-intercept form.

1.   Through $(-1, 2)$ with slope -4.   1. _____

2.

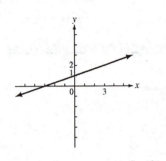

                                                2. _____

3.

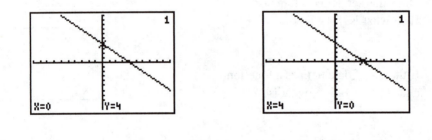

                                                3. _____

4.   Write the equation $y = -\dfrac{4}{3}x + 6$ in standard form.   4. _____

__In Exercises 5-6, graph the linear inequality.__

5. $x - 4y \leq 8$

5.

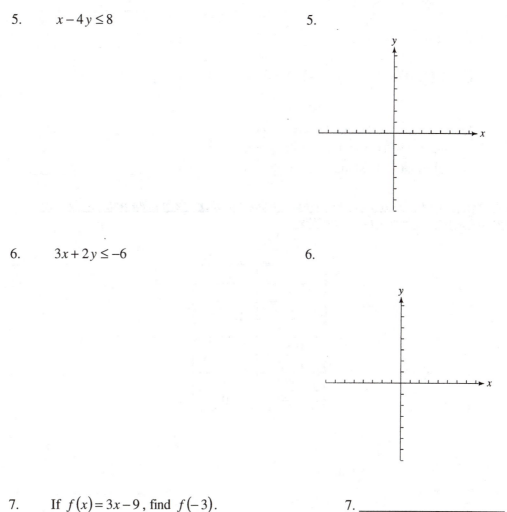

6. $3x + 2y \leq -6$

6.

7.   If $f(x) = 3x - 9$, find $f(-3)$.

7. _____

__In Exercises 8-11, decide whether each of the following represents a function. If it does, give the domain and the range.__

8.

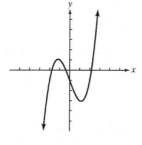

8. _____

9. $y = x^2 + 2$ 9. _____

10. $x = -3$ 10. _____

11. $\{(1,2), (2,3), (3,4), (4,5), (6,7), (7,8)\}$ 11. _____

12. Suppose that f is a function whose graph
    is a line. If $f(5) = 3$ and the y-intercept is
    $(0, -3)$, what is the slope of the line?   12. _____

*__For Exercises 13-16, the table gives data on the number of children nationally who
contracted HIV between 1983 and 1990.__*

1983	25
1984	76
1985	160
1986	203
1987	399
1988	618
1989	644
1990	801

13. Does this set define a function?
    If so what is $f(1987)$?   13. _____

14. If the set of data defines a function f
    and $f(x) = 801$, what is x?   14. _____

15. The data points are closely approximated
    by the line $y = 113.6x - 225326.1$.
    What is the slope of this line?   15. _____

16. What does the slope of the line tell you
    about the annual change in the number
    of children who contract HIV each year?   16. _____

CHAPTER SEVEN, FORM C   **BEGINNING ALGEBRA**

NAME_____   SECTION_____

For Exercises 1-3, write an equation in slope-intercept form.

1.   Through $(-4, -2)$ with slope -1.   1. _____

2.

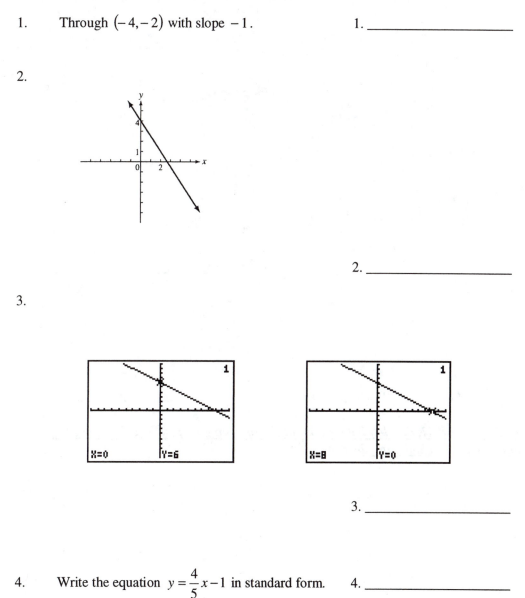

          2. _____

3.

          3. _____

4.   Write the equation $y = \dfrac{4}{5}x - 1$ in standard form.   4. _____

In Exercises 5-6, graph the linear inequality.

5. $2x - 5y \leq 10$

5.

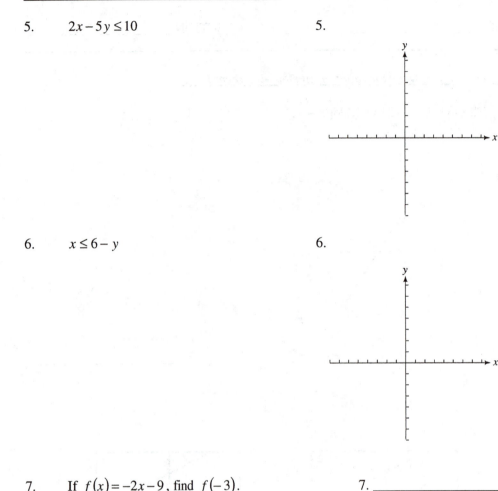

6. $x \leq 6 - y$

6.

7.   If $f(x) = -2x - 9$, find $f(-3)$.

7. _____

In Exercises 8-11, decide whether each of the following represents a function. If it does, give the domain and the range.

8.

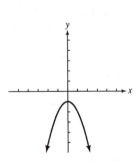

8. _____

9. $y = 2x^2 - 1$ 9. _____

10. $y = -3$ 10. _____

11. $\{(1,2), (2,1), (3,4), (4,3), (6,7), (7,6)\}$ 11. _____

12.   Suppose that f is a function whose graph
       is a line. If $f(3) = -5$ and the y-intercept
       is $(0, -3)$, what is the slope of the line?   12. _____

***For Exercises 13-16, the table gives data on the number (in millions) of fish caught
nationally between 1985 and 1992.***

1985	15.7
1986	11.9
1987	7.1
1988	6.0
1989	5.2
1990	6.0
1991	4.0
1992	3.1

13.   Does this set define a function?
       If so what is $f(1989)$?   13. _____

14.   If the set of data defines a function f
       and $f(x) = 6.0$, what is x?   14. _____

15.   The data points are closely approximated
       by the line $y = -1.58x + 3148.99$.
       What is the slope of this line?   15. _____

16.   What does the slope of the line tell you
       about the annual change in the number
       of fish caught each year?   16. _____

CHAPTER SEVEN, FORM D **BEGINNING ALGEBRA**

NAME_____ SECTION_____

_____

For Exercises 1-3, write an equation in slope-intercept form.

1. Through $(-2, -3)$ with slope -1. 1. _____

2.

2. _____

3.

3. _____

4. Write the equation $y = -\dfrac{1}{5}x + 9$ in standard form. 4. _____

In Exercises 5-6, graph the linear inequality.

5. $3x + 4y \leq 6$

5.

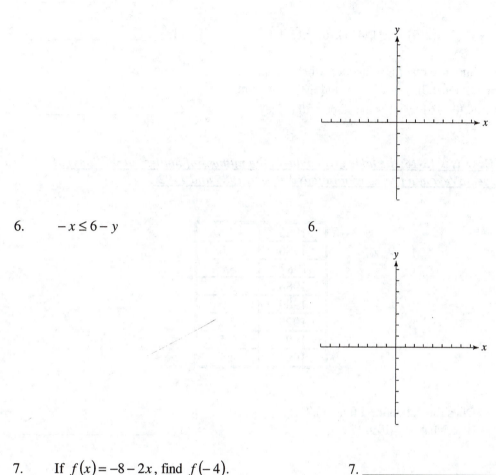

6. $-x \leq 6 - y$

6.

7.   If $f(x) = -8 - 2x$, find $f(-4)$.

7. _____

In Exercises 8-11, decide whether each of the following represents a function. If it does, give the domain and the range.

8.

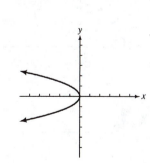

8. _____

9. $x = 2y^2 + 3$ 9. _____

10. $2y = -4$ 10. _____

11. $\{(1,0), (2,0), (3,4), (4,4), (6,1), (7,1)\}$ 11. _____

12.   Suppose that f is a function whose graph
      is a line. If $f(-2) = -5$ and the y-intercept
      is $(0,-6)$, what is the slope of the line?   12. _____

For Exercises 13-16, the table gives data on the number of bottles (in millions) of VitaMeataVegamin produced annually between 1985 and 1992.

1985	3.5
1986	3.0
1987	3.6
1988	3.9
1989	4.3
1990	3.9
1991	3.5
1992	4.1

13.   Does this set define a function?
      If so what is $f(1992)$?   13. _____

14.   If the set of data defines a function f
      and $f(x) = 3.9$, what is x?   14. _____

15.   The data points are closely approximated
      by the line $y = 0.08x - 155.25$.
      What is the slope of this line?   15. _____

16.   What does the slope of the line tell you
      about the number of bottles
      VitaMeataVegamin produced each year?   16. _____

CHAPTER SEVEN, FORM E   BEGINNING ALGEBRA

NAME_____   SECTION_____

For Exercises 1-3, write an equation in slope-intercept form.

1.   Through $(-2,-3)$ with slope -4.

    (a) $y = -4x - 14$ (b) $y = -4x - 11$

    (c) $y = -4x - 10$ (d) $y = -4x - 5$ 1. _____

2. $m = -3;$ $y-\text{intercept } (0,-2)$

    (a) $y = -3x - 2$ (b) $y = -3x + 2$

    (c) $y = \dfrac{3}{2}x$ (d) $y = -\dfrac{3}{2}x$ 2. _____

3.   Through $(4,-4)$ and $(-2,2)$.

    (a) $y = 4x - 1$ (b) $y = -2x + 4$

    (c) $y = -x + 4$ (d) $y = -x$ 3. _____

4.   Write the equation $y = -\dfrac{1}{5}x + 9$ in standard form.

    (a) $x + 5y = 45$ (b) $x + 5y = 9$

    (c) $x - 5y = 45$ (d) $x - 5y = 9$ 4. _____

In Exercises 5-6, graph the linear inequality.

5. $x + 2y \le 4$ 5. _____

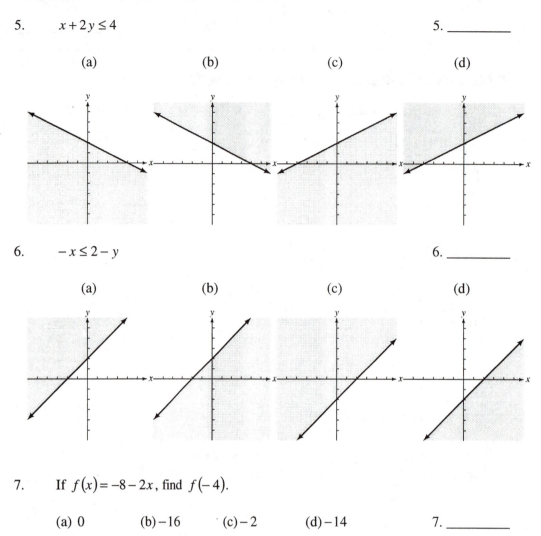

| (a) | (b) | (c) | (d) |

6. $-x \le 2 - y$ 6. _____

| (a) | (b) | (c) | (d) |

7.   If $f(x) = -8 - 2x$, find $f(-4)$.

(a) 0   (b) -16 (c) -2 (d) -14 7. _____

In Exercises 8-10, decide whether each of the following represents a function.

8.   Which one of the following equations
      is *not* that of a function?

(a) $3x - y = 2$ (b) $3x - y = 0$
(c) $3x - y^2 = 2$ (d) $3x^2 - y = 2$ 8. _____

9.   Which one of the following sets of
      ordered pairs is *not* that of a function?

      (a) $\{(1,1),\ (2,1),(3,1),(4,1)\}$
      (b) $\{(1,1),\ (2,2),(3,3),(4,5)\}$
      (c) $\{(1,0),\ (1,1),(2,0),(2,2)\}$
      (d) $\{(1,2),(3,4),(5,6),(8,9)\}$ 9. _____

10.   Which one of the following graphs   10. _____
      is *not* that of a function?

      (a)   (b)   (c)   (d)

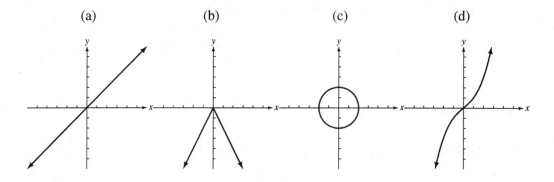

11.   Find the range of $y = |x| - 4$.

      (a) {all real numbers}
      (b) {all real numbers between –4 and 4}
      (c) {all real numbers greater than or equal to –4}
      (d) {all real numbers less than or equal to 4}   11. _____

12.   Suppose that f is a function whose graph
      is a line. If $f(-2)=-5$ and the y-intercept
      is $(0,-6)$, what is the slope of the line?

      (a) $-\dfrac{1}{2}$ (b) $-\dfrac{11}{2}$ (c) $-\dfrac{4}{5}$ (d) $\dfrac{4}{5}$ 12. _____

For Exercises 13-16, the table gives data on the number of bottles (in millions) of VitaMeataVegamin produced annually between 1985 and 1992.

1985	3.5	1989	4.3
1986	3.0	1990	3.9
1987	3.6	1991	3.5
1988	3.9	1992	4.1

13.   If this set defines a function, then what is $f(1992)$?

(a) 3.5   (b) 4.1   (c) 4.3   (d) Not a function   13. _____

14.   If $f(x) = 3.5$, what is x?

(a) 1988   (b) 1989   (c) 1990   (d) 1991   14. _____

15.   The data points are closely approximated by the line $y = 0.08x - 155.25$. What is the slope of this line?

(a) 0.08   (b) –155.25
(c) 155.25   (d) –0.08   15. _____

16.   What does the slope of the line tell you about the number of bottles VitaMeataVegamin produced each year?

(a) Production is constant.
(b) Production is increasing slightly each year.
(c) Production is decreasing slightly each year.
(d) The slope gives us no information about production.   16. _____

CHAPTER SEVEN, FORM F   **BEGINNING ALGEBRA**

NAME_____   SECTION_____

For Exercises 1-3, write an equation in slope-intercept form.

1.   Through $(-3,-4)$ with slope -5.

(a) $y = -5x - 19$ (b) $y = -5x + 11$
(c) $y = -5x + 19$ (d) $y = -5x - 11$ 1. _____

2. $m = -\dfrac{2}{3}$; y-intercept $(0,-4)$

(a) $y = -2x + 12$ (b) $y = -\dfrac{2}{3}x + 4$

(c) $y = -\dfrac{2}{3}x - 4$ (d) $y = -\dfrac{2}{3}x - \dfrac{4}{3}$ 2. _____

3.   Through $(-4,5)$ and $(-7,2)$.

(a) $y = -x - 5$ (b) $y = x + 9$
(c) $y = \dfrac{3}{11}x - \dfrac{1}{11}$ (d) $y = \dfrac{3}{11}x + \dfrac{43}{11}$ 3. _____

4.   Write the equation $y = -\dfrac{3}{5}x - 1$ in standard form.
(a) $3x + 5y = -5$ (b) $3x + 5y = -1$
(c) $3x - 5y = -5$ (d) $3x + y = -5$ 4. _____

In Exercises 5-6, graph the linear inequality.

5. $x - 2y \le 6$ 5. _____

| (a) | (b) | (c) | (d) |

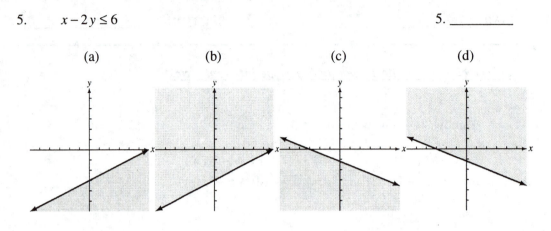

6. $-2x \le 3 - y$ 6. _____

| (a) | (b) | (c) | (d) |

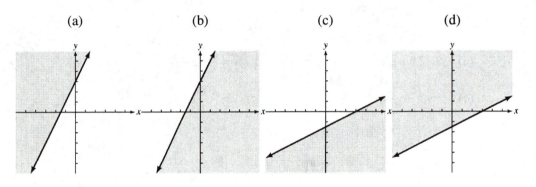

7.   If $f(x) = -4 - 2x$, find $f(-2)$.

(a) 0   (b) -8 (c) -2 (d) -26 7. _____

In Exercises 8-10, decide whether each of the following represents a function.

8.   Which one of the following equations
      is *not* that of a function?

(a) $3x = 2$ (b) $2x - 3y = 0$
(c) $3x - 2y = 0$ (d) $3y = 2$ 8. _____

9.   Which one of the following sets of
        ordered pairs is *not* that of a function?

        (a) $\{(1,1),\ (2,2),(3,3),(4,0)\}$
        (b) $\{(1,2),(3,4),(5,6),(8,8)\}$
        (c) $\{(1,1),\ (2,1),(3,1),(4,1)\}$
        (d) $\{(1,0),\ (1,1),(2,0),(2,2)\}$ 9. _____

10.   Which one of the following graphs   10. _____
        is *not* that of a function?

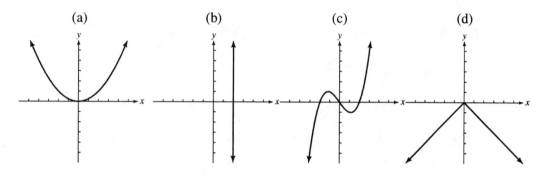

        (a)   (b)   (c)   (d)

11.   Find the range of $y = 5 - |x|$.

        (a) {all real numbers}
        (b) {all real numbers between –5 and 5}
        (c) {all real numbers greater than or equal to –5}
        (d) {all real numbers less than or equal to 5}   11. _____

12.   Suppose that f is a function whose graph
        is a line. If $f(-3) = -4$ and the y-intercept
        is $(0,2)$, what is the slope of the line?

        (a) $\dfrac{1}{4}$ (b) 2   (c) $\dfrac{5}{4}$ (d) $\dfrac{2}{3}$ 12. _____

For Exercises 13-16, the table gives data on the number of children's lemonade stands set up nationally between 1985 and 1990.

1985	29400	1988	29050
1986	29250	1989	28800
1987	29150	1990	27959

13.   If this set defines a function, then what is $f(1990)$?

(a) 29050   (b) 28800   (c) 27959   (d) Not a function   13. _____

14.   If $f(x) = 29150$, what is x?

(a) 1987   (b) 1988   (c) 1989   (d) 1990   14. _____

15.   The data points are closely approximated by the line
$y = -236.4x + 498730.1$. What is the slope of this line?

(a) 236.4   (b) –236.4
(c) 498730.1   (d) 498730.1   15. _____

16.   What does the slope of the line tell you about the number
of children's lemonade stands set up each year?

(a) The number of lemonade stands is increasing each year.
(b) The number of lemonade stands is decreasing each year.
(c) Children don't like lemonade anymore.
(d) The slope gives us no information about production.   16. _____

CHAPTER EIGHT, FORM A

BEGINNING ALGEBRA

NAME_____

SECTION_____

1. Decide whether each ordered pair is a solution of the system
$$x + 2y = -1$$
$$3x - y = 4$$

    (a) $(1, -1)$

    (b) $(2, 2)$

    (c) $(3, -2)$

1. (a) _____

1. (b) _____

1. (c) _____

2. Solve this system by graphing.

$$3x - 2y = 5$$
$$-x + 4y = 5$$

2.

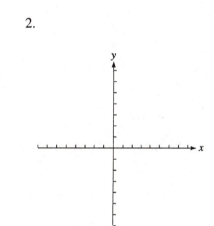

For Exercises 3-4, solve each system by substitution.

3. $4x + 3y = 1$

      $x = 1 - y$

3. _____

4. $7x - 5y = 4$

      $y = 3x - 4$

4. _____

For Exercises 5-10, solve each system by elimination.

5. $x + 4y = 7$

      $3x + 7y = 6$

5. _____

6.　　$-2x + 9y = -7$

　　　　$x - 3y =\ 5$　　　　　　　　　　6. _____

7.　　$-3x + 4y =\ 5$

　　　　$12x - 16y = 18$　　　　　　　　7. _____

8.　　$7x - 8y = 1$

　　　　$-3x +\ y = 2$　　　　　　　　　8. _____

9.　　$5x +\ 2y = 12$

　　　　$3x\ -4y =\ 2$　　　　　　　　　9. _____

10.　　$\dfrac{2}{3}x + \dfrac{1}{4}y = 25$

　　　　$\dfrac{1}{6}x - \dfrac{3}{4}y = -23$　　　　　　10. _____

11.　　Solve the system by any method.

　　　　$y - 2 =\ 5x$
　　　　$10x = 2y - 4$　　　　　　　　11. _____

12.　　Suppose that the graph of a system of two linear
　　　　equations consists of two lines that have different
　　　　slopes and different y-intercepts. How many
　　　　solutions does the system have?　　　　12. _____

For Exercises 13-17, write a system of two equations and solve each problem. Use the six-step method.

13.　　On the opening night of Star Wars®, the Crown
　　　　Theater sold 762 movie tickets. An adult's ticket
　　　　costs $7.00 and a child's ticket costs $4.50. In all,
　　　　$4959 was taken in. How many of each kind of
　　　　ticket were sold?　　　　　　　　13. _____

14.   A chemist has one solution that is 65% saline
       and a second that is 20% saline. How many
       liters of each should be mixed together to get
       120 liters of a solution that is 50% saline?   14. _____

15.   The distance by air from London, England to
       New York City is 945 miles less than the distance
       from Berlin, Germany to Chicago. Together the
       two distances total 7883 miles. How far is it
       between Berlin and Chicago?   15. _____

16.   Two cars leave from Merrillville, Indiana, at the
       same time and travel in the same direction. One
       car travels one and three-quarter times as fast as
       the other. After 3.5 hours, they are 105 miles apart.
       What is the speed of each car?   16. _____

17.   The perimeter of a rectangular garden is 76 feet.
       The length is six feet less than three times the
       width. Find the dimensions of the garden.   17. _____

For Exercises 18-19, graph the solution set of each system of inequalities.

18. $3x + 2y \geq 12$
       $x - y < -2$

18.

19. $y \le -\dfrac{1}{2}x+1$

         $y \ge \quad x -1$ 19.

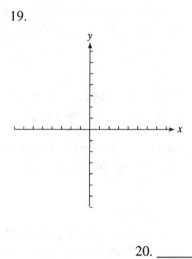

20.   Which one of the following is a calculator   20. _____
         representation of the system given here?

$$4x - 3y \le 12$$
$$x + y \ge 3$$

(a)   (b)

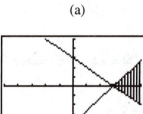   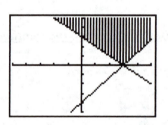

(c)   (d)

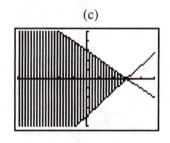   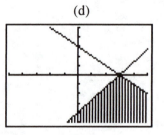

CHAPTER EIGHT, FORM B   BEGINNING ALGEBRA

NAME_____   SECTION_____

1. Decide whether each ordered pair $x - 2y = -1$
    is a solution of the system $3x + y = 4$

    (a)   (1, 3)   1. (a) _____
    (b)   (1, 1)   1. (b) _____
    (c)   (3, 2)   1. (c) _____

2. Solve this system by graphing.

    $$3x - 2y = 12$$
    $$-x + 4y = -14 \qquad\qquad 2.$$

For Exercises 3-4, solve each system by substitution.

3. $2x - 3y = 13$
    $y = 3 - 3x$ 3. _____

4. $2x - y = -2$
    $x = 3y - 1$ 4. _____

For Exercises 5-10, solve each system by elimination.

5. $2x + 5y = -13$
    $2x - 3y = -5$ 5. _____

6. $3x + 4y = 34$
        $3x - 5y = -29$ 6. _____

7. $-4x + 5y = 3$
        $12x - 15y = 9$ 7. _____

8. $4x - 3y = 0$
        $5x - 4y = -2$ 8. _____

9. $2x - y = 16$
        $3x + 2y = 3$ 9. _____

10. $\dfrac{1}{2}x - \dfrac{1}{4}y = 1$

        $\dfrac{1}{3}x + y = 3$ 10. _____

11.   Solve the system by any method.

        $2 - x = 5y$
        $10y = -2x + 4$ 11. _____

12.   Suppose that the graph of a system of two
        linear equations consists of two lines that
        have different slopes and the same y-intercept.
        How many solutions does the system have?   12. _____

For Exercises 13-17, write a system of two equations and solve each problem. Use the six-step method.

13.   The number of calories in one serving of
        whole-milk yogurt is 210 calories less than
        the number of calories in one serving of
        creamed cottage cheese. Together, they
        contain 510 calories. How many calories
        are in each?   13. _____

14.   On the opening night of the Broadway musical
       Rent®, the Aerie Krown Theater sold out. An
       adult's ticket costs $27.00 and a child's ticket
       costs $20.50. In all, $12616 was taken in. If
       the theater seats 500 people, how many of each
       kind of ticket were sold?   14. _____

15.   Peanuts that sell for $3.00 per pound are to be
       mixed with cashews that sell for $6.00 per pound
       to get 48 pounds of mixed nuts that will sell for
       $4.00 per pound. How many pounds of each type
       of nut should be used?   15. _____

16.   Two cars leave from Merrillville, Indiana, at the
       same time and travel in the same direction. One
       car travels one and one-half times as fast as the
       other. After 4 hours and 45 minutes, they are 95
       miles apart. What is the speed of each car?   16. _____

17.   The perimeter of a rectangular garden is 70 meters.
       The length is seven meters less than twice the width.
       Find the dimensions of the garden.   17. _____

For Exercises 18-19, graph the solution set of each system of inequalities.

18. $x + y \geq 2$
       $x + y \leq 4$ 18.

19. $2x + y \geq 6$
        $4x - y \leq 6$ 19.

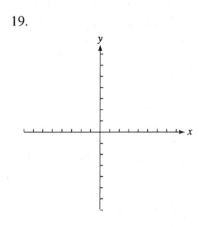

20.   Which one of the following is a calculator   20. _____
        representation of the system given here?

$$2x - 3y \leq 6$$
$$3x + 2y \geq 6$$

(a)   (b)

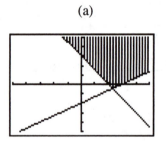   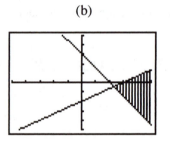

(c)   (d)

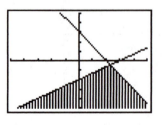   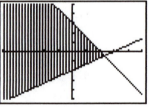

CHAPTER EIGHT, FORM C   **BEGINNING ALGEBRA**

NAME_____   SECTION_____

1.   Decide whether each ordered pair $3x - 5y = 2$
      is a solution of the system $-6x + \ y = 5$

      (a) $(-2, -7)$ 1. (a) _____
      (b) $(4, 2)$ 1. (b) _____
      (c) $(-1, -1)$ 1. (c) _____

2.   Solve this system by graphing.

                $2x + 5y = -9$
                $4x - \ y = 15$ 2.

For Exercises 3-4, solve each system by substitution.

3. $2x + 3y = 12$
      $y = 2x + 4$ 3. _____

4. $-2x + y = 5$
      $x = -2y - 5$ 4. _____

For Exercises 5-10, solve each system by elimination.

5. $x - 3y = -1$
      $2x - y = -2$ 5. _____

6. $4x + 3y = 10$
    $2x + 5y = -2$

6. _____

7. $4x + 8y = 3$
    $x + 2y = 9$

7. _____

8. $3x - 2y = 8$
    $2x - 3y = 4$

8. _____

9. $4x + 5y = 27$
    $3x - 4y = -3$

9. _____

10. $2x - \dfrac{3}{2}y = \dfrac{5}{2}$

    $\dfrac{1}{4}x + \dfrac{5}{4}y = \dfrac{15}{2}$

10. _____

11. Solve the system by any method.

    $4 = 3x - y$
    $9x = 3y + 12$

11. _____

12. Suppose that the graph of a system of two linear
    equations consists of two lines that have the same
    slope and the same y-intercept. How many solutions
    does the system have?

12. _____

For Exercises 13-17, write a system of two equations and solve each problem. Use the six-step method.

13. A 35% antifreeze solution is to be mixed with a
    60% antifreeze solution to obtain 20 liters of a
    40% antifreeze solution. How many liters of each
    solution should be used?

13. _____

14. Two cars leave from Alexandria, Virginia, at the
    same time and travel in opposite directions. One
    car travels 15 mph faster than the other. After
    3 hours, they are 405 miles apart. What is the
    speed of each car?

14. _____

15. Coffee that sells for $2.50 per pound is to be
    mixed with coffee that sells for $2.10 per pound
    to get 50 pounds of a blended coffee that will
    sell for $2.38 per pound. How many pounds of
    each type of coffee should be used?   15. _____

16. On the opening night of the Broadway musical
    Fame®, the Mercedes Theater sold out. An
    adult's ticket costs $22.50 and a child's ticket
    costs $16.00. In all, $15471 was taken in. If
    the theater seats 750 people, how many of
    each kind of ticket were sold?   16. _____

17. The perimeter of a rectangular garden is 134
    meters. The length is four meters more than
    twice the width. Find the dimensions of the
    garden.   17. _____

For Exercises 18-19, graph the solution set of each system of inequalities.

18. $x - y \leq 6$
    $x + 2y \geq 2$ 18.

19. $x + y \geq 2$
        $x - y \geq 3$ 19.

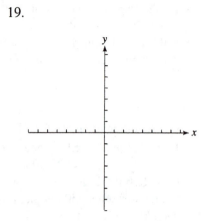

20.   Which one of the following is a calculator   20. _____
        representation of the system given here?

$$3x - y \geq 12$$
$$5x + 3y \geq 12$$

(a)   (b)

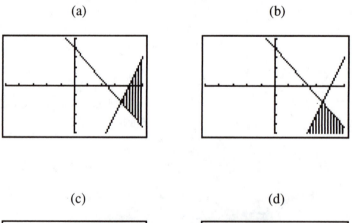

(c)   (d)

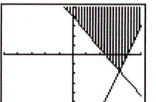   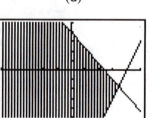

CHAPTER EIGHT, FORM D   **BEGINNING ALGEBRA**

NAME_____   SECTION_____

---

1.   Decide whether each ordered pair $-4x - 2y = 2$
     is a solution of the system $-6x + y = 11$

     (a) $(-1, 5)$ 1. (a) _____
     (b) $(-3, -7)$ 1. (b) _____
     (c) $(-3, -5)$ 1. (c) _____

2.   Solve this system by graphing.

$$2x + 5y = -5$$
$$4x - y = -21$$ 2.

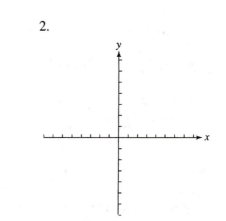

For Exercises 3-4, solve each system by substitution.

3. $4x + 5y = 2$
     $y = 3x - 11$ 3. _____

4. $x = 9 - 3y$
     $3x + 2y = 13$ 4. _____

For Exercises 5-10, solve each system by elimination.

5. $9x + 4y = 5$
     $3x - 2y = 5$ 5. _____

6. $3x + 4y = -6$
      $2x - 3y = -4$ 6. _____

7. $2x + 4y = -24$
      $x + 2y = -20$ 7. _____

8. $3x - 2y = 0$
      $9x + 4y = 5$ 8. _____

9. $x - y = -2$
      $2x - y = 7$ 9. _____

10. $\dfrac{1}{2}x + \dfrac{3}{10}y = \dfrac{1}{2}$

      $\dfrac{2}{5}x - \dfrac{1}{10}y = \dfrac{2}{5}$ 10. _____

11.   Solve the system by any method.

      $x = -2y - 5$
      $3x + 6y = -15$ 11. _____

12.   Suppose that the graph of a system of two linear
      equations consists of two lines that have different
      slopes and different y-intercepts. How many
      solutions does the system have?   12. _____

For Exercises 13-17, Write a system of two equations and solve each problem. Use the six-step method.

13.   A 40% antiseptic solution is to be mixed with
      100% antiseptic solution to obtain 8 liters of
      a 55% antiseptic solution. How many liters
      of each solution should be used?   13. _____

14.   An airplane can fly downwind a distance of
      600 miles in 2 hours. The return trip against
      the same wind takes 3 hours. Find the speed
      of the wind.   14. _____

15.   At Dairy King, ice cream cones cost $0.75
        and sundaes cost $1.95. One day the receipts
        for a total of 298 cones and sundaes were
        $422.70. How many sundaes were sold?   15. _____

16.   On the opening night of the Broadway musical
        Grease®, the Trident Theater sold 800 total
        tickets. An adult's ticket costs $22.00 and a
        child's ticket costs $14.50. In all, $16640
        was taken in. How adult tickets were sold?   16. _____

17.   The perimeter of a rectangular garden is
        200 yards. The length is eight yards more
        than three times the width. Find the
        dimensions of the garden.   17. _____

For Exercises 18-19, graph the solution set of each system of inequalities.

18. $5x + 2y > 3$
        $2x - 3y \le 4$ 18.

19. $4x + 3 < y$
       $3y \geq 2$ 19.

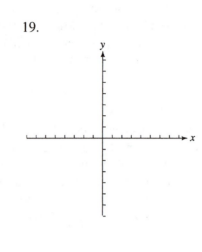

20.   Which one of the following is a calculator   20. _____
       representation of the system given here?

$$5x + 2y \leq \;\; 2$$
$$-3x + 4y \geq -4$$

(a)   (b)

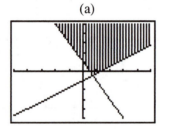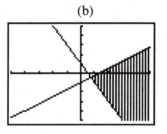

(c)   (d)

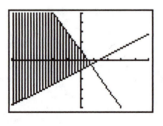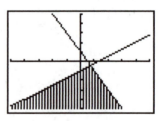

CHAPTER EIGHT, FORM E   **BEGINNING ALGEBRA**

NAME_____   SECTION_____

---

1.   Decide which ordered pair $-4x-2y=-6$
      is a solution of the system $-6x+\;\;y=11$

      (a) $(-1, 5)$ (b) $(-3,-7)^-$ (c) $(-3,-5)$ (d) $(-2, 3)$ 1. _____

2.   Solve this system by graphing. State the y-value of the solution.

                    $6x-3y=-3$ 2.
                    $4x-\;y=-5$

      (a) –2   (b) 2   (c) –3   (d) 3   2. _____

For Exercises 3-4, solve each system by substitution. State the x-value of the solution.

3. $4x+5y=2$
      $y=3x-11$

      (a) 3   (b) 5   (c) 0   (d) $\dfrac{13}{19}$ 3. _____

4. $x=2y+1$
      $2x-4y=5$

      (a) 3   (b) 0
      (c) Infinite number of solutions   (d) No solution   4. _____

For Exercises 5-10, solve each system by elimination. State the x-value of the solution.

5. $9x + 4y = 5$
     $3x - 2y = 5$

   (a) –1   (b) 1
   (c) Infinite number of solutions   (d) No solution   5. _____

6. $5x - 6y = 9$
     $3x + 6y = 7$

   (a) 2   (b) $\dfrac{1}{2}$ (c) 6   (d) $\dfrac{1}{6}$ 6. _____

7. $2x + 4y = -24$
     $x + 2y = -20$

   (a) –2   (b) –4
   (c) Infinite number of solutions   (d) No solution   7. _____

8. $3x - 2y = 0$
     $9x + 4y = 5$

   (a) $\dfrac{1}{3}$ (b) 3
   (c) –3   (d) No solution   8. _____

9. $x - y = -2$
     $2x - y = 7$

   (a) 4   (b) –4
   (c) 9   (d) –3   9. _____

10. $\frac{1}{2}x + \frac{3}{10}y = \frac{1}{2}$

       $\frac{2}{5}x - \frac{1}{10}y = \frac{2}{5}$

       (a) 10   (b) 1
       (c) Infinite number of solutions (d) 0   10. _____

11.   Solve the system by any method.

       $5x - 4y - 8x - 2 = 6x + 3y - 3$
       $4x - y = -2y - 8$

       (a) –3   (b) 4   (c) 6   (d) 7   11. _____

12.   If the graph of a system of two linear equations
       consists of two lines that have the same
       slope and different y-intercepts, then

       (a) the equations are dependent
       (b) the system has infinite solutions
       (c) the system is inconsistent
       (d) the system has only one solution   12. _____

For Exercises 13-17, Write a system of two equations and solve each problem. Use the six-step method.

13.   A 40% acid solution is to be mixed with a 60% acid
       solution to obtain 8 liters of a 55% acid solution. How
       many liters of the 60% acid solution should be used?

       (a) 2   (b) 6   (c) 1.2   (d) 6.8   13. _____

14.   At Bulbs-R-Us, halogen light bulbs cost $14.50 each and
       fluorescent bulbs cost $19.00 each. One day the receipts
       for a total of 376 bulbs were $6469.00. How many
       halogen bulbs were sold?

       (a) 150   (b) 170   (c) 206   (d) 226   14. _____

15.   An airplane can fly downwind a distance of 600 miles
        in 2 hours. The return trip against the same wind takes
        3 hours. Find the speed of the wind.

        (a) 550 mph   (b) 250 mph   (c) 50 mph   (d) 25 mph   15. _____

16.   Derek invested $5000 in two bank accounts. One account
        pays 6% simple interest and the other pays 5% simple
        interest. If he earns $283 annually in interest from the two
        accounts, how much did he originally invest at 6%?

        (a) $1700   (b) $2500   (c) $3300   (d) $3400   16. _____

17.   The perimeter of a rectangular garden is 200 yards.
        The length is eight yards more than three times the
        width. Find the length of the garden.

        (a) 23 yards   (b) 77 yards   (c) 27 yards   (d) 73 yards   17. _____

For Exercises 18-19, graph the solution set of each system of inequalities.

18. $x - y \geq 4$
        $2x + y \leq 2$ 18. _____

(a)   (b)   (c)   (d)

19. $x - y \leq 5$
        $x + 2y \geq 2$ 19. _____

(a)	(b)	(c)	(d)

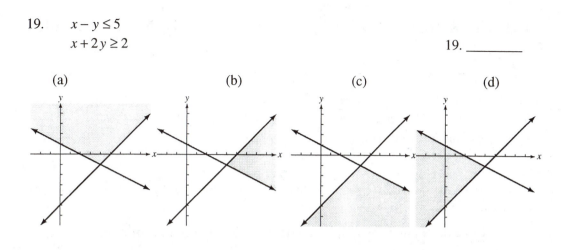

20.   Which one of the following is a calculator representation of the system given here?

$$5x + 2y \geq 2$$
$$3x - 2y \leq 6$$ 20. _____

(a)	(b)

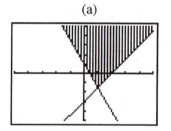 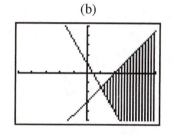

(c)	(d)

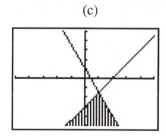 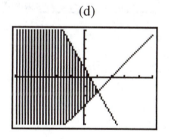

CHAPTER EIGHT, FORM F   BEGINNING ALGEBRA

NAME_____   SECTION_____

---

1.   Decide which ordered pair $x - 2y = -5$
      is a solution of the system $3x + \ y = \ 6$

   (a) $(2, 0)$ (b) $(1,3)$ (c) $\left(0, \dfrac{5}{2}\right)$ (d) $(0, 6)$ 1. _____

2.   Solve this system by graphing. State the y-value of the solution.

$$4x + 7y = -1$$
$$-2x + \ y = \ 5$$

2.

   (a) –2   (b) 2   (c) –1   (d) 1   2. _____

For Exercises 3-4, solve each system by substitution. State the x-value of the solution.

3. $-2x + y = 8$
      $y = 3x + 7$
      (a) 2   (b) –3   (c) 15   (d) 1
                                                              3. _____

4. $x = 3y - 2$
      $2x - 5y = -1$

      (a) 3   (b) 7
      (c) No solution   (d) 1   4. _____

For Exercises 5-10, solve each system by elimination. State the x-value of the solution.

5. $4x + 9y = 2$
      $2x - 3y = 16$

(a) -2   (b) 5
(c) Infinite number of solutions   (d) No solution   5. _____

6. $3x + 5y = 11$
      $-3x + 7y = 13$

(a) 2   (b) $\dfrac{1}{2}$ (c) 3   (d) $\dfrac{1}{3}$ 6. _____

7. $3x - 2y = -5$
      $-6x + 8y = -22$

(a) -7   (b) 7
(c) –8   (d) 8   7. _____

8. $2x + 4y = -10$
      $x + 2y = -20$

(a) 4   (b) 2
(c) Infinite number of solutions   (d) No solution   8. _____

9. $x - 2y = 13$
      $2x - y = -19$

(a) –5   (b) –17
(c) Infinite number of solutions   (d) No solution   9. _____

10.
$$\frac{2}{3}x + \frac{1}{4}y = 18 \qquad \frac{1}{6}x - \frac{3}{8}y = -6$$

(a) 18   (b) 24
(c) Infinite number of solutions   (d) No solution   10. _____

11.   Solve the system by any method.

$$10x - y = 7x + y + 2$$
$$-2x + 3y = 2x - 6y + 29$$

(a) –4   (b) 4   (c) 5   (d) –5   11. _____

12.   If the graph of a system of two linear equations
consists of two lines that have the same
slope and the same y-intercepts, then

(a) the equations are dependent
(b) the equations are independent
(c) the system is inconsistent
(d) the system has only one solution   12. _____

For Exercises 13-17, write a system of two equations and solve each problem. Use the six-step method.

13.   A 75% acid solution is to be mixed with a 50% acid
solution to obtain 12 liters of a 60% acid solution. How
many liters of the 75% acid solution should be used?

(a) 4.8   (b) 7.2   (c) 9   (d) 3   13. _____

14.   At the Hobby Horse, model airplanes cost $29.95 each
and model rockets cost $19.95 each. One day the receipts
for a total of 47 planes and rockets were $1217.65. How
many rockets were sold?

(a) 33   (b) 14   (c) 28   (d) 19   14. _____

15. An airplane can fly downwind a distance of 600 miles
    in 2 hours. The return trip against the same wind takes
    3 hours. Find the speed of the wind.

    (a) 550 mph   (b) 250 mph   (c) 50 mph   (d) 25 mph   15. _____

16. Vonda invested $4000 in two bank accounts. One account
    pays 7% simple interest and the other pays 5% simple
    interest. If he earns $250 annually in interest from the two
    accounts, how much did he originally invest at 5%?

    (a) $1500   (b) $2500   (c) $800   (d) $3200   16. _____

17. The perimeter of a rectangular garden is 250 feet.
    The length is four feet less than twice the
    width. Find the width of the garden.

    (a) 100 feet   (b) 43 feet   (c) 82 feet   (d) 25 feet   17. _____

For Exercises 18-19, graph the solution set of each system of inequalities.

18. $x + y \geq 4$
    $x - 2y \leq -2$ 18. _____

    (a)   (b)   (c)   (d)

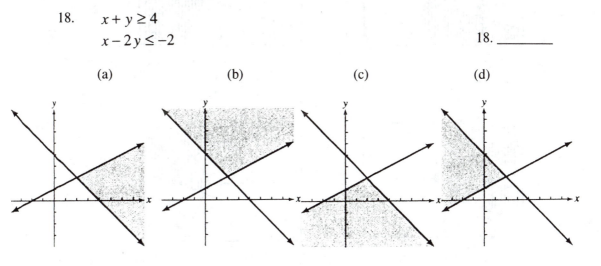

19. $x - y \leq 3$

     $2x + y \geq 2$

19. _____

(a)   (b)   (c)   (d)

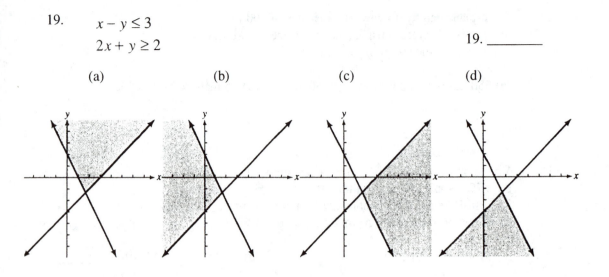

20.   Which one of the following is a calculator representation of the system given here?

$$3x + y \geq 2$$
$$x - 2y \leq 3$$

20. _____

(a)   (b)

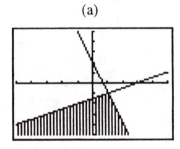
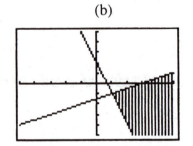

(c)   (d)

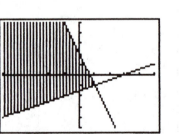
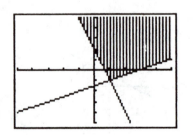

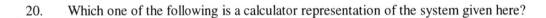

CHAPTER NINE, FORM A

BEGINNING ALGEBRA

NAME_____   SECTION_____

On this test assume that all variables represent positive real numbers.

1.   Find all square roots of 225.   1. _____

2.   Consider $\sqrt{300}$

    (a).   Determine whether it is
           rational or irrational.   2. (a) _____

    (b).   Find a decimal approximation
           to the nearest thousandth.   2. (b) _____

3.   Find the distance between $(4,-3)$ and $(2,5)$.   3. _____

For Exercises 4-14, simplify where possible.

4. $\sqrt[4]{81}$ 4. _____

5. $\sqrt{98}$ 5. _____

6. $\sqrt{\dfrac{200}{169}}$ 6. _____

7. $\sqrt[3]{54}$ 7. _____

8. $\dfrac{64\sqrt{35}}{4\sqrt{7}}$ 8. _____

9. $3\sqrt{18}+4\sqrt{50}$ 9. _____

10. $5\sqrt{20p} - 2\sqrt{45p} - \sqrt{80p}$ 10. _____

11. $\sqrt[3]{24x^5 y^6}$ 11. _____

12. $\left(6 + \sqrt{2}\right)\left(6 - \sqrt{2}\right)$ 12. _____

13. $\left(2 + \sqrt{5}\right)\left(4\sqrt{3} + 1\right)$ 13. _____

14. $\left(3 - \sqrt{2}\right)^2$ 14. _____

15. The hypotenuse of a right triangle measures
     12 feet, and one leg measures 5 feet. Find
     the length of the other leg.

     (a)   Give its length in simplified
           radical form.   15. (a) _____

     (b)   Round your answer to the
           nearest thousandth.   15. (b) _____

In Exercises 16-19, rationalize each denominator.

16. $\dfrac{10\sqrt{6}}{\sqrt{5}}$ 16. _____

17. $\sqrt{\dfrac{10}{3f}}$ 17. _____

18. $\dfrac{6}{\sqrt[3]{2}}$ 18. _____

19. $\dfrac{-5}{\sqrt{6}-4}$

19. _____

For Exercises 20-21, solve each equation.

20. $\sqrt{5y+1} = y+1$

20. _____

21. $2\sqrt{z}-2 = z-10$

21. _____

For Exercises 22-24, simplify each expression.

22. $-8^{2/3}$

22. _____

23. $4^{1/3} \cdot 4^{5/3}$

23. _____

24. $\dfrac{\left(3^6\right)^{1/5}}{3^{11/5}}$

24. _____

25. State whether the following statement is true:

$$\sqrt{16}+\sqrt{9} = \sqrt{16+9} = \sqrt{25}$$

Justify your answer.

25. _____
_____
_____
_____

CHAPTER NINE, FORM B   BEGINNING ALGEBRA

NAME_____   SECTION_____

On this test assume that all variables represent positive real numbers.

1.   Find all square roots of 2500.   1. _____

2.   Consider $\sqrt{162}$

    (a).   Determine whether it is
             rational or irrational.   2. (a) _____

    (b).   Find a decimal approximation
             to the nearest thousandth.   2. (b) _____

3.   Find the distance between $(2,-3)$ and $(5,-4)$.   3. _____

For Exercises 4-14, simplify where possible.

4. $-\sqrt[3]{27}$ 4. _____

5. $-\sqrt{200}$ 5. _____

6. $\sqrt{45}\cdot\sqrt{20}$ 6. _____

7. $\sqrt[3]{-88}$ 7. _____

8. $\dfrac{\sqrt{90}}{18\sqrt{10}}$ 8. _____

9. $5\sqrt{28}-3\sqrt{7}-2\sqrt{63}$ 9. _____

10. $5\sqrt{45p} - 7\sqrt{20p}$

10. _____

11. $\sqrt[4]{64x^{10}y^{7}}$

11. _____

12. $\left(5 + \sqrt{3}\right)\left(5 - \sqrt{3}\right)$

12. _____

13. $\left(2 + \sqrt{6}\right)\left(2\sqrt{6} + 1\right)$

13. _____

14. $\left(\sqrt{11} - \sqrt{10}\right)^{2}$

14. _____

15.   The hypotenuse of a right triangle measures
       22 feet, and one leg measures 10 feet. Find
       the length of the other leg.

       (a)   Give its length in simplified
              radical form.

15. (a) _____

       (b)   Round your answer to the
              nearest thousandth.

15. (b) _____

In Exercises 16-19, rationalize each denominator.

16. $\dfrac{-3}{\sqrt{6}}$

16. _____

17. $\dfrac{12}{\sqrt{6v}}$

17. _____

18. $\dfrac{-12}{\sqrt[3]{4}}$

18. _____

19. $\dfrac{5}{2-\sqrt{6}}$

19. _____

For Exercises 20-21, solve each equation.

20. $3\sqrt{z} = \sqrt{8z+16}$

20. _____

21. $k+3 = \sqrt{k+5}$

21. _____

For Exercises 22-24, simplify each expression.

22. $27^{4/3}$

22. _____

23. $3^{5/6} \cdot 3^{7/6}$

23. _____

24. $\dfrac{\left(5^5\right)^{2/3}}{\left(5^{1/3}\right)^4}$

24. _____

25.   State whether the following statement
       is true:

$$\sqrt{144} + \sqrt{25} = \sqrt{144+25} = \sqrt{169}$$

Justify your answer.

25. _____

_____

_____

_____

CHAPTER NINE, FORM C   BEGINNING ALGEBRA

NAME_____   SECTION_____

On this test assume that all variables represent positive real numbers.

1.   Find all square roots of $\dfrac{121}{36}$.   1. _____

2.   Consider $\sqrt{150}$

    (a).   Determine whether it is
           rational or irrational.   2. (a) _____

    (b).   Find a decimal approximation
           to the nearest thousandth.   2. (b) _____

3.   Find the distance between $(-2,-1)$ and $(4,-9)$.   3. _____

For Exercises 4-14, simplify where possible.

4. $-\sqrt[5]{32}$ 4. _____

5. $-\sqrt{52}$ 5. _____

6. $\sqrt{\dfrac{125}{256}}$ 6. _____

7. $\sqrt[3]{135}$ 7. _____

8. $\dfrac{9\sqrt{35}}{\sqrt{21}}$ 8. _____

9. $2\sqrt{99} - \sqrt{176}$

9. _____

10. $3\sqrt{54p} - 8\sqrt{150p}$

10. _____

11. $\sqrt[3]{16x^4 y^6 z^2}$

11. _____

12. $\left(\sqrt{6} - 8\right)^2$

12. _____

13. $\left(2 + \sqrt{3}\right)\left(3\sqrt{6} - 1\right)$

13. _____

14. $\left(8 + \sqrt{7}\right)\left(8 - \sqrt{7}\right)$

14. _____

15.   Two cars leave an intersection at the same time. One travels due north at 56 miles per hour and The other travels due west at 64 miles per hour. How far apart are they after 3-1/4 hours?

(a)   Give the distance in simplified radical form.

15. (a) _____

(b)   Round your answer to the nearest thousandth.

15. (b) _____

In Exercises 16-19, rationalize each denominator.

16. $\dfrac{7\sqrt{5}}{\sqrt{35}}$

16. _____

17. $\sqrt{\dfrac{11}{6t}}$

17. _____

18. $\dfrac{\sqrt{10}}{4+\sqrt{6}}$

18. _____

19. $\sqrt[4]{\dfrac{2}{27}}$

19. _____

For Exercises 20-21, solve each equation.

20. $5-\sqrt{z}=1$

20. _____

21. $k-3=\sqrt{2k+9}$

21. _____

For Exercises 22-24, simplify each expression.

22. $-(25)^{3/2}$

22. _____

23. $64^{1/6}\cdot 64^{1/2}$

23. _____

24. $\dfrac{\left(6^{3}\right)^{1/4}}{\left(6^{1/4}\right)^{7}}$

24. _____

25.   What is wrong with the following "solution"?

$$\sqrt{2x+1}+5=0$$
$$\sqrt{2x+1}=-5 \qquad \text{Subtract 5}$$
$$2x+1=25 \qquad \text{Square both sides}$$
$$2x=24 \qquad \text{Subtract 1}$$
$$x=12 \qquad \text{Divide by 2}$$

The solution is 12.

25. _____
_____
_____
_____

CHAPTER NINE, FORM D   **BEGINNING ALGEBRA**

NAME_____   SECTION_____

_____

On this test assume that all variables represent positive real numbers.

1.   Find all square roots of $\dfrac{400}{169}$.   1. _____

2.   Consider $\sqrt{800}$

    (a).   Determine whether it is
        rational or irrational.   2. (a) _____

    (b).   Find a decimal approximation
        to the nearest thousandth.   2. (b) _____

3.   Find the distance between $(-5,-3)$ and $(8,-1)$.   3. _____

For Exercises 4-14, simplify where possible.

4. $-\sqrt{160}$ 4. _____

5. $\sqrt{\dfrac{3}{32}} \cdot \sqrt{\dfrac{3}{2}}$ 5. _____

6. $\dfrac{8\sqrt{45}}{3\sqrt{5}}$ 6. _____

7. $\sqrt[3]{54}$ 7. _____

8. $-2\sqrt{48} - 3\sqrt{75}$ 8. _____

9. $\sqrt[4]{32a^7b^5}$ 9. _____

10. $3\sqrt{360p} - 4\sqrt{490p}$ 10. _____

11. $-\sqrt[3]{-8}$ 11. _____

12. $\left(3 - \sqrt{10}\right)^2$ 12. _____

13. $\left(2\sqrt{7} - 1\right)\left(\sqrt{3} + 1\right)$ 13. _____

14. $\left(\sqrt{11} + 4\right)\left(\sqrt{11} - 4\right)$ 14. _____

15. Two cars leave an intersection at the same time.
     One travels due north at 39 miles per hour and
     The other travels due west at 60 miles per hour.
     How far apart are they after 2-1/3 hours?

     (a)   Give the distance in simplified
           radical form.   15. (a) _____

     (b)   Round your answer to the
           nearest thousandth.   15. (b) _____

In Exercises 16-19, rationalize each denominator.

16. $\dfrac{12\sqrt{2}}{\sqrt{3}}$ 16. _____

17. $\sqrt{\dfrac{3}{7t}}$ 17. _____

18. $\dfrac{\sqrt{2}-7}{2-\sqrt{7}}$

18. _____

19. $\sqrt[3]{\dfrac{2}{25}}$

19. _____

For Exercises 20-21, solve each equation.

20. $\sqrt{2x+1}+x=7$

20. _____

21. $\sqrt{2k+9}+3=2$

21. _____

For Exercises 22-24, simplify each expression.

22. $-\left(\dfrac{8}{27}\right)^{4/3}$

22. _____

23. $19^{4/9}\cdot19^{14/9}$

23. _____

24. $\dfrac{\left(5^{3}\right)^{1/5}}{\left(5^{1/5}\right)^{7}}$

24. _____

25. A student in your algebra class states that it is impossible to evaluate $(-16)^{3/4}$ as a real number. Do you agree with her? Justify your answer.

25. _____
_____
_____
_____

CHAPTER NINE, FORM E   **BEGINNING ALGEBRA**

NAME_____   SECTION_____

On this test assume that all variables represent positive real numbers.

For Exercises 1-3, choose the best answer.

1. Find all square roots of 576.

    (a) 24   (b) 26   (c) –24, 24   (d) –26, 26   1._____

2. Which of the following numbers is irrational?

    (a) $\sqrt{27}$ (b) $-\sqrt{36}$ (c) $\sqrt{\dfrac{1}{49}}$ (d) $\sqrt{121}$ 2._____

3. Find the distance between $(-5,-3)$ and $(8,-1)$.

    (a) $\sqrt{13}$ (b) $\sqrt{185}$ (c) $\sqrt{173}$ (d) 5   3._____

For Exercises 4-13, simplify where possible.

4. $\sqrt{392}$

    (a) $14\sqrt{2}$ (b) $49\sqrt{8}$ (c) $8\sqrt{49}$ (d) $2\sqrt{14}$ 4._____

5. $\sqrt{\dfrac{147}{64}}$

    (a) $\dfrac{3\sqrt{7}}{8}$ (b) $\dfrac{7\sqrt{3}}{8}$ (c) $\dfrac{3\sqrt{7}}{4}$ (d) $\dfrac{7\sqrt{3}}{4}$ 5._____

6. $\dfrac{56\sqrt{24}}{14\sqrt{6}}$

    (a) $8\sqrt{6}$ (b) $4\sqrt{6}$ (c) 8   (d) 16   6._____

7. $\sqrt[3]{-16}$

    (a) $-2\sqrt{2}$ (b) $-2\sqrt[3]{2}$ (c) -4 (d) Not a real number 7._____

8. $\sqrt{96} - 4\sqrt{24}$

   (a) $-6\sqrt{6}$ (b) $4\sqrt{6}$ (c) $-4\sqrt{6}$ (d) $-32\sqrt{6}$ 8. _____

9. $\sqrt{32m} - 3\sqrt{50m} + 4\sqrt{8m}$

   (a) $3m\sqrt{2}$ (b) $12m\sqrt{2}$
   (c) $-11\sqrt{2m}$ (d) $-3\sqrt{2m}$ 9. _____

10. $\sqrt[4]{64x^3 y^7 z^4}$

   (a) $2xyz\sqrt[4]{4y^3}$ (b) $4xyz\sqrt[4]{4y^3}$
   (c) $2yz\sqrt[4]{4x^3 y^3}$ (d) $4yz\sqrt[4]{4x^3 y^3}$ 10. _____

11. $\left(5\sqrt{5} + 3\right)\left(5\sqrt{5} - 3\right)$

   (a) 116   (b) 134   (c) 2   (d) $25\sqrt{5} - 3$ 11. _____

12. $\left(\sqrt{11} - \sqrt{10}\right)^2$

   (a) $1 - 2\sqrt{110}$ (b) $21 - 2\sqrt{110}$
   (c) $21 - 2\sqrt{21}$ (d) 1   12. _____

13. $\left(3\sqrt{3} + 7\right)\left(2\sqrt{2} - 1\right)$

   (a) $6\sqrt{6} - 3\sqrt{3} + 14\sqrt{2} - 7$ (b) $6\sqrt{5} - 3\sqrt{3} + 14\sqrt{2} - 7$
   (c) $3\sqrt{3} + \sqrt{7} + 2\sqrt{2} - 1$ (d) $36 - 14\sqrt{2}$ 13. _____

14.   Two cars leave a gas station at the same time. One travels
     due south at 42 miles per hour and the other travels due west
     at 36 miles per hour. How far apart are they after 1-1/2 hours?
     (Round your answer to the nearest thousandth, if necessary.)

   (a) 82.976 miles   (b) 55.317 miles
   (c) 78 miles   (d) 117 miles   14. _____

For Exercises 15-18, rationalize the denominator.

15. $\dfrac{10}{\sqrt{3}}$

(a) $\dfrac{\sqrt{3}}{10}$ (b) $\dfrac{10}{\sqrt{3}}$ (c) $\dfrac{10\sqrt{3}}{3}$ (d) $3\sqrt{3}$ 15. _____

16. $\dfrac{6a}{\sqrt{ab}}$

(a) $\dfrac{6\sqrt{ab}}{ab}$ (b) $6\sqrt{ab}$ (c) $6a\sqrt{ab}$ (d) $\dfrac{6\sqrt{ab}}{b}$ 16. _____

17. $\dfrac{24}{\sqrt[3]{36}}$

(a) $4\sqrt[3]{36}$ (b) $4\sqrt[3]{6}$ (c) 4 (d) $\dfrac{2}{3}$ 17. _____

18. $\dfrac{33}{6-\sqrt{14}}$

(a) $\dfrac{33\left(6+\sqrt{14}\right)}{8}$ (b) $\dfrac{3\left(6+\sqrt{14}\right)}{2}$

(c) $12+2\sqrt{14}$ (d) $\dfrac{33\left(6+\sqrt{14}\right)}{50}$ 18. _____

For Exercises 19-20, solve the equation.

19. $4\sqrt{p}+3=p-2$

(a) 1 , 25 (b) 25 (c) 1 (d) No solution 19. _____

20. $\sqrt{z-2}+3=0$

(a) 11 (b) 5 (c) -1 (d) No solution 20. _____

For Exercises 21-24, simplify the expression.

21. $\left(\dfrac{64}{27}\right)^{2/3}$

    (a) $\dfrac{16}{3}$ (b) $\dfrac{4}{9}$ (c) $\dfrac{4}{3}$ (d) $\dfrac{16}{9}$ 21._____

22. $\left(-\dfrac{243}{32}\right)^{3/5}$

    (a) $-\dfrac{27}{8}$ (b) $-\dfrac{405}{32}$ (c) $\dfrac{3}{2}$ (d) $\dfrac{27}{8}$ 22._____

23. $11^{2/5} \cdot 11^{3/5}$

    (a) $121^{6/25}$ (b) 1 (c) 11 (d) $11^{6/25}$ 23._____

24. $\dfrac{\left(4^{1/5}\right)^3}{4^{8/5}}$

    (a) $4^{8/5}$ (b) 1 (c) 4 (d) $\dfrac{1}{4}$ 24._____

25. When is $\sqrt{x}$ not a real number?

    (a) When $x > 0$
    (b) When $x < 0$
    (c) It is always a real number
    (d) When x is a fraction   25._____

CHAPTER NINE, FORM F   BEGINNING ALGEBRA

NAME_____   SECTION_____

On this test assume that all variables represent positive real numbers.

For Exercises 1-3, choose the best answer.

1.   Find all square roots of 289.

    (a) 23   (b) 17   (c) –23, 23   (d) –17, 17   1._____

2.   Which of the following numbers is irrational?

    (a) $\sqrt{\dfrac{108}{3}}$ (b) $-\sqrt{49}$ (c) $\sqrt{\dfrac{64}{2}}$ (d) $\sqrt{441}$ 2._____

3.   Find the distance between $(-3,-5)$ and $(8,-1)$.

    (a) $\sqrt{137}$ (b) $\sqrt{157}$ (c) $\sqrt{41}$ (d) $\sqrt{61}$ 3._____

For Exercises 4-13, simplify where possible.

4. $\sqrt{360}$

    (a) $6\sqrt{10}$ (b) $10\sqrt{6}$ (c) 60   (d) $6+\sqrt{10}$ 4._____

5. $\sqrt{\dfrac{108}{121}}$

    (a) $\dfrac{6\sqrt{3}}{11}$ (b) $\dfrac{3\sqrt{6}}{11}$ (c) $\dfrac{27\sqrt{2}}{11}$ (d) $\dfrac{2\sqrt{27}}{11}$ 5._____

6. $\dfrac{\sqrt{125}}{5\sqrt{5}}$

    (a) $\sqrt{5}$ (b) $\dfrac{\sqrt{5}}{5}$ (c) 1   (d) $\dfrac{2\sqrt{5}}{5}$ 6._____

7. $\sqrt[5]{96}$

    (a) $3\sqrt[5]{2}$ (b) $2\sqrt[5]{3}$ (c) $3\sqrt{2}$ (d) $2\sqrt{3}$ 7._____

8. $\sqrt{3} + 5\sqrt{2}$

    (a) $6\sqrt{6}$ (b) $6\sqrt{2}$

    (c) Cannot be simplified further   (d) $5\sqrt{5}$ 8. _____

9. $\sqrt{150m} - \sqrt{96m}$

    (a) $m\sqrt{6}$ (b) $6\sqrt{m}$

    (c) $\sqrt{6m}$ (d) $6\sqrt{6m}$ 9. _____

10. $\sqrt[3]{-16x^7 y^{12} z^{15}}$

    (a) $-2x^2 y^4 z^5 \sqrt[3]{2x}$ (b) $2x^3 y^4 z^5 \sqrt[3]{2x}$

    (c) $-4x^2 y^4 z^5$ (d) Not a real number   10. _____

11. $\left(3\sqrt{5} - 6\right)\left(3\sqrt{5} + 6\right)$

    (a) 9   (b) -21 (c) 33   (d) $9\sqrt{5} - 36$ 11. _____

12. $\left(2 + 3\sqrt{5}\right)^2$

    (a) $47 + 12\sqrt{5}$ (b) $49 + 12\sqrt{5}$

    (c) $49 + 6\sqrt{5}$ (d) 49   12. _____

13. $\left(6\sqrt{11} - 5\right)\left(\sqrt{7} - 3\right)$

    (a) $6\sqrt{77} - 6\sqrt{33} - 5\sqrt{7} + 15$ (b) $24\sqrt{11} - 5\sqrt{7} + 15$

    (c) $6\sqrt{77} - 18\sqrt{11} - 5\sqrt{7} + 15$ (d) $24\sqrt{11} - \sqrt{35} + 15$ 13. _____

14.   The legs of a right triangle measure 6 feet and 10 feet. find the measure of the hypotenuse. (Round your answer to the nearest thousandth, if necessary.)

    (a) 8 feet   (b) 11.662 feet

    (c) 16 feet   (d) 18.974 feet   14. _____

For Exercises 15-18, rationalize the denominator.

15. $\dfrac{11\sqrt{10}}{\sqrt{22}}$

   (a) $\sqrt{55}$ (b) $\sqrt{110}$ (c) $11\sqrt{10}$ (d) $10\sqrt{11}$ 15. _____

16. $\sqrt{\dfrac{2b}{a}}$

   (a) $\dfrac{\sqrt{2ab}}{a}$ (b) $\dfrac{2b\sqrt{a}}{a}$ (c) $\dfrac{\sqrt{2ab}}{a^2}$ (d) $\dfrac{\sqrt{2b}}{a}$ 16. _____

17. $\dfrac{-9}{\sqrt[3]{9}}$

   (a) $-3\sqrt[3]{3}$ (b) $-27\sqrt[3]{3}$ (c) 3 (d) -1 17. _____

18. $\dfrac{12}{3+\sqrt{7}}$

   (a) $6\left(3+\sqrt{7}\right)$ (b) $6\left(3-\sqrt{7}\right)$
   (c) $2\left(3+\sqrt{7}\right)$ (d) $2\left(3-\sqrt{7}\right)$ 18. _____

For Exercises 19-20, solve the equation.

19. $3\sqrt{p}+2=p+4$

   (a) 1 , 4 (b) 4 (c) 1 (d) No solution   19. _____

20. $\sqrt{z+13}-1=z$

   (a) -4 , 3 (b) -4 (c) 3 (d) No solution   20. _____

For Exercises 21-24, simplify the expression.

21. $\left(\dfrac{64}{125}\right)^{4/3}$

(a) $\dfrac{4}{5}$ (b) $\dfrac{48}{125}$ (c) $\dfrac{256}{625}$ (d) $\dfrac{16}{25}$ 21. _____

22. $\left(-\dfrac{243}{32}\right)^{2/5}$

(a) $-\dfrac{3}{2}$ (b) $\dfrac{9}{4}$ (c) $\dfrac{3}{2}$ (d) $-\dfrac{9}{4}$ 22. _____

23. $-\left(100^{3/2}\right)$

(a) -1000 (b) 1000 (c) $-\dfrac{1}{1000}$ (d) $\dfrac{1}{1000}$ 23. _____

24. $\dfrac{\left(4^{1/4}\right)^{3}}{4^{11/4}}$

(a) 2 (b) -2 (c) $\dfrac{1}{2}$ (d) $\dfrac{1}{16}$ 24. _____

25. When is $\sqrt[5]{x}$ not a real number?

(a) When $x > 0$
(b) When $x < 0$
(c) It is always a real number
(d) When x is a fraction   25. _____

CHAPTER TEN, FORM A   BEGINNING ALGEBRA

NAME_____   SECTION_____

Items marked * require knowledge of complex numbers.

For Exercises 1-3, solve by using the square root property.

1. $z^2 = 75$ 1. _____

2. $(n-3)^2 = 49$ 2. _____

3. $(5x-2)^2 = 18$ 3. _____

For Exercises 4-5, solve by completing the square.

4. $x^2 + 12x = -9$ 4. _____

5. $2t^2 - 10t = 28$ 5. _____

6.   (a)   Find the discriminant for the quadratic
             equation $9x^2 - 24x + 16 = 0$.   6. (a) _____

      (b)   Based on your answer to part (a), how many
             real solutions does the equation have?   6. (b) _____

For Exercises 7-10, solve by using the quadratic formula.

7. $6y^2 - y - 1 = 0$ 7. _____

8. $2t^2 = 8t + 2$ 8. _____

*9. $5x^2 + 4x + 8 = 0$ 9. _____

*10. $4 + \dfrac{2}{x^2} = \dfrac{1}{x}$

10. _____

For Exercises 11-14, solve by the method of your choice.

11. $t^2 + 36 = -12t$

11. _____

12. $(2x - 1)^2 = 27$

12. _____

*13. $g^2 - 4g + 6 = 0$

13. _____

14. $(3q + 4)(q - 1) = 3$

14. _____

15.   A model rocket is launched into the air from ground level on Earth with an initial velocity of 96 feet per second. Its height s in feet is given by the formula

$$s = -16t^2 + 96t .$$

After how many seconds will the rocket reach a height of 144 feet?

15. _____

16.   Use the formula from Kepler's third law of planetary motion, $P^2 = a^3$, to determine P, in years, for Mercury, if $a = 0.387$ AU.

16. _____

In Exercises 17-20, perform the indicated operations.

*17. $-(3 - i) + (-4 + i) + (8 - 2i)$

17. _____

*18. $(4 - 5i)(-3 + 2i)$

18. _____

*19. $(6-7i)(6+7i)$ 19. _____

*20. $\dfrac{6+5i}{7-3i}$ 20. _____

For Exercises 21-23, sketch each graph and identify the vertex.

21. $y = -x^2 + 4x + 2$ 21. Vertex: _____

22. $y = x^2 - 5x + 6$ 22. Vertex: _____

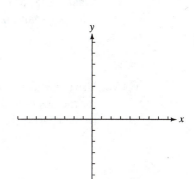

23. $y = -x^2 - 4x - 1$

23. Vertex: _____

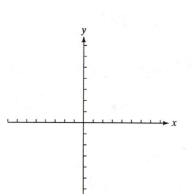

24.   Complete the following:

(a)   Determine the number of real solutions of
        $-x^2 - 4x - 1 = 0$ by looking at the graph.   24. (a) _____

(b)   Use the quadratic formula to find the
        exact values of the real solutions.
        Give the solution set.   24. (b) _____

(c)   Use a calculator to find approximations
        for the answers. Round your answers
        to the nearest thousandth.   24. (c) _____

25.   Find two numbers whose sum is 900 and
        whose product is a maximum.   25. _____

CHAPTER TEN, FORM B   **BEGINNING ALGEBRA**

NAME_____   SECTION_____

*Items marked * require knowledge of complex numbers.*

For Exercises 1-3, solve by using the square root property.

1. $r^2 = 98$ 1. _____

2. $(d - 3)^2 = 100$ 2. _____

3. $(2x - 3)^2 = 48$ 3. _____

For Exercises 4-5, solve by completing the square.

4. $w^2 + 10w = -9$ 4. _____

5. $2t^2 - 10t = 5$ 5. _____

6.   (a)   Find the discriminant for the quadratic
               equation $6x^2 + 5x - 6 = 0$.   6. (a) _____

       (b)   Based on your answer to part (a), how many
               real solutions does the equation have?   6. (b) _____

For Exercises 7-10, solve by using the quadratic formula.

7. $15y^2 - 2y = 10$ 7. _____

*8. $7p^2 - 2p + 3 = 0$ 8. _____

9. $t^2 - 8t = -3$ 9. _____

*10. $3 + \dfrac{2}{x} = -\dfrac{5}{x^2}$ 10. _____

For Exercises 11-14, solve by the method of your choice.

11. $f^2 + 27 = -12f$ 11. _____

12. $(2x - 7)^2 = 20$ 12. _____

13. $4g^2 = 3g + 2$ 13. _____

14. $(q + 3)(2q - 1) = -5$ 14. _____

15.   A model rocket is launched into the air from ground
       Level on Earth with an initial velocity of 84 feet per
       second. Its height s in feet is given by the formula

$$s = -16t^2 + 84t .$$

       After how many seconds will the rocket
       reach a height of 108 feet?   15. _____

16.   Use the formula from Kepler's third law of
       planetary motion, $P^2 = a^3$, to determine P,
       in years, for Neptune, if $a = 30.1$ AU.   16. _____

In Exercises 17-20, perform the indicated operations.

*17. $-(5 + i) - (-4 - 9i) + (8 - 3i)$ 17. _____

*18. $(-1 - 5i)(3 + 11i)$ 18. _____

*19. $(4 - 13i)(4 + 13i)$ 19. _____

*20. $\dfrac{-2+3i}{1-9i}$ 20. _____

For Exercises 21-23, sketch each graph and identify the vertex.

21. $y = x^2 + 2x - 3$ 21. Vertex: _____

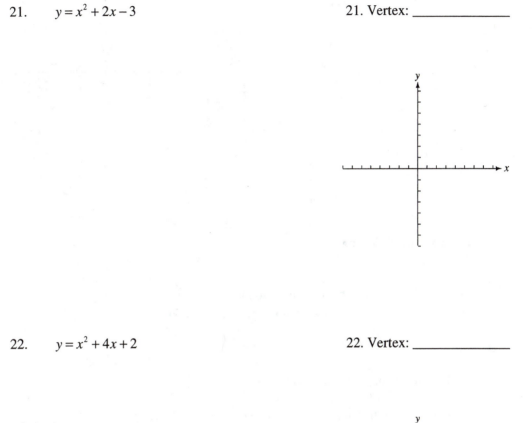

22. $y = x^2 + 4x + 2$ 22. Vertex: _____

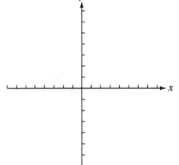

23. $y = -x^2 + 4x + 1$ 23. Vertex: _____

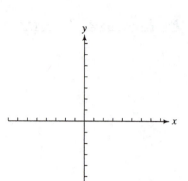

24.   Complete the following:

(a) Determine the number of real solutions of
$-x^2 + 4x + 1 = 0$ by looking at the graph.   24. (a) _____

(b) Use the quadratic formula to find the
exact values of the real solutions.
Give the solution set.   24. (b) _____

(c) Use a calculator to find approximations for the
answers. Round your answers to the nearest
thousandth.   24. (c) _____

25.   Find two numbers whose sum is 150 and
whose product is a maximum.   25. _____

CHAPTER TEN, FORM C   BEGINNING ALGEBRA

NAME_____   SECTION_____

*Items marked * require knowledge of complex numbers.*

For Exercises 1-3, solve by using the square root property.

1. $w^2 = 108$ 1. _____

2. $(x-4)^2 = 121$ 2. _____

3. $(5x-1)^2 = 45$ 3. _____

For Exercises 4-5, solve by completing the square.

4. $w^2 + 14w = 15$ 4. _____

5. $3t^2 - 8t = 16$ 5. _____

6.   (a)   Find the discriminant for the quadratic
             equation $x^2 + 2x + 6 = 0$.   6. (a) _____

      (b)   Based on your answer to part (a), how many
             real solutions does the equation have?   6. (b) _____

For Exercises 7-10, solve by using the quadratic formula.

*7. $5y^2 + 4y = -8$ 7. _____

8. $p^2 + 9p + 5 = 0$ 8. _____

9. $t^2 - 7t = -5$ 9. _____

*10. $5 - \dfrac{1}{x} = -\dfrac{2}{x^2}$ 10. _____

For Exercises 11-14, solve by the method of your choice.

11. $f^2 - 27 = 6f$ 11. _____

12. $(8x - 1)^2 = 12$ 12. _____

13. $2g^2 = 2 - 3g$ 13. _____

14. $(q + 1)(2q - 3) = 2$ 14. _____

15.   A twin-engine model rocket is launched into the air from ground level on Earth with an initial velocity of 176 feet per second. Its height s in feet is given by the formula

$$s = -16t^2 + 176t.$$

After how many seconds will the rocket reach a height of 480 feet?   15. _____

16.   Use the formula from Kepler's third law of planetary motion, $P^2 = a^3$, to determine P, in years, for Venus, if $a = 0.723$ AU.   16. _____

In Exercises 17-20, perform the indicated operations.

*17. $(11 - 5i)(-5 + 11i)$ 17. _____

*18. $-(6 - 5i) - (4 - 2i) + (-8 - 3i)$ 18. _____

*19. $\dfrac{-3 + 2i}{4 - 7i}$ 19. _____

*20. $(7-6i)(7+6i)$

20. _____

For Exercises 21-23, sketch each graph and identify the vertex.

21. $y = x^2 + 2x + 3$

21. Vertex: _____

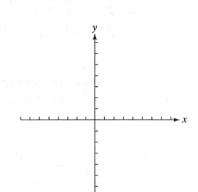

22. $y = x^2 - 5x - 6$

22. Vertex: _____

23. $y = x^2 + 2$ 23. Vertex: _____

$$ $$

24.   Complete the following:

(a) Determine the number of real solutions of
$x^2 + 2 = 0$ by looking at the graph.   24. (a) _____

(b) Use the quadratic formula to find the
exact values of the real solutions. Give
the solution set.   24. (b) _____

(c) Use a calculator to find approximations for the
answers. Round your answers to the nearest
thousandth.   24. (c) _____

25.   Find two numbers whose sum is 70 and
whose product is a maximum.   25. _____

CHAPTER TEN, FORM D   BEGINNING ALGEBRA

NAME_____   SECTION_____

---

*Items marked * require knowledge of complex numbers.*

For Exercises 1-3, solve by using the square root property.

1. $g^2 = 162$ 1. _____

2. $(x+9)^2 = 36$ 2. _____

3. $(2x-9)^2 = 75$ 3. _____

For Exercises 4-5, solve by completing the square.

4. $w^2 - 8w = -16$ 4. _____

5. $3t^2 + 6t = 12$ 5. _____

6.   (a)   Find the discriminant for the quadratic
             equation $3x^2 + 2x - 6 = 0$.   6. (a) _____

      (b)   Based on your answer to part (a), how many
             real solutions does the equation have?   6. (b) _____

For Exercises 7-10, solve by using the quadratic formula.

*7. $y^2 = 4y - 6$ 7. _____

8. $p^2 + 4 = 6p$ 8. _____

9. $7 - \dfrac{2}{x} = \dfrac{2}{x^2}$ 9. _____

*10. $x^2 - 6x + 12 = 0$ 10. _____

For Exercises 11-14, solve by the method of your choice.

11. $16f^2 + 9 = 24f$ 11. _____

12. $(3y + 2)^2 = 24$ 12. _____

13. $g^2 = 21 - 4g$ 13. _____

14. $(q + 1)(3q + 4) = 3$ 14. _____

15.   A twin-engine model rocket is launched into the air
      from ground level on Earth with an initial velocity of
      224 feet per second. Its height s in feet is given by
      the formula

$$s = -16t^2 + 224t .$$

After how many seconds will the rocket
reach a height of 784 feet?   15. _____

16.   Use the formula from Kepler's third law of
      planetary motion, $P^2 = a^3$, to determine P,
      in years, for Uranus, if $a = 19.2$ AU.   16. _____

In Exercises 17-20, perform the indicated operations.

*17. $(1 - 4i)(-5 + 9i)$ 17. _____

*18. $(6 - 4i) - (7 - i) - (-9 - 2i)$ 18. _____

*19. $\dfrac{-1 + 9i}{5 - 3i}$ 19. _____

*20. $(9-8i)(9+8i)$ 20. _____

For Exercises 21-23, sketch each graph and identify the vertex.

21. $y=-x^2+6x-13$ 21. Vertex: _____

22. $y=x^2-3$ 22. Vertex: _____

23. $y = (x-2)^2$ 23. Vertex: _____

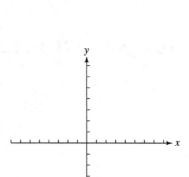

24.   Complete the following

        (a) Determine the number of real solutions of
             $(x-2)^2 = 0$ by looking at the graph.   24. (a) _____

        (b) Use the quadratic formula to find
            the exact values of the real solutions.
            Give the solution set.   24. (b) _____

        (c) Use a calculator to find approximations for the
             answers. Round your answers to the nearest
             thousandth.   24. (c) _____

25.   Find two numbers whose sum is 144 and
        whose product is a maximum.   25. _____

CHAPTER TEN, FORM E   BEGINNING ALGEBRA

NAME_____   SECTION_____

For Exercises 1-3, solve by using the square root property.

1. $x^2 = 75$

(a) $-3\sqrt{5}$, $3\sqrt{5}$ (b) $3\sqrt{5}$

(c) $-5\sqrt{3}$, $5\sqrt{3}$ (d) $5\sqrt{3}$ 1._____

2. $(b-9)^2 = 16$

(a) -4 , 4 (b) 13 , 5

(c) -13 , -5 (d) -1 , 7 2._____

3. $(2t-5)^2 = 24$

(a) $5+2\sqrt{3}$, $5-2\sqrt{3}$ (b) $5+\sqrt{6}$, $5-\sqrt{6}$

(c) $\dfrac{5+2\sqrt{6}}{2}$, $\dfrac{5-2\sqrt{6}}{2}$ (d) $-\dfrac{19}{2}$, $\dfrac{29}{2}$ 3._____

For Exercises 4-5, solve by completing the square.

4. $x^2 + 4x = 6$

(a) $-2+2\sqrt{5}$, $-2-2\sqrt{5}$ (b) $-2+\sqrt{10}$, $-2-\sqrt{10}$

(c) $2+2\sqrt{5}$, $2-2\sqrt{5}$ (d) $2+\sqrt{10}$, $2-\sqrt{10}$ 4._____

5. $15x^2 + 7x = 2$

(a) $\dfrac{-7+\sqrt{71}}{30}$, $\dfrac{-7-\sqrt{71}}{30}$ (b) $\dfrac{2}{3}$, $-\dfrac{1}{5}$

(c) $\dfrac{-7+i\sqrt{71}}{30}$, $\dfrac{-7-i\sqrt{71}}{30}$ (d) $-\dfrac{2}{3}$, $\dfrac{1}{5}$ 5._____

6.   When using the quadratic formula to solve a certain quadratic
     equation, the discriminant is –36. Describe the solution(s) of
     this equation.

     (a)   There are no real number solutions
     (b)   There is exactly one rational number solution
     (c)   There is exactly one irrational number solution
     (d)   There are exactly two real number solutions   6. _____

For Exercises 7-10, solve by using the quadratic formula.

7. $x^2 + 8x = 2$

     (a) $-4 + \sqrt{14}$, $-4 - \sqrt{14}$ (b) $4 + \sqrt{14}$, $4 - \sqrt{14}$
     (c) $-4 + 3\sqrt{2}$, $-4 - 3\sqrt{2}$ (d) $4 + 3\sqrt{2}$, $4 - 3\sqrt{2}$ 7. _____

8. $2z^2 = 4z + 3$

     (a) $\dfrac{2 + \sqrt{10}}{2}$, $\dfrac{2 - \sqrt{10}}{2}$ (b) $\dfrac{-2 + \sqrt{10}}{4}$, $\dfrac{-2 - \sqrt{10}}{4}$

     (c) $\dfrac{-2 + \sqrt{10}}{2}$, $\dfrac{-2 - \sqrt{10}}{2}$ (d) $1 + \sqrt{10}$, $1 - \sqrt{10}$ 8. _____

9. $3x^2 + 4x + 12 = 0$

     (a) $-2 + 4i\sqrt{2}$, $-2 - 4i\sqrt{2}$ (b) $-6 + 4i\sqrt{2}$, $-6 - 4i\sqrt{2}$

     (c) $\dfrac{2 + 4i\sqrt{2}}{3}$, $\dfrac{2 - 4i\sqrt{2}}{3}$ (d) $\dfrac{-2 + 4i\sqrt{2}}{3}$, $\dfrac{-2 - 4i\sqrt{2}}{3}$ 9. _____

10. $x^2 + \dfrac{7}{6}x - \dfrac{5}{6} = 0$

     (a) $\dfrac{5}{3}$, $-\dfrac{1}{2}$ (b) 3 , -10

     (c) $\dfrac{-7 + i\sqrt{71}}{12}$, $\dfrac{-7 - i\sqrt{71}}{12}$ (d) $-\dfrac{5}{3}$, $\dfrac{1}{2}$ 10. _____

For Exercises 11-14, solve by the method of your choice.

11. $t^2 - 12t = -32$

   (a) $-4, -8$ (b) $-4, 8$ (c) $4, -8$ (d) $4, 8$ 11. _____

12. $4y^2 = 4y - 1$

   (a) $-2, 2$ (b) $\dfrac{1}{4}$ (c) $\dfrac{1}{2}$ (d) $-\dfrac{1}{2}, \dfrac{1}{2}$ 12. _____

13. $(3x - 1)(x + 3) = 1$

   (a) $\dfrac{-4 + 2\sqrt{7}}{3}, \dfrac{-4 - 2\sqrt{7}}{3}$ (b) $-3, \dfrac{4}{3}$

   (c) $\dfrac{-4 + \sqrt{7}}{6}, \dfrac{-4 - \sqrt{7}}{6}$ (d) $-3, \dfrac{1}{3}$ 13. _____

14. $6r^2 - 5r = 6$

   (a) $\dfrac{5 + \sqrt{119}}{12}, \dfrac{5 - \sqrt{119}}{12}$ (b) $-\dfrac{3}{2}, \dfrac{2}{3}$

   (c) $\dfrac{5 + 13i}{12}, \dfrac{5 - 13i}{12}$ (d) $\dfrac{3}{2}, -\dfrac{2}{3}$ 14. _____

15.   A twin-engine model rocket is launched into the air
       from a height of 70 feet with an initial velocity of
       96 feet per second. Its height s in feet is given by
       the formula

$$s = -16t^2 + 96t + 70.$$

   At what times will the rocket be 58 feet above
   the ground?

   (a)   7 seconds   (b) 1 and 8 seconds
   (c)   3 seconds   (d) 2 and 4 seconds   15. _____

16.   Find the value of x in this right triangle.

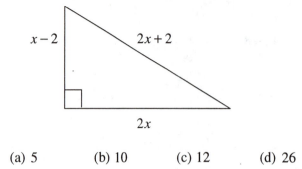

(a) 5   (b) 10   (c) 12   (d) 26   16. _____

17.   Which one of these equations has
        exactly one real number solution?

(a) $(3z-2)^2 = 16$ (b) $(4z-1)^2 = 0$

(c) $(3z+2)^2 = 16$ (d) $(4z-1)^2 + 8 = 0$ 17. _____

For Exercises 18-21, perform the indicated operation.

18. $-(3-2i)-(6+4i)+(11-15i)$

(a) $2-9i$ (b) $2-13i$ (c) $2-17i$ (d) $-20-17i$ 18. _____

19. $(3+2i)(4-i)$

(a) $14+5i$ (b) $14-5i$ (c) $10+5i$ (d) $10-5i$ 19. _____

20. $(3+7i)(3-7i)$

(a) 2   (b) 16   (c) 58   (d) -40 20. _____

21. $\dfrac{8-i}{3-2i}$

(a) $2-i$ (b) $2+i$ (c) $-2+i$ (d) $-2-i$ 21. _____

For Exercises 22-23, sketch the graph of the equation and identify the vertex.

22. $y = -x^2 + 4x + 2$ 22. _____

    (a)   (b)   (c)   (d)

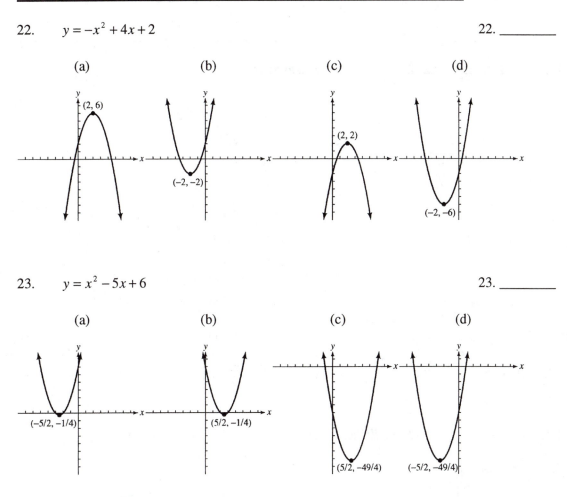

23. $y = x^2 - 5x + 6$ 23. _____

    (a)   (b)   (c)   (d)

24.   Graph $y = -2x^2 + 7x + 3$ in the standard window
      of a graphing calculator to determine the number of
      solutions of the quadratic equation

$$-2x^2 + 7x + 3 = 0.$$

    (a) 0   (b) 1   (c) 2   (d) 3   24. _____

25.   Which of the following equations should be used to
      find two numbers whose sum is 144 and whose product
      is a maximum?

    (a) $M(x) = x(144 - x)$ (b) $M(x) = x(144 + x)$
    (c) $M(x) = 144 - 2x$ (d) $M(x) = 144 + 2x$ 25. _____

CHAPTER TEN, FORM F   **BEGINNING ALGEBRA**

NAME_____   SECTION_____

For Exercises 1-3, solve by using the square root property.

1. $x^2 = 98$

   (a) $-2\sqrt{7}$, $2\sqrt{7}$ (b) $2\sqrt{7}$
   (c) $-7\sqrt{2}$, $7\sqrt{2}$ (d) $7\sqrt{2}$ 1._____

2. $(b-16)^2 = 25$

   (a) -5 , 5 (b) 21 , 11
   (c) 9 , -1 (d) -9 , 1 2._____

3. $(2k-3)^2 = 32$

   (a) $3+2\sqrt{2}$, $3-2\sqrt{2}$ (b) 7 , -1
   (c) $\dfrac{3+4\sqrt{2}}{2}$, $\dfrac{3-4\sqrt{2}}{2}$ (d) $-\dfrac{3}{2}$, $\dfrac{9}{2}$ 3._____

For Exercises 4-5, solve by completing the square.

4. $r^2 + 4r = 11$

   (a) $-2+\sqrt{15}$, $-2-\sqrt{15}$ (b) $-4+\sqrt{15}$, $-4-\sqrt{15}$
   (c) $2+\sqrt{15}$, $2-\sqrt{15}$ (d) $4+\sqrt{15}$, $4-\sqrt{15}$ 4._____

5. $7x^2 - 2x = -1$

   (a) $\dfrac{-1+i\sqrt{6}}{14}$, $\dfrac{-1-i\sqrt{6}}{14}$ (b) $\dfrac{-1+i\sqrt{6}}{7}$, $\dfrac{-1-i\sqrt{6}}{7}$
   (c) $\dfrac{1+i\sqrt{6}}{14}$, $\dfrac{1-i\sqrt{6}}{14}$ (d) $\dfrac{1+i\sqrt{6}}{7}$, $\dfrac{1-i\sqrt{6}}{7}$ 5._____

6.   When using the quadratic formula to solve a certain quadratic
      equation, the discriminant is 0. Describe the solution(s) of
      this equation.

      (a)   There are no real number solutions
      (b)   There is exactly one rational number solution
      (c)   There is exactly one irrational number solution
      (d)   There are exactly two real number solutions   6. _____

__For Exercises 7-10, solve by using the quadratic formula.__

7. $15y^2 - 4y = 1$

      (a) $\dfrac{2+\sqrt{19}}{15}$, $\dfrac{2-\sqrt{19}}{15}$ (b) $\dfrac{2+i\sqrt{11}}{15}$, $\dfrac{2-i\sqrt{11}}{15}$

      (c) $\dfrac{-2+\sqrt{19}}{15}$, $\dfrac{-2-\sqrt{19}}{15}$ (d) $\dfrac{-2+i\sqrt{11}}{15}$, $\dfrac{-2-i\sqrt{11}}{15}$ 7. _____

8. $1 = \dfrac{5}{x} - \dfrac{6}{x^2}$

      (a) $\dfrac{-5+7i}{2}$, $\dfrac{-5-7i}{2}$ (b) 3 , 2

      (c) -3 , -2 (d) 1 , -6 8. _____

9. $x^2 = 8x - 3$

      (a) $4+\sqrt{13}$, $4-\sqrt{13}$ (b) $-4+\sqrt{13}$, $-4-\sqrt{13}$

      (c) $4+\sqrt{19}$, $4-\sqrt{19}$ (d) $-4+\sqrt{19}$, $-4-\sqrt{19}$ 9. _____

10. $2x^2 + 5x + 2 = 0$

      (a) -2 , $-\dfrac{1}{2}$ (b) $\dfrac{1}{2}$, 2

      (c) $\dfrac{-5+3i}{4}$, $\dfrac{-5-3i}{4}$ (d) $\dfrac{5+3i}{4}$, $\dfrac{5-3i}{4}$ 10. _____

For Exercises 11-14, solve by the method of your choice.

11. $t^2 + 8t = -7$

    (a) -7 , 1 (b) $-4 + \sqrt{23}$, $-4 - \sqrt{23}$

    (c) $4 + \sqrt{23}$, $4 - \sqrt{23}$ (d) -7 , -1 11. _____

12. $9y^2 + 1 = 6y$

    (a) -3 , 3 (b) $\dfrac{1}{3}$ (c) $\dfrac{1}{9}$ (d) $-\dfrac{1}{3}$, $\dfrac{1}{3}$ 12. _____

13. $(3w - 1)(w + 2) = 1$

    (a) $\dfrac{-5 + \sqrt{61}}{2}$, $\dfrac{-5 - \sqrt{61}}{2}$ (b) $\dfrac{-5 + \sqrt{11}}{2}$, $\dfrac{-5 - \sqrt{11}}{2}$

    (c) $\dfrac{-5 + \sqrt{61}}{6}$, $\dfrac{-5 - \sqrt{61}}{6}$ (d) $\dfrac{-5 + \sqrt{11}}{6}$, $\dfrac{-5 - \sqrt{11}}{6}$ 13. _____

14. $(3q + 1)^2 = 12$

    (a) $\dfrac{-1 + 3\sqrt{2}}{3}$, $\dfrac{-1 - 3\sqrt{2}}{3}$ (b) $-\dfrac{1}{3} + 2\sqrt{3}$, $-\dfrac{1}{3} - 2\sqrt{3}$

    (c) $\dfrac{-1 + 2\sqrt{3}}{3}$, $\dfrac{-1 - 2\sqrt{3}}{3}$ (d) $\dfrac{1 + 3\sqrt{2}}{3}$, $\dfrac{1 - 3\sqrt{2}}{3}$ 14. _____

15.   A twin-engine model rocket is launched into the air from ground level with an initial velocity of 128 feet per second. Its height s in feet is given by the formula

$$s = -16t^2 + 128t .$$

At what times will the rocket be 192 feet above the ground?

    (a)   12 seconds   (b) 2 and 6 seconds

    (c)   8 seconds   (d) 12 and 16 seconds   15. _____

16.   Find the value of x in this right triangle.

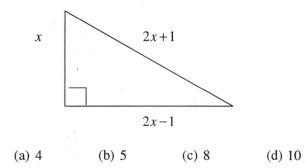

    (a) 4   (b) 5   (c) 8   (d) 10   16. _____

17.   Which one of these equations has
exactly one real number solution?

    (a) $(2z-3)^2 = 0$ (b) $(3z+2)^2 = 9$

    (c) $(3z-2)^2 = 9$ (d) $(4z-5)^2 +9 = 0$ 17. _____

For Exercises 18-21, perform the indicated operation.

18. $-(2-3i)-(4+6i)+(15-11i)$

    (a) $9-8i$ (b) $9+8i$ (c) $9-20i$ (d) $9-14i$ 18. _____

19. $(3-i)(4-3i)$

    (a) $15-13i$ (b) $15-5i$ (c) $9-13i$ (d) $9-5i$ 19. _____

20. $(1+9i)(1-9i)$

    (a) 82   (b) 80   (c) 10   (d) -8 20. _____

21. $\dfrac{9+7i}{2+3i}$

    (a) $3-i$ (b) $3+i$ (c) $-3+i$ (d) $-3-i$ 21. _____

For Exercises 22-23, sketch the graph of the equation and identify the vertex.

22.　　$y = x^2 + 2x - 3$　　　　　　　　　　　　　　　　22. _____

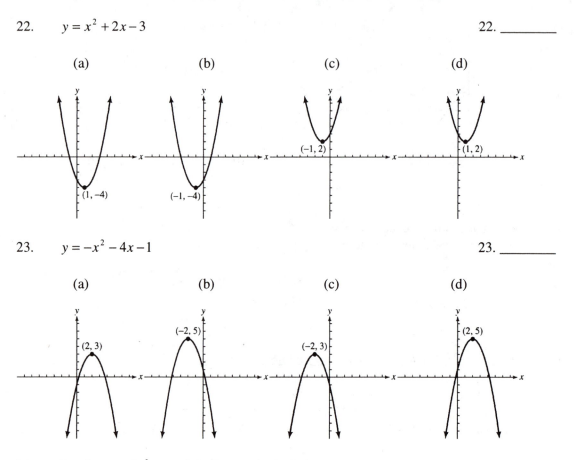

    (a)　　　　　　　(b)　　　　　　　(c)　　　　　　　(d)

23.　　$y = -x^2 - 4x - 1$　　　　　　　　　　　　　　23. _____

    (a)　　　　　　　(b)　　　　　　　(c)　　　　　　　(d)

24.　　Graph $y = -3x^2 + x - 2$ in the standard window
　　　　of a graphing calculator to determine the number of
　　　　solutions of the quadratic equation

$$-3x^2 + x - 2 = 0$$

　　　　(a) 0　　　　　(b) 1　　　　　(c) 2　　　　　(d) 3　　　　　24. _____

25.　　Which of the following equations should be used to
　　　　find two numbers whose sum is 200 and whose product
　　　　is a maximum?

　　　　(a) $M(x) = 200 - 2x$　　　　　　(b) $M(x) = 200 - 2x$
　　　　(c) $M(x) = x(200 + x)$　　　　　(d) $M(x) = x(200 - x)$　　　25. _____

FINAL
EXAMINATIONS

FINAL EXAM, FORM A   **BEGINNING ALGEBRA**

NAME_____   SECTION_____

For Exercises 1-6, perform the indicated operations.

1. $\dfrac{40 - 1^3 - (2 \cdot 8)}{2^5 - 3 \cdot 3}$ 1. _____

2. $-\left|-9\right| - \left|7 - 8\right|$ 2. _____

3. $-11 - (-2)(7) + 5 - (3 - 7)$ 3. _____

4. $-2w - 7 - 13w + 8$ 4. _____

5. $8x - 18 - 2x + 3 - 11x - 6 - 3$ 5. _____

6. $4(y - 3) - (y - 4)$ 6. _____

For Exercises 7-9, solve each equation.

7. $2z - 5 = 3$ 7. _____

8. $4a - 9a + 2a - 7 = 17$ 8. _____

9. $-8(x - 4) - 3(2 - x) = -9$ 9. _____

For Exercises 10-11, solve each problem.

10.   Find the measure of each angle.

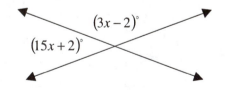

10. _____

11.   The perimeter of a football field is 340 yards. The
      length of the field is 20 yards more than twice the
      width. What are the dimensions of the field?   11. _____

12. Solve the formula $A = p + prt$ for p.

12. _____

For Exercises 13-14, solve each inequality and graph the solution set.

13. $-4t \le 16$

13.

14. $-9p + 6(p-5) > -4p - 5(2p-5)$

14.

15. Graph the equation $2x - 5y = -10$.

15.

16. Find the slope of the line passing through the points $(-2, 3)$ and $(-4, -7)$.

16. _____

17. Subtract:

$$\left(-8ab^2 + 4a - 2b^2\right) - \left(2ab^2 - 3a^2b + 2a - b\right)$$

17. _____

18. Create a table of values and graph $y = 2x^2 - 1$.

18.

For Exercises 19-21, simplify each expression. Write answers with positive exponents.

19. $\left(3^{-2}x^3\right)^{-2}$

19. _____

20. $\left(\dfrac{c^{-5}d^3}{c^{-7}d^{-3}}\right)^5$

20. _____

21. $\left(\dfrac{5}{y^{-3}}\right)^{-3}$

21. _____

For Exercises 22-24, perform each indicated operation.

22. $(3x-4)(5x+1)$

22. _____

23. $(2k-3)(3k^2-2k-2)$

23. _____

24. $\dfrac{2x^3-7x^2-14x-11}{x-5}$

24. _____

25. (a) The number of possible hands in seven-card stud poker is approximately 134,000,000. Write this number in scientific notation.

25. (a) _____

(b) Americans drink approximately 270000000 gallons of fruit juice every 30 days. How much is consumed in one day? Write your answer in scientific notation.

25. (b) _____

For Exercises 26-29, factor the polynomial.

26. $25a^3-100a^2b$

26. _____

27. $12y^2+5y-3$

27. _____

28. t^4-81

28. _____

29. $36p^2-84pq+49q^2$

29. _____

30. Solve by factoring:

$$x^2-5x-84=0$$

30. _____

31.   If an object is dropped in Earth's atmosphere,
      the distance d in feet it falls in t seconds is given
      by the formula $d = 16t^2$. How long will it take
      for an object to fall 300 feet?   31. _____

32.   Solve the quadratic inequality:

      $$x^2 - 7x + 12 < 0$$ 32. _____

33.   Write the rational expression

      $$\frac{g^2 - 36}{g^2 - g - 42}$$

      in lowest terms. For what value(s) of g is
      the original rational expression undefined?   33. _____

For Exercises 34-37, perform each operation and write the answer in lowest terms.

34. $\dfrac{4}{b-5} \div \dfrac{8}{3b-15}$ 34. _____

35. $\dfrac{3}{w-5} + \dfrac{5}{w}$ 35. _____

36. $\dfrac{1}{a^2-4} - \dfrac{a+2}{a^2-4a-12}$ 36. _____

37. $\dfrac{\dfrac{1}{y} - \dfrac{1}{x}}{\dfrac{1}{x} + \dfrac{1}{y}}$ 37. _____

38.   Solve:

      $$\frac{12}{x} - \frac{8}{x-4} = -2$$ 38. _____

39.   Brad's car travels 12 mph faster than
      Dave's car. In the same time that Brad's
      car travels 300 miles, Dave travels 240 miles.
      Find the speed of each of their cars.   39. _____

40. Two views of the same line are shown, and the displays at the bottom indicate the coordinates of two points on the line.

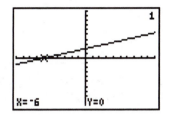 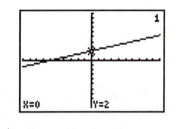

(a) Find the slope of the line.

(b) Find the equation of the line in the form $y = mx + b$.

(c) Explain why the graph indicates that when the equation is written in $y = mx + b$ form, the value of b must be positive.

40. (a) _____

40. (b) _____

40. (c) _____

41. Write an equation of a line with slope -3 and y-intercept $(0, -2)$. Give it in the form $Ax + By = C$.

41. _____

42. Graph the inequality $5x - 4y < -20$.

42.

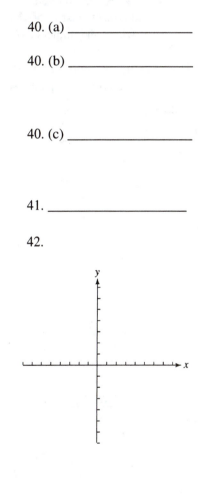

43. If $f(x) = x^3 - 2x^2 - 3x + 4$, find $f(-3)$.

43. _____

For Exercises 44-45, solve each system of equations.

44. $4x - 5y = -4$
        $6x + 10y = 1$ 44. _____

45. $3x + y = -1$
        $-6x - 2y = 2$ 45. _____

46.   Chunky Chicken charges $6.00 for a
        chicken dinner and $3.50 for an order
        of buffalo wings. After filling 28 orders
        for wings and dinners, Chunky Chicken
        had collected $140.50. How many orders
        of wings and dinners did they sell?   46. _____

47.   Graph the solution set of
        the system of inequalities:

                $3x - y \geq 6$
                $2x + y < 4$ 47.

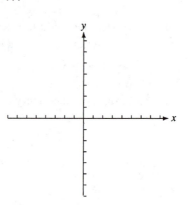

For Exercises 48-54, simplify each expression as much as possible.

48. $\sqrt{121}$ 48. _____

49. $\dfrac{3\sqrt{5}}{\sqrt{6}}$ 49. _____

50. $\sqrt[3]{\dfrac{7}{9}}$

50. _____

51. $2\sqrt{6} - 2\sqrt{24} + \sqrt{216}$

51. _____

52. $\sqrt[3]{54x^6 y^4} + \sqrt[3]{16x^6 y^4}$

52. _____

53. $x - 5 = \sqrt{x + 7}$

53. _____

54. $27^{4/3}$

54. _____

For Exercises 55-58, solve each quadratic equation using the method indicated. Give only real solutions.

55. $(2x - 3)^2 = 18$ (square root property)

55. _____

56. $x^2 + 6 = -4x$ (completing the square)

56. _____

57. $x(12x + 11) = 5$ (quadratic formula)

57. _____

58. $6x(x - 2) - 5x(x - 1) = -10$ (any method)

58. _____

59. A Little League baseball diamond is actually a square 60 feet on a side. How far is it from home plate to second base?

59. _____

60. Write in standard form:

(a) $(-3 - 9i) - (-2 + 7i) + 4 + 11i$

60. (a) _____

(b) $\dfrac{-14 - 6i}{7 + 6i}$

60. (b) _____

61.   Find the complex solutions of $x^2 + x + 1 = 0$.   61. _____

62.   Graph the quadratic function $f(x) = 4 - x^2$
      and identify the vertex. Give the domain
      and the range.

62.

63.   If the graph of the parabola $y = ax^2 + bx + c$
      opens downward and has two negative
      x-values for its x-intercepts, will the value
      for its y-intercept be positive or negative?   63. _____

FINAL EXAM, FORM B

BEGINNING ALGEBRA

NAME_____   SECTION_____

For Exercises 1-6, perform the indicated operations.

1. $\dfrac{40-1^3-(2\cdot 8)}{2^5-3\cdot 3}$

    (a) $\dfrac{23}{87}$ (b) $\dfrac{27}{26}$ (c) $\dfrac{21}{23}$ (d) 1 1._____

2. $-\left|-9\right|-\left|7-8\right|$

    (a) -10 (b) 10 (c) -8 (d) 8 2._____

3. $-11-(-2)(7)+5-(3-7)$

    (a) -24 (b) -11 (c) 4 (d) 12 3._____

4. $-2w-7-13w+8$

    (a) $-11w-15$ (b) $-15w-15$
    (c) $-11w+1$ (d) $-15w+1$ 4._____

5. $8x-18-2x+3-11x-6-3$

    (a) $-5x-24$ (b) $5x-24$
    (c) $-5x-18$ (d) $-29x$ 5._____

6. $4(y-3)-(y-4)$

    (a) $3y-8$ (b) $3y-16$
    (c) $3y+8$ (d) -12 6._____

For Exercises 7-9, solve each equation.

7. $2z-5=3$

    (a) $-\dfrac{1}{5}$ (b) -1 (c) 1 (d) 4 7._____

8. $4a - 9a + 2a - 7 = -17$

   (a) -8 (b) 8 (c) $-\dfrac{24}{7}$ (d) $\dfrac{10}{3}$ 8. _____

9. $-8(x - 4) - 3(2 - x) = -9$

   (a) 7 (b) $-\dfrac{17}{5}$ (c) $-\dfrac{27}{11}$ (d) $\dfrac{35}{11}$ 9. _____

For Exercises 10-11, solve each problem.

10.   Find the measure of each angle.

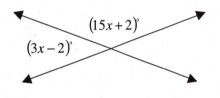

   (a) $152°, 28°$ (b) $152°, 152°$
   (c) $28°, 28°$ (d) $120°, 60°$ 10. _____

11.   The perimeter of a football field is 340 yards. The
       length of the field is 20 yards more than twice the
       width. What are the dimensions of the field?

   (a) 100 x 70   (b) 120 x 50   (c) 200 x 140   (d) 240 x 100   11. _____

12.   Solve the formula $A = p + prt$ for p.

   (a) $p = A - prt$ (b) $p = \dfrac{A}{2rt}$

   (c) $p = \dfrac{A}{1 + rt}$ (d) $p = \dfrac{A - p}{rt}$ 12. _____

For Exercises 13-14, solve each inequality.

13. $-4t \le 16$

   (a) $t \ge -4$ (b) $t \le -4$ (c) $t \ge 4$ (d) $t \le 4$ 13. _____

14. $9(4-w)+5w-12 \geq 3w-26+1$

     (a) $w \leq 1$ (b) $w \geq 1$ (c) $w \leq 7$ (d) $w \geq -7$ 14. _____

15.   Graph the equation $2x-5y=-10$.   15. _____

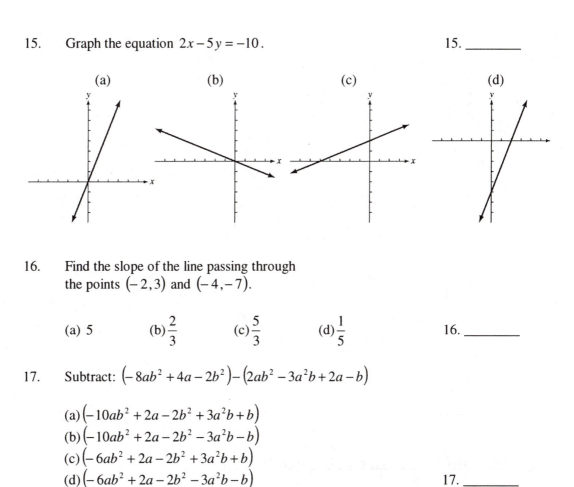

     (a)   (b)   (c)   (d)

16.   Find the slope of the line passing through
the points $(-2,3)$ and $(-4,-7)$.

     (a) 5   (b) $\dfrac{2}{3}$ (c) $\dfrac{5}{3}$ (d) $\dfrac{1}{5}$ 16. _____

17.   Subtract: $\left(-8ab^2+4a-2b^2\right)-\left(2ab^2-3a^2b+2a-b\right)$

     (a) $\left(-10ab^2+2a-2b^2+3a^2b+b\right)$
     (b) $\left(-10ab^2+2a-2b^2-3a^2b-b\right)$
     (c) $\left(-6ab^2+2a-2b^2+3a^2b+b\right)$
     (d) $\left(-6ab^2+2a-2b^2-3a^2b-b\right)$ 17. _____

18.   Graph $y=2x^2-4$.   18. _____

     (a)   (b)   (c)   (d)

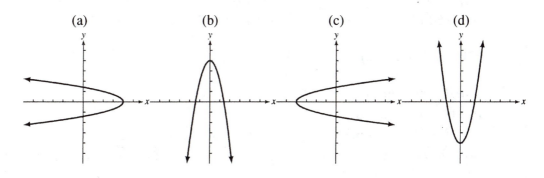

For Exercises 19-21, simplify each expression. Write answers with positive exponents.

19. $\left(3^{-2} x^3\right)^{-2}$

   (a) $\dfrac{3^4}{x^6}$ (b) $\dfrac{x}{3^4}$ (c) $3^4 x^6$ (d) $\dfrac{1}{3^4 x^6}$ 19. _____

20. $\left(\dfrac{c^{-5} d^3}{c^{-7} d^{-3}}\right)^5$

   (a) $\dfrac{d^{30}}{c^{10}}$ (b) $\dfrac{1}{c^{10}}$

   (c) $c^{10} d^{30}$ (d) $c^2 d^6$ 20. _____

21. $\left(\dfrac{5}{y^{-3}}\right)^{-3}$

   (a) $\dfrac{y^9}{5^3}$ (b) $\dfrac{5^3}{y^9}$

   (c) $\dfrac{1}{5^3 y^9}$ (d) $5^3 y^9$ 21. _____

For Exercises 22-24, perform each indicated operation.

22. $(3x - 4)(5x + 1)$

   (a) $15x^2 + 23x - 4$ (b) $15x^2 - 23x - 4$
   (c) $15x^2 + 17x - 4$ (d) $15x^2 - 17x - 4$ 22. _____

23. $(2k - 3)(3k^2 - 2k - 2)$

   (a) $6k^3 - 13k^2 + 2k + 6$
   (b) $6k^3 - 5k^2 - 10k + 6$
   (c) $6k^3 - 13k^2 - 2k + 6$
   (d) $6k^3 - 5k^2 - 2k + 6$ 23. _____

24. $\dfrac{2x^3 - 7x^2 - 14x - 11}{x - 5}$

(a) $2x^2 - 17x - 99 + \dfrac{484}{x - 5}$

(b) $2x^2 - 17x - 99 - \dfrac{506}{x - 5}$

(c) $2x^2 + 3x + 1 - \dfrac{6}{x - 5}$

(d) $2x^2 + 3x + 1 - \dfrac{16}{x - 5}$ 24. _____

25. Americans drink approximately 270000000 gallons of fruit juice every 30 days. How many gallons are consumed in one day? Write your answer in scientific notation.

(a) 9.0×10^6 (b) 9.0×10^{-6}

(c) 8.1×10^9 (d) 8.1×10^{-9} 25. _____

For Exercises 26-29, factor the polynomial completely.

26. $25a^3 - 100a^2 b$

(a) $25a^2(a - 4b)$ (b) $25(a - 2b)(a + 2b)$

(c) $25(a^2 - b)(a + 4b)$ (d) $25(a^2 - 4b)(a + b)$ 26. _____

27. $12y^2 + 5y - 3$

(a) $(2y - 3)(6y + 1)$ (b) $(4y + 3)(3y - 1)$

(c) $(12y - 1)(y + 3)$ (d) $(6y - 3)(2y + 1)$ 27. _____

28. $t^4 - 81$

(a) $(t^2 + 9)(t - 3)(t - 3)$ (b) $(t^2 - 9)(t^2 - 9)$

(c) $(t^2 + 9)(t + 3)(t - 3)$ (d) $(t - 9)(t^3 + 9)$ 28. _____

29. $36p^2 - 84pq + 49q^2$

(a) $(12p - 7q)(3p - 7q)$ (b) $(36p - q)(p - 49q)$

(c) $(6p - 7q)(6p - 7q)$ (d) $(18p - 7q)(2p - 7q)$ 29. _____

30. Solve by factoring: $x^2 - 5x - 84 = 0$

    (a) $-12, 7$ (b) $12, -7$ (c) $-21, 4$ (d) $21, -4$ 30. _____

31. If an object is dropped in Earth's atmosphere, the distance d in feet it falls in t seconds is given by the formula $d = 16t^2$. How long will it take for an object to fall 900 feet?

    (a) 4.68 seconds   (b) 7.5 seconds
    (c) 21.9 seconds   (d) 56.25 seconds   31. _____

32. Solve the quadratic inequality:

    $$x^2 - 7x + 12 < 0$$

    (a) $3 < x < 4$ (b) $x < 3$ and $x > 4$
    (c) $x < -3$ and $x > -4$ (d) $x < 3$ or $x > 4$ 32. _____

33. Write the following rational expression in lowest terms.

    $$\frac{g^2 - 36}{g^2 - g - 42}$$

    (a) $\dfrac{g + 6}{g + 7}$ (b) $\dfrac{g - 6}{g - 7}$ (c) $\dfrac{g - 6}{g + 7}$ (d) Not reducible   33. _____

For Exercises 34-37, perform each operation and write the answer in lowest terms.

34. $\dfrac{4}{b - 5} \div \dfrac{8}{4b - 20}$

    (a) 1   (b) 2   (c) $\dfrac{1}{2}$ (d) $\dfrac{8}{(b - 5)^2}$ 34. _____

35. $\dfrac{3}{w - 5} + \dfrac{5}{w}$

    (a) $\dfrac{8}{2w - 5}$ (b) $\dfrac{8w - 25}{w(w - 5)}$

    (c) $\dfrac{3w}{5(w - 5)}$ (d) $\dfrac{8w - 15}{w(w - 5)}$ 35. _____

36. $\dfrac{1}{a^2-4} - \dfrac{a+2}{a^2-4a-12}$

(a) $\dfrac{1}{(a+2)(a-2)(a-6)}$

(b) $\dfrac{a+1}{4(a+4)}$

(c) $\dfrac{-a^2+a-2}{(a+2)(a-2)(a-6)}$

(d) $\dfrac{-1(a+1)}{a-6}$

36. _____

37. $\dfrac{\dfrac{1}{y}-\dfrac{1}{x}}{\dfrac{1}{x}+\dfrac{1}{y}}$

(a) $\dfrac{1}{xy}$ (b) $\dfrac{x-y}{x+y}$ (c) $\dfrac{xy}{x+y}$ (d) -1

37. _____

38.   Solve:

$$\dfrac{12}{x} - \dfrac{8}{x-4} = -2$$

(a) $-6, 4$ (b) $6, -4$ (c) -6 (d) No solution

38. _____

39.   Brad's car travels 12 mph faster than Dave's car. In the same time that Brad's car travels 300 miles, Dave travels 240 miles. Find the speed of each of their cars.

(a) 36 mph, 48 mph

(b) 48 mph, 60 mph

(c) 60 mph, 72 mph

(d) 68 mph, 80 mph

39. _____

40. Two views of the same line are shown, and the displays at the bottom indicate the coordinates of two points on the line. Find the equation of the line in the form $y = mx + b$.

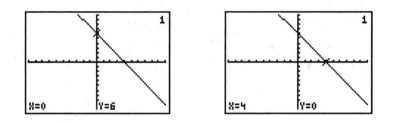

(a) $y = -\dfrac{2}{3}x + 4$ (b) $y = -\dfrac{3}{2}x + 4$

(c) $y = -\dfrac{2}{3}x + 6$ (d) $y = -\dfrac{3}{2}x + 6$ 40. _____

41. Write an equation of a line with slope –3 and y-intercept $(0, -2)$. Give it in the form $Ax + By = C$.

(a) $3x + y = -2$ (b) $2x + y = -3$

(c) $3x - 2y = 0$ (d) $2x - 3y = 0$ 41. _____

42. Graph the inequality $3x - 4y < -12$.   42. _____

   (a)   (b)   (c)   (d)

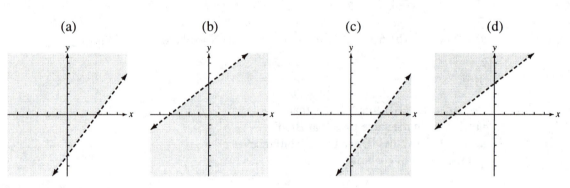

43. If $f(x) = x^3 - 2x^2 - 3x + 4$, find $f(-2)$.

(a) -18 (b) -6 (c) -2 (d) 10 43. _____

44.   Chunky Chicken charges $6.00 for a chicken dinner and $3.50 for an order
       of buffalo wings. After filling 28 orders for wings and dinners, Chunky Chicken
       had collected $140.50. How many orders of wings and dinners did they sell?

       (a) 11 orders of wings and 17 dinners
       (b) 17 orders of wings and 11 dinners
       (c) 12 orders of wings and 15 dinners
       (d) 15 orders of wings and 12 dinners   44. _____

For Exercises 45-46, solve the systems of linear equations.

45. $4x - 5y = -4$
       $6x + 10y = 1$

       (a) $(-1, 0)$ (b) $\left(-\dfrac{1}{2}, \dfrac{2}{5}\right)$

       (c) All real numbers   (d) No solution   45. _____

46. $3x + y = -1$
          $-6x - 2y = 2$

       (a) $(0, -1)$ (b) $(1, -4)$
       (c) All real numbers   (d) No solution   46. _____

47.   Graph the solution set of the system of inequalities: 47. _____

                     $3x - y \geq 6$
                     $2x + y < 4$

       (a)   (b)   (c)   (d)

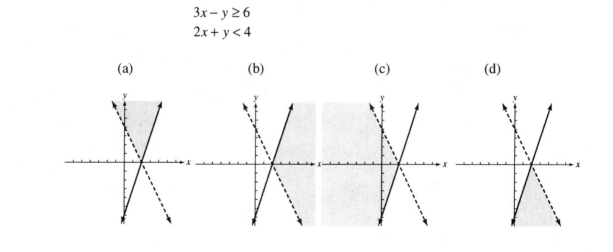

For Exercises 48-54, simplify each expression as much as possible.

48. $\sqrt{289}$

    (a) 17   (b) 13   (c) $17, -17$ (d) $13, -13$ 48._____

49. $\dfrac{3\sqrt{5}}{\sqrt{6}}$

    (a) $\dfrac{\sqrt{30}}{2}$ (b) $\dfrac{5\sqrt{6}}{2}$ (c) $3\sqrt{5}$ (d) $30\sqrt{3}$ 49._____

50. $\sqrt[3]{\dfrac{7}{9}}$

    (a) $\dfrac{\sqrt[3]{21}}{27}$ (b) $\dfrac{\sqrt[3]{21}}{3}$ (c) $\dfrac{\sqrt[3]{7}}{3}$ (d) $\sqrt[3]{7}$ 50._____

51. $5\sqrt{24} - 2\sqrt{6} + \sqrt{54}$

    (a) $5\sqrt{6}$ (b) $11\sqrt{6}$ (c) $3\sqrt{72}$ (d) $3\sqrt{84}$ 51._____

52. $\sqrt[3]{54x^6 y^4} + \sqrt[3]{16x^6 y^4}$

    (a) $2x^2 y \sqrt[3]{10y}$ (b) $7x^2 y \sqrt[3]{2y}$
    (c) $x^2 y \sqrt[3]{70y}$ (d) $5x^2 y \sqrt[3]{2y}$ 52._____

53. $x - 5 = \sqrt{x + 7}$

    (a) 2   (b) 9   (c) $2, 9$ (d) No solution   53._____

54. $64^{4/3}$

    (a) 16   (b) 256   (c) 1024   (d) 4096   54 ._____

For Exercises 55-58, solve each quadratic equation. Give only real solutions.

55. $(2x-3)^2 = 18$

(a) $\dfrac{3+3\sqrt{2}}{2}, \dfrac{3-3\sqrt{2}}{2}$ (b) $\dfrac{3+2\sqrt{3}}{2}, \dfrac{3-2\sqrt{3}}{2}$

(c) $\dfrac{2+3\sqrt{2}}{3}, \dfrac{2-3\sqrt{2}}{3}$ (d) $\dfrac{2+2\sqrt{3}}{3}, \dfrac{2-2\sqrt{3}}{3}$

55. _____

56. $x^2 + 4 = -6x$

(a) $5+\sqrt{3}, 5-\sqrt{3}$ (b) $3+\sqrt{5}, 3-\sqrt{5}$

(c) $-5+\sqrt{3}, -5-\sqrt{3}$ (d) $-3+\sqrt{5}, -3-\sqrt{5}$

56. _____

57. $x(12x+11)=5$

(a) $-\dfrac{5}{4}, \dfrac{1}{3}$ (b) $\dfrac{5}{4}, -\dfrac{1}{3}$

(c) $-\dfrac{4}{5}, 3$ (d) $\dfrac{4}{5}, -3$

57. _____

58. $6x(x-2)-5x(x-1)=-10$

(a) $5, -2$ (b) $-5, -2$

(c) $5, 2$ (d) $-5, 2$

58. _____

59.   A Little League baseball diamond is actually a square 60 feet on a side. How far is it for the catcher to throw the ball from home plate to second base (rounded to the nearest foot)?

(a) 120 feet   (b) 90 feet   (c) 85 feet   (d) 75 feet

59. _____

For Exercises 60-61, write the answer in standard form.

60. $(-3-9i)-(-2+7i)+(4+11i)$

   (a) $-1-5i$ (b) $-1+9i$ (c) $3-5i$ (d) $3+9i$ 60. _____

61. $\dfrac{3+2i}{4-3i}$

   (a) $\dfrac{6}{25}+\dfrac{9}{25}i$ (b) $\dfrac{18}{25}+\dfrac{17}{25}i$

   (c) $\dfrac{1}{5}$ (d) $\dfrac{6}{25}+\dfrac{17}{25}i$ 61. _____

62.   Find the complex solutions of $x^2+x+1=0$.

   (a) $\dfrac{-1+i\sqrt{5}}{2}, \dfrac{-1-i\sqrt{5}}{2}$ (b) $\dfrac{-1+i\sqrt{3}}{2}, \dfrac{-1-i\sqrt{3}}{2}$

   (c) $\dfrac{1+i\sqrt{5}}{2}, \dfrac{1-i\sqrt{5}}{2}$ (d) $\dfrac{1+i\sqrt{3}}{2}, \dfrac{1-i\sqrt{3}}{2}$ 62. _____

63.   Find the vertex of the parabola $f(x)=-x^2+8x-13$.

   (a) $(-4,3)$ (b) $(4,3)$ (c) $(4,-3)$ (d) $(-4,-29)$ 63. _____

ANSWERS
TO
CHAPTER TEST FORMS

Chapter 1, Form A

1. $\dfrac{24}{13}$

2. $\dfrac{11}{6}$

3. $\dfrac{10}{7}$

4. False

5.

6. −9

7. 3

8. Positive

9. 7

10. $-7\dfrac{1}{12}$

11. −10

12. 10

13. 49

14. Undefined

15. 7

16. −4

17. −3

18. −29

19. −7

20. 1302 feet

21. B

22. D

23. A

24. E

25. C

26. $21x$

27. $15x - 3$

28. $-6a + 8b - 4c$

29. $^{+}274$

30. -350 million bushels

31. $\dfrac{4}{5}$

32. 4.1 million

33. Answers will vary;

$$20 \div 5 = 4, 5 \div 20 = \dfrac{1}{4}$$

is one example.

Chapter 1, Form B

1. $\dfrac{25}{14}$

2. $\dfrac{7}{5}$

3. $\dfrac{15}{7}$

4. False

5.

6. $-|-9|$

7. 14

8. Negative

9. 0

10. $-5\dfrac{11}{20}$

11. 2

12. 12

13. 75

14. $\dfrac{2}{3}$

15. 8

16. -5

17. -3

18. -127

19. $\dfrac{21}{8}$

20. $-4°F$

21. C

22. D

23. E

24. A

25. B

26. $20y$

27. $29x - 38$

28. $-6r + 15s - 21t$

29. -213 thousand

30. 450 million bushels

31. $\dfrac{29}{50}$

32. 1.17 million

33. Answers will vary;

$5 - 3 = 2, 3 - 5 = -2$

is one example.

Chapter 1, Form C

1. $\dfrac{13}{44}$

2. $\dfrac{13}{8}$ or $1\dfrac{5}{8}$

3. $\dfrac{49}{55}$

4. True

5.

6. -0.998

7. -12

8. Negative

9. -19

10. $-1\dfrac{1}{30}$

11. -2

12. 23

13. 76

14. 2

15. 6

16. -2

17. -3

18. 8

19. $\dfrac{21}{10}$

20. $76°\,F$

21. E

22. D

23. A

24. C

25. B

26. $31a - 14$

27. $16t$

28. $5(4r - 3s)$

29. $^{+}4.8$ million

30. -350 million bushels

31. $\dfrac{93}{100}$

32. 7500

33. 0

Chapter 1, Form D

1. $\dfrac{33}{46}$

2. $\dfrac{41}{24}$

3. $\dfrac{5}{4}$

4. True

5.

6. -0.676

7. 10

8. Positive

9. -26

10. $-1\dfrac{7}{8}$

11. 16

12. 95

13. 54

14. 1

15. 4

16. 3

17. -5

18. -44

19. 2

20. $-19°\mathrm{F}$

21. D

22. A

23. B

24. E

25. C

26. $26m - 7$

27. $-3c$

28. $4(5r - 3t)$

29. $^{+}32.5$ million

30. -250 million bushels

31. $\dfrac{19}{20}$

32. 6900

33. $\dfrac{0}{4} \cdot \dfrac{4}{0} \neq 1$

Chapter 1, Form E

1. C
2. B
3. A
4. A
5. B
6. C
7. B
8. D
9. D
10. A
11. B
12. D
13. C
14. A
15. D
16. B
17. D
18. C
19. A
20. B
21. C
22. B
23. C
24. E
25. D
26. B
27. A
28. A
29. A
30. D
31. C
32. C
33. B

Chapter 1, Form F

1. D
2. A
3. B
4. B
5. A
6. C
7. A
8. B
9. B
10. C
11. D
12. D
13. D
14. A
15. B
16. D
17. D
18. A
19. C
20. A
21. D
22. B
23. E
24. C
25. A
26. D
27. B
28. D
29. B
30. C
31. A
32. B
33. A

Chapter 2, Form A

1. $x = 1$

2. $d = -3$

3. $x = 7.2$

4. All real numbers

5. $y = -37$

6. $p = -16$

7. $w = -80$

8. -5

9. 12 feet

10. (a) $b = \dfrac{2A - Bh}{h}$

10. (b) $b = 15$

11. Negative

12. $138°, 42°$

13. $125°, 125°$

14. $55°$

15. $x = -20$

16. \$16.38

17. two 100-sheet rolls at \$0.79

18. 5 gallons

19. 1080 miles

20. 28

21. \$900

22. $w \le 3$

23. $0 \le y \le 3$

24. $-6 < x < 2$

25. 90 or higher

Chapter 2, Form B

1. $x = -3$

2. $d = 1$

3. $a = 12.7$

4. No solution

5. $x = -25$

6. All real numbers

7. $k = 90$

8. -10

9. 17 inches

10. (a) $m = \dfrac{y - b}{x}$

10. (b) $m = 4$

11. Positive

12. $106^\circ, 74^\circ$

13. $118^\circ, 118^\circ$

14. 30°

15. $x = -16$

16. $21.09

17. 12 for $4.68

18. 240 gallons

19. 9 hours

20. 13

21. $1300

22. $p < -3$

23. $-1 \le y < 2$

24. $5 < t < 10$

25. 5 hours or more

Chapter 2, Form C

1. $x = 1$

2. $a = -10$

3. $y = -14.2$

4. All real numbers

5. $n = -\dfrac{25}{2}$

6. No solution

7. $x = 40$

8. 7

9. 1480 points

10. (a) $h = \dfrac{2A}{b}$

10. (b) $h = 7$

11. Answers will vary

12. $123°, 57°$

13. $150°, 150°$

14. $37°$

15. $s = 48$

16. $33.48

17. 5 for $2.85

18. 75 ounces

19. 2 hours

20. 24

21. $1250

22. $q > 5$

23. $1 \le r < 4$

24. $12 < t < 16$

25. $348 or more

Chapter 2, Form D

1. $a = 10$

2. $z = 0$

3. No solution

4. $m = 2.3$

5. $n = -\dfrac{12}{5}$

6. All real numbers

7. $x = 30$

8. 80

9. 92 costumes

10. (a) $b = \dfrac{2A - Bh}{h}$

10. (b) $b = 6$

11. Answers will vary

12. $138°, 42°$

13. $39°, 39°$

14. $43°$

15. $s = 17$

16. $29.59

17. Four 16-ounce cans

18. 80 gallons

19. 7 hours

20. 48

21. $2000

22. $b \le 1$

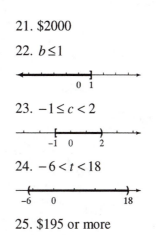

23. $-1 \le c < 2$

24. $-6 < t < 18$

25. $195 or more

Chapter 2, Form E

1. B
2. A
3. A
4. D
5. A
6. C
7. D
8. D
9. C
10. A
11. A
12. D
13. B
14. D
15. B
16. B
17. B
18. D
19. C
20. B
21. B
22. A
23. A
24. C
25. D

Chapter 2, Form F

1. B
2. B
3. A
4. B
5. C
6. A
7. D
8. C
9. A
10. A
11. C
12. B
13. A
14. A
15. C
16. B
17. C
18. C
19. C
20. B
21. C
22. D
23. A
24. A
25. C

Chapter 3, Form A

1. A

2. 50%

3. $-1, -13, 4$

4. $-\dfrac{3}{2}, \dfrac{3}{2}, \dfrac{1}{2}$

5. No

6. $y = 0$

7. x-intercept: $\left(\dfrac{3}{2}, 0\right)$

    y-intercept: $(0, -3)$

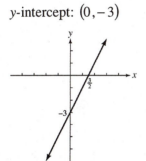

8. x-intercept: None

   y-intercept: $(0, -4)$

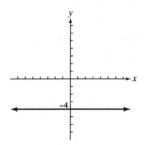

9. No. The domain element 6 corresponds

with two unique range elements, 7 and 8.

10. Inputs: $\{1, 2, 3, 4, 6\}$

   Outputs: $\{2, 3, 4, 5, 7, 8\}$

11. $m = -\dfrac{8}{5}$

12. $m = 2$

13. $m = 2$

14. $m = \dfrac{3}{2}$

15. $m = -3$

16. $m = \dfrac{1}{3}$

Chapter 3, Form B

1. A

2. 100%

3. $-5, -2, -20$

4. $-3, 4, 2$

5. No

6. $x = 0$

7. x-intercept: $(-8, 0)$

   y-intercept: $(0, 4)$

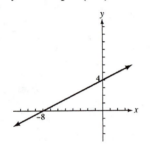

8. x-intercept: None

   y-intercept: $(0, -3)$

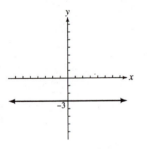

9. Yes. Every element in the domain corresponds with one unique element in the range.

10. Inputs: $\{1, 2, 3, 4, 6, 8\}$

    Outputs: $\{2, 3, 4, 5, 7, 8\}$

11. $m = -8$

12. $m = -\dfrac{7}{3}$

13. $m = \dfrac{3}{2}$

14. $m = \dfrac{2}{3}$

15. $m = -\dfrac{2}{3}$

16. $m = \dfrac{3}{2}$

Chapter 3, Form C

1. B

2. 20%

3. $-\dfrac{5}{2}, -1, -35$

4. $-\dfrac{9}{2}, -5, -3$

5. No

6. x

7. x-intercept: $\left(\dfrac{4}{5}, 0\right)$

   y-intercept: $(0, -2)$

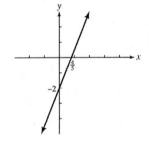

8. x-intercept: $(5, 0)$

   y-intercept: None

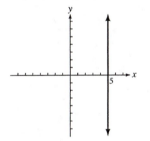

9. No. The domain element 4 corresponds with two unique range elements, 5 and 6.

10. Inputs: $\{1,2,3,4,6\}$

    Outputs: $\{2,3,4,5,6,7\}$

11. $m = -11$

12. $m = \dfrac{2}{3}$

13. $m = \dfrac{2}{5}$

14. $m = 3$

15. $m = \dfrac{11}{9}$

16. $m = -\dfrac{9}{11}$

Chapter 3, Form D

1. B

2. 50%

3. $-2, 13, -\dfrac{17}{5}$

4. $-3, -8, -4$

5. Yes

6. y

7. x-intercept: $(6, 0)$

   y-intercept: $(0, -4)$

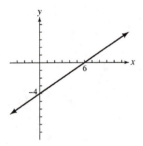

8. x-intercept: $(-2, 0)$

   y-intercept: None

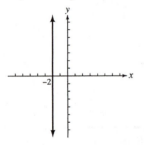

9. Yes. Every element in the domain corresponds with one unique element in the range.

10. Inputs: $\{1, 2, 3, 4, 5, 6\}$

    Outputs: $\{2, 3, 4, 5, 6, 11\}$

11. Undefined

12. $m = -\dfrac{2}{3}$

13. $m = -\dfrac{7}{3}$

14. $m = -1$

15. $m = -\dfrac{1}{9}$

16. $m = 9$

Chapter 3, Form E

1. C
2. D
3. D
4. B
5. D
6. A
7. D
8. A
9. B
10. B
11. A
12. B
13. C
14. B
15. A
16. A

Chapter 3, Form F

1. C
2. D
3. A
4. D
5. C
6. B
7. A
8. A
9. B
10. C
11. D
12. A
13. D
14. B
15. A
16. D

Chapter 4, Form A

1. $5y^2 + 2y - 1$; Degree = 2; Trinomial

2. $2a^5 - 2a^4$; Degree = 5; Binomial

3.

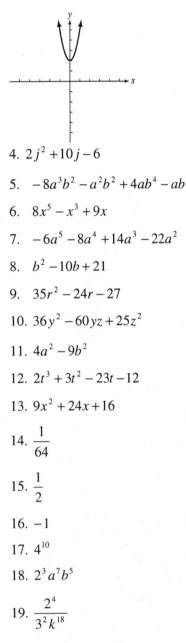

4. $2j^2 + 10j - 6$

5. $-8a^3b^2 - a^2b^2 + 4ab^4 - ab$

6. $8x^5 - x^3 + 9x$

7. $-6a^5 - 8a^4 + 14a^3 - 22a^2$

8. $b^2 - 10b + 21$

9. $35r^2 - 24r - 27$

10. $36y^2 - 60yz + 25z^2$

11. $4a^2 - 9b^2$

12. $2t^3 + 3t^2 - 23t - 12$

13. $9x^2 + 24x + 16$

14. $\dfrac{1}{64}$

15. $\dfrac{1}{2}$

16. -1

17. 4^{10}

18. $2^3 a^7 b^5$

19. $\dfrac{2^4}{3^2 k^{18}}$

20. Disagree.

$-3^4 = -81, \text{but} \left(-3\right)^4 = {}^+ 81.$

21. $-r^2 + 3r - 2 - \dfrac{4}{r}$

22. $3a + 4$

23. $2g^3 + 2g^2 + 6g + 6 - \dfrac{6}{g-1}$

24. (a) 2.6×10^7

24. (b) 0.0000045

24. (c) 250

25. 13 pounds

Chapter 4, Form B

1. $2y + 4$; Degree = 1; Binomial

2. $12a^6 - 4a^3$; Degree = 6; Binomial

3.

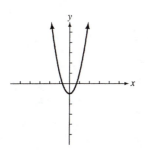

4. $-4p^2 + 15$

5. $6a^3b^2 + 8ab$

6. $16x^5 - 13x^2 + 14$

7. $-20c^2 + 25c^3 - 15c^4$

8. $z^2 + 12z + 27$

9. $10m^2 - 7m - 12$

10. $49x^2 - y^2$

11. $16 + 64d + 64d^2$

12. $3z^3 + 13z^2 + 11z - 14$

13. $4x^2 + 11x - 20$

14. $\dfrac{1}{64}$

15. 1

16. $\dfrac{5}{8}$

17. 9^5

18. $3^2 a^{12} b^8$

19. $3^3 xy^3 z$

20. Disagree. $(-3)^0 = 1$, not -1.

21. $2xy^2 - y + 3$

22. $3v + 1$

23. $2h^2 + 5h - 3 - \dfrac{2}{3h - 5}$

24. (a) 5.8×10^6

24. (b) 0.0091

24. (c) 0.031

25. $161,000,000

Chapter 4, Form C

1. $7a^3 - 10a^2$; Degree = 3; Binomial

2. $-3x^4 + 4x^3 + x^2 - 2x + 3$;

   Degree = 4; None of these

3.

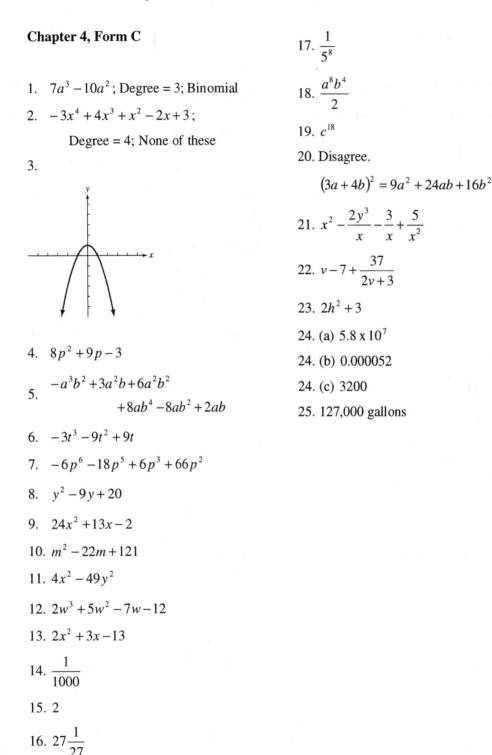

4. $8p^2 + 9p - 3$

5. $\begin{aligned} -a^3b^2 + 3a^2b + 6a^2b^2 \\ + 8ab^4 - 8ab^2 + 2ab \end{aligned}$

6. $-3t^3 - 9t^2 + 9t$

7. $-6p^6 - 18p^5 + 6p^3 + 66p^2$

8. $y^2 - 9y + 20$

9. $24x^2 + 13x - 2$

10. $m^2 - 22m + 121$

11. $4x^2 - 49y^2$

12. $2w^3 + 5w^2 - 7w - 12$

13. $2x^2 + 3x - 13$

14. $\dfrac{1}{1000}$

15. 2

16. $27\dfrac{1}{27}$

17. $\dfrac{1}{5^8}$

18. $\dfrac{a^8 b^4}{2}$

19. c^{18}

20. Disagree.

   $(3a + 4b)^2 = 9a^2 + 24ab + 16b^2$

21. $x^2 - \dfrac{2y^3}{x} - \dfrac{3}{x} + \dfrac{5}{x^2}$

22. $v - 7 + \dfrac{37}{2v + 3}$

23. $2h^2 + 3$

24. (a) 5.8×10^7

24. (b) 0.000052

24. (c) 3200

25. 127,000 gallons

Chapter 4, Form D

1. $4a^3 - 3a^2 - a$; Degree = 3; Trinomial

2. $10x^3 + 1$; Degree = 3; Binomial

3.

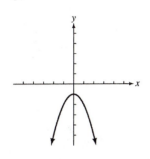

4. $6p + 11$

5. $8ab^4 - 9ab$

6. $3t^3 + 9t^2 - 9t$

7. $-6p^4 + 8p^3 + 10p^2$

8. $b^2 + b - 20$

9. $24x^2 + 55x - 24$

10. $49m^2 - 84mn + 36n^2$

11. $81p^2 - 4q^2$

12. $6x^3 - 17x^2 + 22x - 35$

13. $4y^2 + 8y - 8$

14. $\dfrac{25}{9}$

15. -2

16. $\dfrac{3}{4}$

17. 4^9

18. $2^3 \cdot 3^2 a^5 b^6$

19. $a^{16}b^{16}$

20. Agree.

$(-5)^3 = -125$, and $-5^3 = -125$.

21. $xy^2 + 2x - 4$

22. $3v^2 - 5v + 6$

23. $3h^2 - 4h + 11 - \dfrac{15}{h+1}$

24. (a) 7.6×10^8

24. (b) 0.000081

24. (c) 12000

25. $32{,}000$ bacteria

Chapter 4, Form E

1. B
2. D
3. B
4. B
5. C
6. D
7. D
8. A
9. C
10. B
11. B
12. C
13. D
14. D
15. B
16. D
17. A
18. C
19. A
20. B
21. A
22. C
23. A
24. C
25. B

Chapter 4, Form F

1. B
2. C
3. A
4. B
5. C
6. A
7. A
8. B
9. A
10. B
11. B
12. C
13. C
14. C
15. A
16. D
17. C
18. A
19. A
20. D
21. C
22. C
23. C
24. B
25. D

Chapter 5, Form A

1. B

2. $(3a+7)(7a+3)$

3. Prime

4. $(c-4)(c-5)$

5. $4w^2(w-4)(2w+1)$

6. $(4x-3)(3x+2)$

7. $(t-3)(t^2+3t+9)$

8. Prime

9. $(5x-3)(3x+8)$

10. $(x^2+9)(x+3)(x-3)$

11. $(2w+5x)(4w^2-10wx+25x^2)$

12. $(w+3z)(x-2y)$

13. $2a(a+5)^2$

14. $(3x-10y)^2$

15. Because the square of any binomial will result in a trinomial.

16. $g=6,-3$

17. $t=6,2$

18. $d=\dfrac{2}{9},-\dfrac{2}{9}$

19. $x=0,\dfrac{1}{2},-\dfrac{1}{2}$

20. The ball will be at a height of 48 feet at 1 second and 3 seconds.

21. (a) $x-7$

21. (b) 12 feet

22. 26 mpg

23. $x<-\dfrac{5}{3}$ or $x>\dfrac{1}{2}$

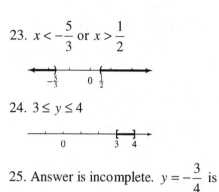

24. $3\le y\le 4$

25. Answer is incomplete. $y=-\dfrac{3}{4}$ is also a correct response.

Chapter 5, Form B

1. C

2. $(x-6)(x-5)$

3. $6y^3(3y-7)$

4. Prime

5. $(4y+5)(y+5)$

6. $(5x+2)(4x-3)$

7. $2(3t+2v)(9t^2-6tv+4v^2)$

8. $(a-1)(b-1)$

9. $-8(z-3)^2$

10. $(5-2x)(25+10x+4x^2)$

11. $(8c-1)^2$

12. $3x(2x-5)(3x+1)$

13. $b^2(c-1)^2$

14. $(9x^2+4)(3x+2)(3x-2)$

15. Because the square of any binomial will result in a trinomial.

16. $h=5,-7$

17. $t=6,3$

18. $d=\dfrac{7}{3},-\dfrac{7}{3}$

19. $x=0,\dfrac{1}{8},-\dfrac{1}{8}$

20. The ball will be at a height of 48 feet at 4 seconds and 5 seconds.

21. (a) x

21. (b) 120 feet

22. 26 mpg

23. $y<-\dfrac{2}{3}$ or $y>\dfrac{5}{2}$

24. $-6\le c\le -5$

25. Answer is incomplete. $c=-\dfrac{1}{4}$ is also a correct response.

Chapter 5, Form C

1. A

2. $(x-3)(x-5)$

3. Prime

4. $5b^2c^2(9b-6b^2c+1)$

5. $(5y+1)(3y-2)$

6. $(2y-5z)(y^2+z^2)$

7. $(4s+3t)^2$

8. $(1+5g)(1-5g+25g^2)$

9. $4(9z^2+16)$

10. $(3r+4)(r-1)$

11. $(2a-b)(4a^2+2ab+b^2)$

12. $6(d+3)(d-3)$

13. $(5d+1)(2d+1)$

14. $(9x^2+16)(3x+4)(3x-4)$

15. Because the square of any binomial will result in a trinomial.

16. $h=8,-5$

17. $x=-\dfrac{1}{2}$

18. $f=\dfrac{8}{5},-\dfrac{8}{5}$

19. $y=0,25$

20. The ball will be at a height of 48 feet at 5 seconds and 8 seconds.

21. (a) $x-7000$

21. (b) 5,000 feet

22. 26 mpg

23. $s<\dfrac{4}{3}$ or $s>\dfrac{5}{2}$

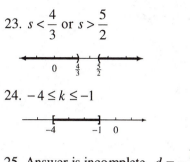

24. $-4 \le k \le -1$

25. Answer is incomplete. $d=-\dfrac{1}{3}$ is also a correct response

Chapter 5, Form D

1. C

2. $(c-7)(c-1)$

3. Prime

4. $5y^2x^2(9y-6y^2x+1)$

5. $(6a-5)(a-2)$

6. $(2b-5d)(b^2+d^2)$

7. $(6s+7t)^2$

8. $(3+5p)(9-15p+25p^2)$

9. $25(q^2+4)$

10. $(3r-2)(r-2)$

11. $(3a-2b)(9a^2+6ab+4b^2)$

12. $3(h+4)(h-4)$

13. $(4d+1)(2d-1)$

14. $2(4x^2+1)(2x+1)(2x-1)$

15. Because the square of any binomial
   will result in a trinomial.

16. $x = \dfrac{3}{2}, -\dfrac{2}{3}$

17. $h = 9, -4$

18. $y = 0$, 81

19. $p = \dfrac{8}{7}, -\dfrac{8}{7}$

20. The ball will be at a height of 48 feet
   at 3 seconds and 4 seconds.

21. (a) $x + 70$

21. (b) Eastbound car: 80 miles;
       Northbound car: 150 miles

22. 20 mpg

23. $\dfrac{1}{3} < x < \dfrac{5}{4}$

24. $d \le -5$ or $d \ge 3$

25. Answer is incomplete. $d = -\dfrac{5}{4}$ is

also a correct response.

Chapter 5, Form E

1. B
2. D
3. A
4. A
5. B
6. B
7. B
8. D
9. D
10. B
11. C
12. B
13. B
14. C
15. D
16. C
17. A
18. A
19. C
20. B
21. A
22. B
23. A
24. B
25. B

Chapter 5, Form F

1. B
2. A
3. A
4. A
5. C
6. C
7. C
8. B
9. C
10. B
11. A
12. B
13. D
14. C
15. D
16. D
17. A
18. A
19. A
20. B
21. C
22. B
23. D
24. C
25. B

Chapter 6, Form A

1. $y \neq 1, 5$

2. (a) $-\dfrac{12}{7}$

2. (b) Undefined

3. Answers will vary. $\dfrac{3x-4}{5x+6}$ is one

    example.

4. $-4ab^4$

5. $\dfrac{x-3}{x+3}$

6. $\dfrac{27}{50}$

7. $\dfrac{4(g-1)}{3(g-3)}$

8. $\dfrac{(a-4)^2}{(2a-3)(a-3)}$

9. $30b^3$

10. $(4x-1)(2x-1)(x+4)$

11. $81x^2$

12. 16

13. 2

14. $\dfrac{4}{a-3}$

15. $\dfrac{2(w+1)}{(2w-5)}$

16. $\dfrac{4r^2-10r+5}{(r+4)(r-2)(r-1)}$

17. $\dfrac{16y^2z}{9}$

18. $-\dfrac{c}{c+2}$

19. $x \neq 2, -2$

20. $x = 2$

21. No solution

22. $D = \dfrac{b}{H} - d$

23. 2 mph

24. $1\dfrac{1}{5}$ hours

25. 192 lbs./sq. ft.

Chapter 6, Form B

1. $x \neq -1, -4$

2. (a) $-\dfrac{8}{21}$

2. (b) $-\dfrac{27}{70}$

3. Answers will vary. $\dfrac{7+8p}{1-5p}$ is one

   example.

4. $-15ab^5$

5. $\dfrac{3x+1}{x-4}$

6. $\dfrac{54}{125}$

7. $\dfrac{8g-3}{3g}$

8. $\dfrac{6x+1}{5x}$

9. $75x^4$

10. $(5x-1)(3x+1)(2x+3)$

11. $40c^6$

12. $24v$

13. 3

14. $\dfrac{37}{5(a-5)}$

15. $\dfrac{2w-9}{w-3}$

16. $\dfrac{-r^2+4r+9}{(r+5)(r-5)(r-1)}$

17. $\dfrac{2z+1}{z}$

18. $\dfrac{4b-3}{a(b+2)}$

19. $r \neq 3, -3$

20. $x = 20$

21. $t = 3, -2$

22. $b = \dfrac{2-HD}{H-1}$

23. Car = 70 mph, Van = 50 mph

24. $2\dfrac{2}{9}$ hours

25. 12 ohms

Chapter 6, Form C

1. $b \neq -1, -2$

2. (a) Undefined

2. (b) $-\dfrac{27}{92}$

3. Answers will vary. $\dfrac{6n-7}{5n+1}$ is one

   example.

4. $-7fg^2$

5. $\dfrac{2p-1}{5p-2}$

6. $\dfrac{7}{8}$

7. $\dfrac{6g(3g+5)}{(g+2)^2}$

8. $\dfrac{2(4x+3)}{x}$

9. $182t^5$

10. $(7x+2)(7x-1)(2x-7)$

11. $-36d^2$

12. $14v$

13. 3

14. $\dfrac{45}{4(a-5)}$

15. $\dfrac{2q-7}{q-4}$

16. $\dfrac{-j^2-4j-10}{(j+3)(j-3)(j-4)}$

17. $\dfrac{4t-1}{t}$

18. $4c^2$

19. $k \neq 6, -6$

20. $p = 96$

21. $t = 6, -4$

22. $t = \dfrac{A-p}{P \cdot r}$

23. Passenger Train = 84 mph,

    Freight Train = 68 mph

24. $1\dfrac{1}{3}$ hours

25. 60 amps

Chapter 6, Form D

1. $d \neq 4, 5$

2. (a) Undefined

2. (b) $\dfrac{1}{7}$

3. Answers will vary. $\dfrac{6s-1}{5s+4}$ is one

   example.

4. $-5g^2h^2$

5. $\dfrac{3p+8}{3p+2}$

6. $\dfrac{81}{25}$

7. $\dfrac{3(g-3)}{2}$

8. $\dfrac{2z-5}{2z-7}$

9. $180t^5$

10. $(3x+4)(3x-4)(2x+3)$

11. $12b^3d^2$

12. $28x$

13. 3

14. $\dfrac{37}{6(a+5)}$

15. $\dfrac{1-8q}{7q-6}$

16. $\dfrac{3j^2+j+2}{(j+3)(j+2)(j-1)}$

17. $\dfrac{9y-1}{y}$

18. $5s^2t^4$

19. $k \neq 5, -5$

20. $p = \dfrac{27}{4}$

21. $v = 5$

22. $G = \dfrac{76-HL}{4L}$

23. 15 mph

24. $2\dfrac{2}{5}$ hours

25. 7.2 amps

Chapter 6, Form E

1. A
2. C
3. D
4. D
5. C
6. C
7. A
8. B
9. B
10. C
11. B
12. D
13. D
14. A
15. C
16. A
17. B
18. C
19. C
20. B
21. A
22. B
23. D
24. B
25. A

Chapter 6, Form F

1. B
2. A
3. B
4. B
5. A
6. A
7. D
8. A
9. A
10. C
11. C
12. D
13. A
14. B
15. D
16. C
17. A
18. C
19. C
20. D
21. C
22. B
23. B
24. B
25. A

Chapter 7, Form A

1. $y = -2x - 7$

2. $y = x - 1$

3. $y = \dfrac{1}{3}x + 2$

4. $3x + 4y = -28$

5.

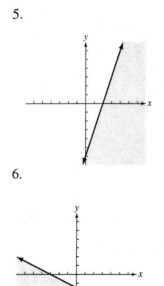

6.

7. 22

8. Not a function

9. Not a function

10. Function; Domain: All real numbers; Range: All real numbers

11. Not a function

12. False: Vertical lines are not functions

13. Yes. $f(1991) = 20063$

14. $x = 1987$

15. $m = 1848.2$

16. The number of automobile leases is increasing annually.

Chapter 7, Form B

1. $y = -4x - 2$

2. $y = \frac{1}{3}x + 1$

3. $y = -x + 4$

4. $4x + 3y = 18$

5.

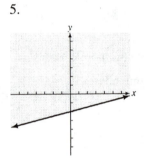

6.

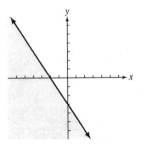

7. -18

8. Function; Domain: All real numbers;
    Range: All real numbers

9. Function; Domain: All real numbers;
    Range: $y \geq 2$

10. Not a function

11. Function; Domain: $\{1,2,3,4,6,7\}$;
    Range: $\{2,3,4,5,7,8\}$

12. $\frac{6}{5}$

13. Yes. $f(1987) = 399$

14. $x = 1990$

15. $m = 113.6$

16. The number of children who contract
    HIV is increasing annually.

Chapter 7, Form C

1. $y = -x - 6$

2. $y = -\dfrac{3}{2}x + 4$

3. $y = -\dfrac{3}{4}x + 6$

4. $4x - 5y = 5$

5.

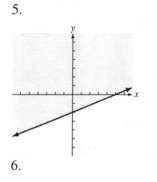

6.

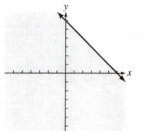

7. -3

8. Function; Domain: All real numbers; Range: $y \le -1$

9. Function; Domain: All real numbers; Range: $y \ge -1$

10. Function; Domain: All real numbers; Range: $y = -3$

11. Function; Domain: $\{1,2,3,4,6,7\}$; Range: $\{1,2,3,4,6,7\}$

12. $-\dfrac{2}{3}$

13. Yes. $f(1989) = 5.2$

14. $x = 1988$ or 1990

15. $m = -1.58$

16. The number of fish caught is decreasing annually.

Chapter 7, Form D

1. $y = -x - 5$

2. $y = \dfrac{3}{2}x - 3$

3. $y = 3x - 6$

4. $x + 5y = 45$

5.

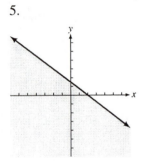

6.

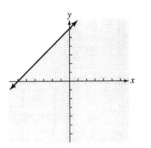

7. 0

8. Not a function

9. Not a function

10. Function; Domain: All real numbers;
    Range: $y = -2$

11. Function; Domain: $\{1, 2, 3, 4, 6, 7\}$
    Range: $\{0, 1, 4\}$

12. $-\dfrac{1}{2}$

13. Yes. $f(1992) = 4.1$

14. $x = 1988 \text{ or } 1990$

15. $m = 0.08$

16. The production of
    VitaMeataVegamin is increasing
    annually.

Chapter 7, Form E

1. B
2. A
3. D
4. A
5. A
6. B
7. A
8. E
9. C
10. C
11. C
12. A
13. B
14. D
15. A
16. B

Chapter 7, Form F

1. A
2. C
3. B
4. A
5. B
6. B
7. A
8. A
9. D
10. B
11. D
12. B
13. C
14. A
15. B
16. B

Chapter 8, Form A

1. (a) Yes

1. (b) No

1. (c) No

2. $(3,2)$

3. $(-2,3)$

4. $(2,2)$

5. $(-5,3)$

6. $(8,1)$

7. No solution

8. $(-1,-1)$

9. $(2,1)$

10. $(24,36)$

11. Infinite solutions

12. One

13. 612 adults, 150 children

14. 80 liters of 65% solution; 40 liters of 20% solution

15. 4,414 miles

16. 40 mph, 70 mph

17. width = 11 feet, length = 27 feet

18.

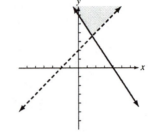

19.

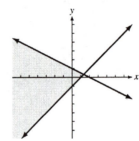

20. B

Chapter 8, Form B

1. (a) No

1. (b) Yes

1. (c) No

2. $(2,-3)$

3. $(2,-3)$

4. $(-1,0)$

5. $(-4,-1)$

6. $(2,7)$

7. No solution

8. $(6,8)$

9. $(5,-6)$

10. $(3,2)$

11. Infinite solutions

12. One

13. Yogurt = 150 calories;   Cottage
    Cheese = 360 calories

14. 364 adults, 136 children

15. 32 lbs. of peanuts; 16 lbs. of cashews

16. 40 mph, 60 mph

17. width = 14 meters,   length
    = 21 meters

18.

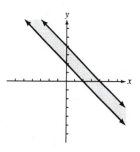

19.

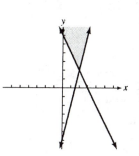

20. A

Chapter 8, Form C

1. (a) No

1. (b) No

1. (c) Yes

2. $(3,-3)$

3. $(0,4)$

4. $(-3,-1)$

5. $(-1,0)$

6. $(4,-2)$

7. No solution

8. $\left(\dfrac{16}{5},\dfrac{4}{5}\right)$

9. $(3,3)$

10. $(5,5)$

11. Infinite solutions

12. Infinite solutions

13. 4 liters of 60% solution;

   16 liters of 35% solution

14. 60 mph, 75 mph

15. 35 lbs. @ $2.50/lb.;

   15 lbs. @ $2.10/lb.

16. 534 adults, 216 children

17. width = 21 meters,

   length = 46 meters

18.

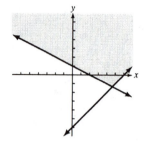

19.

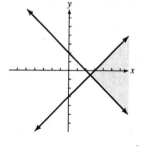

20. A

Chapter 8, Form D

1. (a) No

1. (b) No

1. (c) No

2. $(-5,1)$

3. $(3,-2)$

4. $(3,2)$

5. $(1,-1)$

6. $(-2,0)$

7. No solution

8. $\left(\dfrac{1}{3},\dfrac{1}{2}\right)$

9. $(9,11)$

10. $(1,0)$

11. Infinite solutions

12. one

13. 6 liters of 40% solution;

   2 liters of 100% solution

14. 50 mph

15. 166 sundaes; 132 cones

16. 672 adults; 128 children

17. width = 23 yards, length = 77 yards

18.

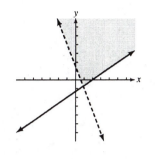

19.

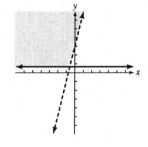

20. C

Chapter 8, Form E

1. A
2. C
3. A
4. D
5. B
6. A
7. D
8. A
9. C
10. B
11. A
12. C
13. B
14. A
15. C
16. C
17. B
18. D
19. A
20. A

Chapter 8, Form F

1. B
2. D
3. D
4. B
5. B
6. D
7. A
8. D
9. B
10. A
11. B
12. A
13. A
14. D
15. C
16. A
17. B
18. B
19. A
20. D

Chapter 9, Form A

1. $-15, 15$

2. (a) Irrational

2. (b) 17.321

3. $d = 2\sqrt{17}$

4. 3

5. $7\sqrt{2}$

6. $\dfrac{10\sqrt{2}}{13}$

7. $3\sqrt[3]{2}$

8. $16\sqrt{5}$

9. $29\sqrt{2}$

10. 0

11. $2xy^2\sqrt[3]{3x^2}$

12. 34

13. $8\sqrt{3} + 4\sqrt{15} + \sqrt{5} + 2$

14. $11 - 6\sqrt{2}$

15. (a) $\sqrt{119}$

15. (b) 10.909

16. $2\sqrt{30}$

17. $\dfrac{\sqrt{30f}}{3f}$

18. $3\sqrt[3]{4}$

19. $\dfrac{\sqrt{6}+4}{2}$

20. $y = 0, 3$

21. $z = 16$

22. -4

23. 16

24. $\dfrac{1}{3}$

25. Statement is false.

$\sqrt{16} + \sqrt{9} = 4 + 3 = 7;$

$\sqrt{16+9} = \sqrt{25} = 5;$

$7 \neq 5$

Chapter 9, Form B

22. 81

23. 9

24. 25

1. $-50, 50$

25. Statement is false.

$$\sqrt{144} + \sqrt{25} = 12 + 5 = 17;$$
$$\sqrt{144 + 25} = \sqrt{169} = 13;$$
$$17 \neq 13$$

2. (a) Irrational

2. (b) 12.728

3. $d = \sqrt{10}$

4. -3

5. $-10\sqrt{2}$

6. 30

7. $-2\sqrt[3]{11}$

8. $\dfrac{1}{6}$

9. $\sqrt{7}$

10. $\sqrt{5p}$

11. $2x^2 y \sqrt[4]{4x^2 y^3}$

12. 22

13. $5\sqrt{6} + 14$

14. $21 - 2\sqrt{110}$

15. (a) $8\sqrt{6}$

15. (b) 19.596

16. $-\dfrac{\sqrt{6}}{2}$

17. $\dfrac{2\sqrt{6v}}{v}$

18. $-6\sqrt[3]{2}$

19. $-\dfrac{5\left(2 + \sqrt{6}\right)}{2}$

20. $z = 16$

21. $k = -1$

Chapter 9, Form C

1. $-\dfrac{11}{6}, \dfrac{11}{6}$

2. (a) Irrational

2. (b) 12.247

3. $d = 10$

4. -2

5. $-2\sqrt{13}$

6. $\dfrac{5\sqrt{5}}{16}$

7. $3\sqrt[3]{5}$

8. $\dfrac{9\sqrt{5}}{\sqrt{3}}$ or $3\sqrt{15}$

9. $2\sqrt{11}$

10. $-31\sqrt{6p}$

11. $2xy^2 \sqrt[3]{2xz^2}$

12. $70 - 16\sqrt{6}$

13. $6\sqrt{6} + 9\sqrt{2} - \sqrt{3} - 2$

14. 57

15. (a) $26\sqrt{113}$

15. (b) 276.384 miles

16. $\sqrt{7}$

17. $\dfrac{\sqrt{66t}}{6t}$

18. $\dfrac{2\sqrt{10} - \sqrt{15}}{5}$

19. $\dfrac{\sqrt[4]{6}}{3}$

20. $z = 16$

21. $k = 8$

22. -125

23. 16

24. $\dfrac{1}{6}$

25. When you subtract 5 from both sides of the equation, you are left with a square root set equal to a negative number. This is impossible.

Chapter 9, Form D

1. $-\dfrac{20}{13}, \dfrac{20}{13}$

2. (a) Irrational

2. (b) 28.284

3. $d = \sqrt{173}$

4. $-4\sqrt{10}$

5. $\dfrac{3}{8}$

6. 8

7. $3\sqrt[3]{2}$

8. $-23\sqrt{3}$

9. $2ab\sqrt[4]{2a^3 b}$

10. $-10\sqrt{10p}$

11. 2

12. $19 - 6\sqrt{10}$

13. $2\sqrt{21} + 2\sqrt{7} - \sqrt{3} - 1$

14. -5

15. (a) $7\sqrt{569}$

15. (b) 166.976

16. $4\sqrt{6}$

17. $\dfrac{\sqrt{21t}}{7t}$

18. $-\dfrac{2\sqrt{2} + \sqrt{14} - 7\sqrt{7} - 14}{3}$

19. $\dfrac{\sqrt[3]{10}}{5}$

20. $x = 4$

21. No solution

22. $-\dfrac{16}{81}$

23. 361

24. $\dfrac{1}{5^{4/5}}$

25. Yes, I agree. To evaluate this expression, I would need to take the fourth root of a negative number, which is impossible under the real numbers.

Chapter 9, Form E

1. C
2. A
3. C
4. A
5. B
6. C
7. B
8. C
9. D
10. C
11. A
12. B
13. A
14. A
15. C
16. D
17. B
18. B
19. B
20. D
21. D
22. A
23. C
24. D
25. B

Chapter 9, Form F

1. D
2. C
3. A
4. A
5. A
6. C
7. B
8. C
9. C
10. A
11. A
12. B
13. C
14. B
15. A
16. A
17. A
18. B
19. A
20. C
21. C
22. B
23. A
24. D
25. C

Chapter 10, Form A

1. $z = \pm 5\sqrt{3}$

2. $n = 10, -4$

3. $x = \dfrac{2 \pm 3\sqrt{2}}{5}$

4. $x = -6 \pm 3\sqrt{3}$

5. $t = 7, -2$

6. (a) 0

6. (b) One real solution

7. $y = \dfrac{1}{2}, -\dfrac{1}{3}$

8. $t = 2 \pm \sqrt{5}$

9. $x = \dfrac{-2 \pm 6i}{5}$

10. $x = \dfrac{1 \pm i\sqrt{31}}{8}$

11. $t = -6$

12. $x = \dfrac{1 \pm 3\sqrt{3}}{2}$

13. $g = 2 \pm i\sqrt{2}$

14. $q = \dfrac{-1 \pm \sqrt{85}}{6}$

15. $t = 3$ seconds

16. $P = 0.24$

17. $1 + 0i$

18. $-2 + 23i$

19. 85

20. $\dfrac{27 + 53i}{58}$

21. Vertex: $(2, 6)$

22. Vertex: $\left(\dfrac{5}{2}, -\dfrac{1}{4}\right)$

23. Vertex: $(-2, 3)$

24. (a) Two real solutions

24. (b) $x = 1, -5$

24. (c) $x = 1.000, -5.000$

25. $450, 450$

Chapter 10, Form B

1. $r = \pm 7\sqrt{2}$

2. $d = 13, -7$

3. $x = \dfrac{3 \pm 4\sqrt{3}}{2}$

4. $w = -1, -9$

5. $t = \dfrac{5 \pm \sqrt{35}}{2}$

6. (a) 169

6. (b) Two real solutions

7. $y = \dfrac{1 \pm \sqrt{151}}{15}$

8. $p = \dfrac{1 \pm 2i\sqrt{5}}{7}$

9. $t = 4 \pm \sqrt{13}$

10. $x = \dfrac{-1 \pm i\sqrt{14}}{3}$

11. $f = -3, -9$

12. $x = \dfrac{7 \pm 2\sqrt{5}}{2}$

13. $g = \dfrac{3 \pm \sqrt{41}}{8}$

14. $q = -\dfrac{1}{2}, -2$

15. 3 seconds

16. $P = 165.14$

17. $7 + 5i$

18. $52 - 26i$

19. 185

20. $\dfrac{-29 - 15i}{82}$

21. Vertex: $(-1, -4)$

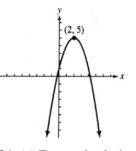

22. Vertex: $(-2, -2)$

23. Vertex: $(2, 5)$

24. (a) Two real solutions

24. (b) $x = -4, -2$

24. (c) $x = -4.000, -2.000$

25. 75 , 75

Chapter 10, Form C

1. $w = \pm 6\sqrt{3}$

2. $x = 15, -7$

3. $x = \dfrac{1 \pm 3\sqrt{5}}{5}$

4. $w = -15, 1$

5. $t = -\dfrac{4}{3}, 4$

6. (a) -20

6. (b) No real solutions

7. $y = \dfrac{-2 \pm 6i}{5}$

8. $p = \dfrac{-9 \pm \sqrt{61}}{2}$

9. $t = \dfrac{7 \pm \sqrt{29}}{2}$

10. $x = \dfrac{1 \pm i\sqrt{39}}{10}$

11. $f = -3, 9$

12. $x = \dfrac{1 \pm 2\sqrt{3}}{8}$

13. $g = \dfrac{1}{2}, -2$

14. $q = \dfrac{1 \pm \sqrt{41}}{4}$

15. 5 seconds

16. $P = 0.61$

17. $0 + 146i$

18. $-18 + 4i$

19. $\dfrac{-2 - i}{5}$

20. 85

21. Vertex: $(-1, 2)$

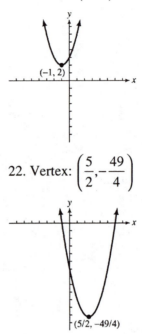

22. Vertex: $\left(\dfrac{5}{2}, -\dfrac{49}{4} \right)$

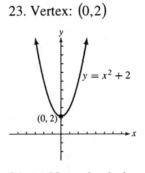

23. Vertex: $(0, 2)$

24. (a) No real solutions

24. (b) $x = -1 \pm \sqrt{5}$

24. (c) $x = 1.236, -3.236$

25. 35 , 35

Chapter 10, Form D

1. $g = \pm 9\sqrt{2}$

2. $x = -15, -3$

3. $x = \dfrac{9 \pm 5\sqrt{3}}{2}$

4. $w = 4$

5. $t = -1 \pm \sqrt{5}$

6. (a) 76

6. (b) Two real solutions

7. $y = 2 \pm i\sqrt{2}$

8. $p = 3 \pm \sqrt{5}$

9. $x = \dfrac{1 \pm \sqrt{15}}{7}$

10. $x = 3 \pm i\sqrt{3}$

11. $f = \dfrac{3}{4}$

12. $y = \dfrac{-2 \pm 2\sqrt{6}}{3}$

13. $g = 3, -7$

14. $q = \dfrac{-7 \pm \sqrt{37}}{6}$

15. 7 seconds

16. $P = 84.13$

17. $31 + 29i$

18. $8 - i$

19. $\dfrac{-16 + 21i}{17}$

20. 145

21. Vertex: $(3, -4)$

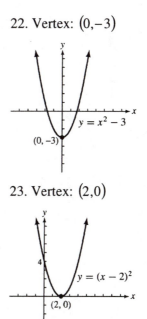

22. Vertex: $(0, -3)$

23. Vertex: $(2, 0)$

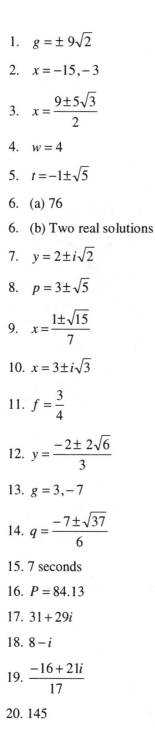

24. (a) One real solution

24. (b) $x = \dfrac{3}{2}$

24. (c) $x = 1.500$

25. 72 , 72

Chapter 10, Form E

1. C
2. B
3. C
4. B
5. D
6. A
7. C
8. A
9. D
10. D
11. D
12. C
13. A
14. D
15. D
16. C
17. B
18. C
19. A
20. C
21. B
22. A
23. B
24. C
25. A

Chapter 10, Form F

1. C
2. B
3. C
4. A
5. D
6. B
7. A
8. B
9. A
10. A
11. D
12. B
13. C
14. C
15. B
16. C
17. A
18. D
19. C
20. A
21. A
22. B
23. C
24. A
25. D

ANSWERS
TO
FINAL EXAMINATIONS

Final Exam,
Form A

1. 1

2. -10

3. 12

4. $-15w+1$

5. $-5x-24$

6. $3y-8$

7. $z=4$

8. $a=-8$

9. $x=7$

10. $152°,28°$

11. width = 50 yards; length = 120
    yards

12. $p=\dfrac{A}{1+rt}$

13. $t\geq-4$

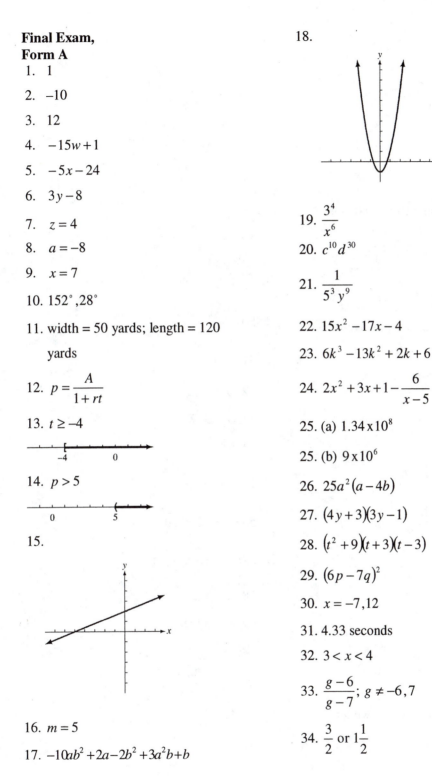

14. $p>5$

15.

16. $m=5$

17. $-10ab^2+2a-2b^2+3a^2b+b$

18.

19. $\dfrac{3^4}{x^6}$

20. $c^{10}d^{30}$

21. $\dfrac{1}{5^3y^9}$

22. $15x^2-17x-4$

23. $6k^3-13k^2+2k+6$

24. $2x^2+3x+1-\dfrac{6}{x-5}$

25. (a) 1.34×10^8

25. (b) 9×10^6

26. $25a^2(a-4b)$

27. $(4y+3)(3y-1)$

28. $(t^2+9)(t+3)(t-3)$

29. $(6p-7q)^2$

30. $x=-7,12$

31. 4.33 seconds

32. $3<x<4$

33. $\dfrac{g-6}{g-7};\ g\neq-6,7$

34. $\dfrac{3}{2}$ or $1\dfrac{1}{2}$

35. $\dfrac{8w-25}{w(w-5)}$

36. $\dfrac{-a^2+a-2}{(a+2)(a-2)(a-6)}$

37. $\dfrac{x-y}{x+y}$

38. $x = 6, -4$

39. Dave = 48 mph; Brad = 60 mph

40. (a) $m = \dfrac{1}{3}$

40. (b) $y = \dfrac{1}{3}x + 2$

40. (c) The value of b is positive because the second picture shows the line crossing the y-axis at +2.

41. $3x + y = -2$

42.

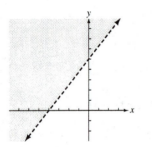

43. –32

44. $\left(-\dfrac{1}{2}, \dfrac{2}{5}\right)$

45. Infinite solutions

46. 17 dinners; 11 orders of wings

47.

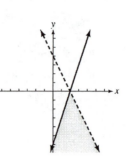

48. 11

49. $\dfrac{\sqrt{30}}{2}$

50. $\dfrac{\sqrt[3]{21}}{3}$

51. $4\sqrt{6}$

52. $5x^2 y\sqrt[3]{2y}$

53. $x = 9$

54. 81

55. $x = \dfrac{3 \pm 3\sqrt{2}}{2}$

56. $x = -2 \pm i\sqrt{2}$

57. $x = -\dfrac{5}{4}, \dfrac{1}{3}$

58. $x = 5, 2$

59. 84.85 feet

60. (a) $3 - 5i$

60. (b) $\dfrac{-134 + 42i}{85}$

61. $x = \dfrac{-1 \pm i\sqrt{3}}{2}$

62.

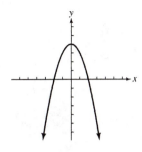

Vertex: $(0, 4)$

Domain: All real numbers

Range: $y \le 4$

63. Negative

Final Exam, Form B

1. D
2. A
3. D
4. D
5. A
6. A
7. D
8. D
9. A
10. A
11. B
12. C
13. A
14. C
15. C
16. A
17. A
18. D
19. A
20. C
21. C
22. D
23. A
24. C
25. A
26. A
27. B
28. C
29. C
30. B
31. B
32. A
33. B
34. B
35. B
36. C
37. B
38. B
39. B
40. D
41. A
42. D
43. B
44. A
45. B
46. C
47. D
48. C
49. A
50. B
51. B
52. D
53. B
54. B
55. A
56. D
57. A
58. C
59. C
60. C
61. D
62. B
63. B

ADDITIONAL
EXERCISES

CHAPTER 1   THE REAL NUMBER SYSTEM

Section 1.1   Fractions

Objective 1   Learn the definition of factor.

Identify the number as *prime, composite,* or *neither.*

1. 19   2. 35   3. 1   4. 51

5. 43   6. 596   7. 3993   8. 61

9. 127   10. 2301   11. 8448   12. 89

Write the number as the product of prime factors.

13. 28   14. 48   15. 64   16. 78

17. 84   18. 104   19. 180   20. 156

21. 196   22. 228   23. 37   24. 546

Objective 2   Write fractions in lowest terms.

Write the fraction in lowest terms.

25. $\dfrac{30}{36}$ 26. $\dfrac{15}{25}$ 27. $\dfrac{25}{60}$ 28. $\dfrac{42}{15}$

29. $\dfrac{48}{150}$ 30. $\dfrac{36}{24}$ 31. $\dfrac{28}{336}$ 32. $\dfrac{180}{216}$

33. $\dfrac{144}{324}$ 34. $\dfrac{252}{882}$ 35. $\dfrac{224}{504}$ 36. $\dfrac{990}{132}$

Objective 3   Multiply and divide fractions.

Find the product and write it in lowest terms.

37. $\dfrac{7}{3} \cdot \dfrac{6}{14}$ 38. $\dfrac{3}{8} \cdot \dfrac{16}{21}$ 39. $\dfrac{7}{6} \cdot \dfrac{18}{14}$ 40. $\dfrac{3}{4} \cdot 20$

41. $\dfrac{15}{11} \cdot \dfrac{22}{10}$

42. $\dfrac{12}{15} \cdot \dfrac{10}{15}$

43. $\dfrac{5}{8} \cdot \dfrac{16}{25}$

44. $\dfrac{6}{7} \cdot \dfrac{28}{21}$

45. $2\dfrac{1}{5} \cdot 3\dfrac{1}{3}$

46. $4\dfrac{3}{8} \cdot 5\dfrac{3}{7}$

47. $8\dfrac{5}{6} \cdot 3\dfrac{1}{5}$

48. $5\dfrac{3}{4} \cdot 3\dfrac{1}{2}$

Find the quotient and write it in lowest terms.

49. $\dfrac{5}{4} \div \dfrac{15}{24}$

50. $\dfrac{3}{8} \div \dfrac{9}{32}$

51. $\dfrac{2}{3} \div \dfrac{14}{15}$

52. $\dfrac{5}{7} \div \dfrac{10}{14}$

53. $\dfrac{12}{13} \div 6$

54. $\dfrac{12}{11} \div \dfrac{36}{99}$

55. $\dfrac{4}{11} \div 32$

56. $\dfrac{16}{15} \div \dfrac{28}{60}$

57. $8\dfrac{7}{10} \div 3\dfrac{5}{8}$

58. $1\dfrac{4}{9} \div 2\dfrac{2}{3}$

59. $9\dfrac{5}{8} \div 3\dfrac{1}{2}$

60. $5\dfrac{1}{3} \div 1\dfrac{1}{3}$

Objective 4   Add and subtract fractions

Find the sum and write it in lowest terms.

61. $\dfrac{3}{10} + \dfrac{1}{10}$

62. $\dfrac{3}{8} + \dfrac{7}{8}$

63. $\dfrac{5}{6} + \dfrac{1}{6}$

64. $\dfrac{5}{12} + \dfrac{5}{18}$

65. $\dfrac{1}{5} + \dfrac{7}{10}$

66. $\dfrac{13}{45} + \dfrac{41}{75}$

67. $\dfrac{13}{8} + \dfrac{3}{4}$

68. $\dfrac{11}{16} + \dfrac{7}{12}$

69. $\dfrac{1}{6} + \dfrac{3}{4} + \dfrac{2}{9}$

70. $3\dfrac{2}{3} + 4\dfrac{5}{6}$

71. $6\dfrac{3}{5} + 5\dfrac{1}{2}$

72. $4\dfrac{3}{4} + 5\dfrac{1}{8}$

Find the difference and write it in lowest terms.

73. $\dfrac{5}{6} - \dfrac{1}{6}$

74. $\dfrac{7}{11} - \dfrac{4}{11}$

75. $\dfrac{11}{12} - \dfrac{5}{12}$

76. $\dfrac{8}{9} - \dfrac{3}{5}$

77. $\dfrac{3}{8} - \dfrac{1}{6}$

78. $\dfrac{7}{15} - \dfrac{3}{10}$

79. $\dfrac{7}{30} - \dfrac{3}{20}$

80. $\dfrac{19}{18} - \dfrac{3}{4}$

81. $\dfrac{31}{16} - \dfrac{11}{9}$

82. $10\dfrac{1}{9} - 2\dfrac{2}{3}$

83. $11\dfrac{1}{6} - 2\dfrac{2}{3}$

84. $3\dfrac{1}{8} - 1\dfrac{5}{6}$

Objective 5   Solve applied problems that involve fractions.

Solve the problem.

85. A triangle has sides of length 1/2 foot, 1 1/4 feet, and 1 1/8 feet. What is the distance around the triangle?

86. Betty bought material to make five different dresses. If the first dress takes 1 5/8 yards, the second dress takes 1 1/3 yards, the third dress takes 1 3/8 yards, the fourth dress takes 1 1/2 yards, and the fifth dress takes 1 7/8 yards, how many yards did she buy?

87. Saul sold 3/5 bushel of potatoes, 2/5 bushel of apples, 3/4 bushel of pears, 1/4 bushel of peppers, and 1 1/4 bushels of tomatoes. How many bushels of fruits and vegetables did he sell?

88. Pete has 12 2/3 cords of firewood for sale. If he sells the firewood in face cord lots (a face cord equals 1/3 of a cord), how many face cords does he have for sale?

89. Petri is planning a party for 28 guests. If he plans 2/3 pound of beef for each person, how many pounds of beef should he buy?

90. If an upholsterer needs 6 3/4 yards of fabric to re-cover a sofa, how many sofas can he re-cover with 54 yards of fabric?

91. Two sockets in a socket wrench set have diameters of 11/16 inch and 5/8 inch. What is the difference between these diameters?

92. A cake recipe calls for 1 2/3 cups of sugar. A caterer has 20 cups of sugar on hand. How many cakes can she make?

93. Marissa bought 8 yards of fabric. She used 1 3/4 yards for a blouse and 3 5/8 yards for a skirt. How many yards were left?

94. Lois Schwartz worked 40 hours during a certain week. She worked 8 ½ hours c
    Monday, 7 ¾ hours on Tuesday, 6 5/8 hours on Wednesday, and 8 7/8 hours on
    Thursday. How many hours did she work on Friday?

Objective 6   Interpret data in a circle graph

Use the circle graph to answer the questions.

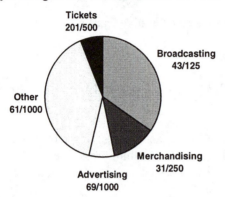

Major League Baseball Revenue in 1998

95. What fractional part of total revenue went to broadcasting?

96. What fractional part of total revenue went to areas other than merchandising?

97. What fractional part of total revenue went to areas other than broadcasting and
    advertising?

98. If the annual revenue was $2.7 billion:
    (a) How much more was spent on tickets than on broadcasting?
    (b) How much was spent on areas other than advertising?
    (c) How much more was spent on advertising than on "other" expenses?

Mixed Exercises

Write the number as the product of prime factors.

99. 63   100. 112   101. 41

102. 1110   103. 720   104. 910

Perform the indicated operation. Write the answer in lowest terms.

105. $\dfrac{4}{15}+\dfrac{3}{5}$

106. $\dfrac{2}{3}-\dfrac{7}{12}$

107. $\dfrac{75}{30}\cdot\dfrac{24}{25}$

108. $\dfrac{2}{9}\div\dfrac{14}{27}$

109. $\dfrac{11}{13}+\dfrac{15}{26}$

110. $\dfrac{7}{8}\cdot\dfrac{16}{21}$

111. $\dfrac{7}{2}\div\dfrac{28}{6}$

112. $\dfrac{33}{45}-\dfrac{7}{18}$

113. $3\dfrac{1}{5}+2\dfrac{11}{12}$

114. $10\dfrac{7}{8}-6\dfrac{5}{6}$

115. $4\dfrac{1}{8}\cdot3\dfrac{1}{3}$

116. $9\dfrac{1}{5}\div4\dfrac{3}{5}$

117. $\dfrac{1}{2}+\dfrac{1}{6}$

118. $3\dfrac{1}{5}+1\dfrac{3}{4}$

119. $2\dfrac{2}{3}+3\dfrac{1}{6}$

120. $5\dfrac{3}{4}-2\dfrac{4}{5}$

121. $\dfrac{3}{4}-\dfrac{4}{9}$

122. $\dfrac{13}{15}-\dfrac{1}{5}$

Solve the problem.

123. A dressmaker bought 20 yards of fabric. She made three different garments from this fabric. The first required 3 2/3 yards, the second 5 1/4 yards, and the third 2 1/2 yards. How much fabric was left?

124. Gena picked 8 2/3 bushels of apples in the morning and 5 ¾ bushels in the afternoon. How many more bushels did she pick in the morning?

Writing/Conceptual Exercises

125. How many prime numbers are there between 10 and 30?

126. A student adds two fractions in the following way:

$$\frac{3}{5}+\frac{5}{6}=\frac{8}{11}.$$

How did the student obtain this answer? Explain to the student why his work is incorrect and how it can be corrected.

127. A student was given the following problem:

$$\frac{5}{9} \cdot \frac{3}{7}.$$

She told you that her first step would be to find the LCD. How would you respond?

128. Which one of the following is the correct way to write 42/60 in lowest terms?

(a) $\dfrac{42}{60} = \dfrac{3 \cdot 14}{3 \cdot 20} = \dfrac{14}{20}$
　　　　　　　　　　　　　　(b) $\dfrac{42}{60} = \dfrac{6 \cdot 7}{3 \cdot 20} = \dfrac{7}{10}$

(c) $\dfrac{42}{60} = \dfrac{30 + 12}{48 + 12} = \dfrac{30}{48} = \dfrac{5}{8}$
　　　　　　(d) $\dfrac{42}{60} = \dfrac{49 - 7}{70 - 10} = \dfrac{7}{10}$

Section 1.2   Exponents, Order of Operations, and Inequality

Objective 1   Use exponents

Find the value of the exponential expression.

129. 9^2 　　　　　　　130. 10^4 　　　　　　131. 8^3 　　　　　　132. 3^6

133. $\left(\dfrac{11}{12}\right)^2$ 　　　　134. $\left(\dfrac{1}{2}\right)^8$ 　　　　135. $\left(\dfrac{3}{7}\right)^3$ 　　　　136. $\left(\dfrac{5}{6}\right)^4$

137. $(.7)^2$ 　　　　　138. $(.02)^3$ 　　　　139. $(.56)^2$ 　　　　140. $(1.9)^3$

141. 6^4 　　　　　　142. 18^2 　　　　　143. 10^5 　　　　　144. 11^3

Objective 2   Use the order of operations rules.

Find the value of each expression.

145. $3 \cdot 8 + 2$ 　　　　　　　　146. $4 + 3 \cdot 7$

147. $5 + 16 \div 2$ 　　　　　　　　148. $8 \cdot 15 - 10^2$

149. $3 \cdot 8 + 8 \cdot 2$ 　　　　　　　150. $25 \div 5 - 3 \cdot 1$

151. $\dfrac{1}{2} \cdot \dfrac{2}{3} + \dfrac{3}{4} \cdot \dfrac{5}{3}$

152. $\left(\dfrac{3}{4}\right)^2 - \left(\dfrac{2}{5}\right)\left(\dfrac{5}{4}\right)$

153. $(1.2)(2.3) - (.4)(.8)$

154. $2^3 + 3^2 - 4(8 - 5)$

Objective 3   Use more than one grouping symbol.

Find the value of the expression.

155. $6[5 + 3(2)]$

156. $6 + 3[7(2) - 9]$

157. $5 + 4[2(6) + 3]$

158. $2[6(5) - 10]$

159. $3 + 7[15 - 6(2)]$

160. $7[2(5) - 4] + 6$

161. $4[3 + 9(7 - 2)]$

162. $90 - 3[8(5 - 2 + 6)]$

163. $5^2[(22 - 8) + 4]$

164. $3^3[(6 + 9) - 2]$

165. $\dfrac{7^2 + 5(3 - 1)}{5^2 - 2(5 + 1)}$

166. $\dfrac{6 \cdot 10 + 9 \cdot 3}{2(4 - 2)}$

167. $\dfrac{5(4^2 - 7)}{3 \cdot 8 - 9}$

168. $\dfrac{10(5 - 2) - 9(6 - 4)}{2(4 - 1) - 2^2}$

Objective 4   Know the meanings of $\neq, <, >, \leq$, and $\geq$.

Tell whether the statement is *true* or *false*.

169. $98 > 99$

170. $45 \leq 45$

171. $3 \cdot 4 \div 2^2 \neq 3$

172. $3.02 \leq 3.2$

173. $2[3(4) + 6(5)] \geq 85$

174. $4[3 \cdot 2 + 5(6)] \leq 150$

175. $\dfrac{3}{8} > \dfrac{5}{16}$

176. $2\dfrac{1}{4} \geq 1\dfrac{3}{4}$

177. $1\frac{1}{2} + 2\frac{3}{4} < 4$

178. $\frac{6 + 4 \cdot 2}{13 - 2 \cdot 3} \geq 2$

179. $3 \geq \frac{4(3+1) - 2(2+1)}{3 \cdot 2 - 1}$

180. $\frac{4 + 3 \cdot 2^2}{5^2 - 3 \cdot 3} < 1$

Objective 5   Translate word statements to symbols.

Write the word statement in symbols.

181. Ten equals thirteen minus three.

182. Fourteen is less than twenty.

183. Twelve is not equal to seven.

184. Twenty-two is greater than nineteen.

185. The difference between twenty and eight is greater than ten.

186. Nineteen is less than thirty-two.

187. Six is greater than or equal to zero.

188. The sum of eight and eighteen is greater than fifteen.

189. Three is greater than the quotient of ten and five.

190. Seventeen is less than the product of three and ten.

Objective 6   Write statements that change the direction of inequality symbols.

Write the statement with the inequality symbol reversed while keeping the same meaning.

191. $5 < 7$

192. $12 \geq 11$

193. $0 < \frac{1}{100}$

194. $\frac{1}{2} > \frac{1}{3}$

195. $\frac{2}{3} < \frac{3}{4}$

196. $\frac{2}{7} \leq \frac{2}{5}$

197. .48 > .47   198. .921 ≤ .922

199. 0 ≤ 1   200. .1 > .01

Objective 7   Interpret data in a bar graph

Use the bar graph to answer the questions. Write your answer using signed numbers.

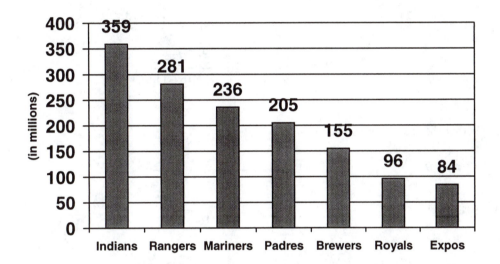

Value of Seven Major League Baseball Teams

201. What is the difference in value between the Mariners and the Royals?

202. What is the difference in value between the Brewers and the Rangers?

203. Which difference in value is greater: the difference between the Indians and the
     Mariners or the difference between the Padres and the Expos?

204. The Indians are worth $60 million less than the Yankees, who are not on this list.
     What is the difference in value between the Yankees and the Expos?

Mixed Exercises

Find the value of the expression.

205. $\left(\dfrac{3}{4}\right)^5$ 206. $(4.5)^3$

207. $3 \cdot 9 + 5 \cdot 8$

208. $6 \cdot 12 - 35 \div 5$

209. $\dfrac{3 \cdot 8 - 4 \cdot 5}{2(3^2 - 1)}$

210. $6[3 + 2(5^2 - 4)]$

211. 8^3

212. $25 \div 5 - 3 \cdot 1$

213. $\left(\dfrac{4}{5}\right)^2$

214. $\left(\dfrac{2}{3}\right)^4$

215. $4 \cdot 13 - 3 \cdot 12$

216. $2 + 3^3 - 5(1 + 2)$

217. $2[(14 + 1) - 12]$

218. $6[4 \cdot 3 - (2 + 7)]$

219. $\dfrac{2 + 3(8 + 1)}{4 \cdot 5 + 9}$

220. $\dfrac{5 - 4(3 - 2)}{4(6 - 1)}$

Tell whether the statement is *true* or *false*.

221. $12 \neq 3 + 9$

222. $5^2 + 12^2 \geq 13^2$

223. $9 \cdot 8 - 6 \cdot 12 < 0$

224. $\dfrac{3 + 2 \cdot 5}{16 - 3 \cdot 4} \geq 3$

225. $\dfrac{11}{6} > \dfrac{7}{3}$

226. $\dfrac{2}{5} + \dfrac{1}{2} = \dfrac{3}{10}$

Write the word statement in symbols.

227. Eight equals thirteen minus five.

228. Fourteen is less than thirty.

229. Ten is not equal to eleven.

230. The difference between thirty and six is greater than five.

Write the statement with the inequality symbol reversed while keeping the same meaning.

231. $101 \geq 99$

232. $.213 < .231$

233. $\dfrac{7}{12} \leq \dfrac{2}{3}$

234. $\dfrac{99}{100} > \dfrac{98}{99}$

Writing/Conceptual Exercises

Decide whether the statement is *true* or *false*. If it is false, explain why.

235. The expression $4 + 6 \cdot 2$ and $(4 + 6) \cdot 2$ have the same meaning.

236. The exponential expression 4^6 means $4 \cdot 4 \cdot 4 \cdot 4 \cdot 4 \cdot 4 \cdot 4$.

237. The statement $3 \le 9$ can also be written as $9 \ge 3$.

238. When evaluating $\left(5^2 + 2^4\right)^3$, what is the *last* exponent that would be applied?

239. Construct a true statement that involves multiplication on the left side, the symbol >, and subtraction on the right side.

240. Construct a false statement that involves division on the left side, the symbol $\le$, and subtraction on the right side. Then tell why the statement is false and how it could be changed to become true.

Section 1.3   Variables, Expressions, and Equations

Objective 1   Define variable, and find the value of an algebraic expression, given the values of the variables.

Find the numerical value if $x = 4$ and $y = 3$.

241. $x + 8$

242. $5x^2$

243. $2x^3$

244. x^4

245. $2x - 2y + 5$

246. $2(7x - 8y)$

247. $\dfrac{x^2}{x+1}$

248. $\dfrac{2x + 3y}{x - y + 1}$

249. $\dfrac{2x - 1}{14}$

250. $\dfrac{x}{5} + \dfrac{y}{4}$

251. $\dfrac{x^2 + 4y^2}{3x^2 + 2y}$

252. $\dfrac{3x + y^2}{2x + 3y}$

Objective 2   Convert phrases from words to algebraic expressions.

Change the word phrase to an algebraic expression. Use x as the variable to represent the number.

253. Eight added to a number

254. Twelve times a number

255. A number subtracted from thirteen

256. The product of 10 and four more than a number

257. The difference between twice a number and 15

258. Seven subtracted from twice a number

259. Ten times a number, added to 11

260. The difference between eight times a number and 7

261. The sum of a number and 4 divided by twice the number

262. Half a number subtracted from two-thirds of the number

Objective 3   Identify solutions of equations.

Decide whether the given number is a solution of the equation.

263. $r - 3 = 11$; 14

264. $x + 6 = 15$; 9

265. $6b + 2(b + 3) = 14$; 1

266. $2x^2 - 1 = 33$; 4

267. $\dfrac{p + 4}{p - 2} = 2$; 8

268. $\dfrac{m + 2}{3m - 10} = 1$; 8

269. $\dfrac{x^2 - 7}{x} = 6$; 7

270. $3y + 5(y - 5) = 22$; 6

271. $(a + 3)^3 = 125$; 2

272. $x^2 + 2x - 1 = 9$; 2

273. $7r + 5r + 8 = 15$; $\dfrac{12}{7}$

274. $\dfrac{p + 5}{3 - p} = \dfrac{17}{7}$; $\dfrac{2}{3}$

Objective 4   Identify solutions of equations from a set of numbers.

Change the word statement to an equation. Use x as the variable. Then find the solution for the equation from the set {0, 2, 4, 6, 8, 10}.

275. The sum of a number and four is ten.

276. A number minus three is equal to five.

277. The sum of three times a number and five is 23.

278. Five more than a number is 11.

279. The product of a number and five is 40.

280. Three times a number is equal to two more than twice the number.

281. 10 divided by a number is three more than the number.

282. 20 divided by a number is 5.

283. A number divided by six is zero.

284. The quotient of a number and ten is one.

Objective 5 Distinguish between an expression and an equation.

Identify as an *expression* or an *equation*.

285. $3x - 4y$

286. $3x = 4y$

287. $4x + 2y = 8$

288. $4x + 2y - 8$

289. $x^2 + 5x + 6$

290. $x^2 + 5x + 6 = 0$

291. $\dfrac{x+3}{5}$

292. $\dfrac{x+3}{5} = 2x$

293. $y = 2x^2 + 3$

294. $2x^2 + 3$

Mixed Exercises

Find the numerical value if $x = 3$ and $y = 5$.

295. $9x - 5y + 7$

296. $3(8x + 4y)$

297. $\dfrac{7x - 2y}{x + y + 3}$

298. $\dfrac{x}{6} + \dfrac{y}{2}$

299. $\dfrac{x^2 + 2y^2}{4x^2 + y}$

300. $\dfrac{5x + y^2}{4x - 2y}$

Change the word phrase to an algebraic expression. Use x as the variable to represent the number.

301. A number subtracted from twenty

302. The product of fifteen and a number

303. A number divided by one hundred

304. The product of twelve and three more than a number

305. Three times a number divided by two less than the number

306. Six more than the product of a number and twelve

Decide whether the given number is a solution of the equation.

307. $2s^2 + 6 = 56$; 5

308. $\dfrac{y + 3}{y - 3} = 4$; 5

309. $\dfrac{q^2 + 5}{q + 15} = 3$; 7

310. $z^2 + 4z - 2 = 30$; 4

311. $(y - 7)^2 = 4$; 10

312. $2(b + 5) - 3b = 3$; 7

Change the word statement to an equation. Use x as the variable. Then find the solution for the equation from the set {1, 3, 5, 7, 9}.

313. The product of a number and 6 is 30.

314. The quotient of 63 and a number is 7.

315. The product of 4 and five more than a number is 32.

316. Twice a number subtracted from 30 is four times the number.

Identify as an *expression* or an *equation*.

317. $2x + 5y = 10$

318. $2x + 5y - 10$

319. $3(r - s) + 2(r + s)$

320. $x^3 + x^2 + x + 1 = 0$

Writing/Conceptual Exercises

321. Why is $4x^3$ not the same as $4x \cdot 4x \cdot 4x$?

322. Explain in your own words why, when evaluating the expression $3y^2$ for $y = 5$, 5 must be squared *before* multiplying by 3. What expression would we be evaluating if we were to multiply 5 by 3 and then square the result?

323. There are many pairs of values m and n for which $m + 4n$ will equal 8. Name two such pairs.

324. In the phrase "Five less than the quotient of a number and 9," does the word *and* signify the operation of addition? Explain.

Section 1.4   Real Numbers and the Number Line

Objective 1   Set up number lines.

Graph the group of numbers on a number line.

325. $-2, -1, 0, 2, 4$

326. $3, 5, -1, -3$

327. $0, 4, 2, -2, -4$

328. $-6, 3, -4, 1$

329. $\frac{1}{2}, 0, -3, -\frac{5}{8}$

330. $0, 1, 2, 3, \frac{9}{2}$

331. $-\frac{3}{2}, 0, \frac{3}{2}, \frac{7}{2}$

332. $-3\frac{1}{2}, -\frac{3}{2}, 0, \frac{1}{2}, 1$

333. $0, 1\frac{1}{2}, 3, 5\frac{1}{2}, 6$

334. $-4.5, -2.3, 1.7, 3.5$

Objective 2   Classify numbers.

List all of the sets among the following to which the number belongs: *natural numbers, whole numbers, integers, rational numbers, irrational numbers, real numbers.*

335. 5

336. -3

337. $\dfrac{4}{5}$

338. 0

339. $-\dfrac{5}{7}$

340. -2.6

341. $-4\dfrac{5}{8}$

342. $\sqrt{3}$

343. 11.45

344. $-\sqrt{15}$

Objective 3   Tell which of two different real numbers is smaller.

Select the smaller of the two given numbers.

345. $-15,\ -14$

346. $0,\ -10$

347. $-.802,\ -.820$

348. $-1.99,\ -2.01$

349. $-\dfrac{2}{3},\ -\dfrac{4}{5}$

350. $\dfrac{5}{12},\ \dfrac{6}{13}$

Decide whether the statement is *true* or *false.*

351. $-100 < -99$

352. $0 > -5$

353. $-12 > -11$

354. $1 < -1$

Objective 4   Find additive inverses and absolute values of real numbers.

Find the additive inverse of the number.

355. 25

356. -15

357. 6.5

358. $\dfrac{5}{8}$

359. $-\dfrac{9}{11}$

360. $-2\dfrac{5}{8}$

361. $10 - 10$

362. -9.99

363. $20 - 13$

Simplify by removing absolute value symbols.

364. $|-8|$

365. $|123|$

366. $|0|$

367. $-|15|$

368. $-|-20|$

369. $|12-5|$

370. $-|99-98|$

371. $|-7.52|$

372. $-|-.7|$

373. $\left|\dfrac{1}{2}-\dfrac{1}{3}\right|$

374. $\left|3\dfrac{1}{2}-2\dfrac{1}{4}\right|$

375. $-\left|7\dfrac{3}{8}-4\dfrac{3}{4}\right|$

Select the smaller of the two given numbers.

376. $|-8|,\ |5|$

377. $-|2|,\ -|-1|$

378. $-|-7|,\ -|10|$

379. $-|-5|,\ -|-8|$

380. $|18-15|,\ |22-13|$

Objective 5   Interpret real number meanings from a table of data.

Use the table below of counties in the state of Indiana to answer the questions.

County Name	1997 Estimated Population	1998 Estimated Population	Numeric Population Change 1997-1998
Allen	311928	314218	2290
Benton	9661	9725	64
Delaware	117520	116828	-692
Elkhart	170668	172310	1642
Henry	48896	48785	-111
Lake	478536	478323	-213
Monroe	116612	115130	-1482
Union	7320	7263	-57

Source: U.S Bureau of Census

381. Which county experienced the greatest population growth?

382. Which county experienced the greatest decline in population?

383. What is the total change in population between Elkhart County and Henry County?

384. What is the total change in population between Union County and Delaware County?

Mixed Exercises

List all of the sets among the following to which the number belongs: *natural numbers, whole numbers, integers, rational numbers, irrational numbers,* and *real numbers.*

385. -7

386. 8

387. $\dfrac{5}{3}$

388. -2.5

389. $\sqrt{49}$

390. $\sqrt{50}$

Decide whether the statement is *true* or *false.*

391. $-2 < 3$

392. $6 > -5$

393. $-8 < -10$

394. $-5 \geq -2$

395. $-7 \leq 5$

396. $4 \geq 0$

Find the additive inverse of the number.

397. $1\dfrac{3}{5}$

398. $-2\dfrac{1}{3}$

399. 14

400. $\dfrac{10}{17}$

401. 45

402. $19 - 8$

Simplify by removing absolute value symbols.

403. $-|-37|$

404. $|17 - 13|$

405. $|15 - 12|$

406. $-|12 - 4|$

407. $-|5 - 1|$

408. $-|12 - 10|$

Select the larger of the two given numbers.

409. $-11, -12$

410. $-1.2, -2.1$

411. $|-6|, 5$

412. $|-20|, |-22|$

413. $-|-10|, -|-9|$

414. $|12 - 15|, |13 - 8|$

Writing/Conceptual Exercises

Decide whether the statement is *true* or *false*.

415. Every whole number is a rational number.

416. Every whole number is positive.

417. Every irrational number is a real number.

418. Some natural numbers are irrational.

Give three numbers that satisfy the given condition.

419. Integers but not whole numbers

420. Real numbers but not irrational numbers

421. Rational numbers but not positive numbers

422. Irrational numbers but not negative numbers

Section 1.5   Addition and Subtraction of Real Numbers

Objective 1   Add two numbers with the same sign.

Find the sum.

423. $8+3$	424. $4+12$
425. $-2+(-5)$	426. $-8+(-6)$
427. $-9+(-9)$	428. $-1+(-11)$
429. $-3+(-7)$	430. $-11+(-19)$
431. $-20+(-27)$	432. $-17+(-18)$
433. $-34+(-53)$	434. $-82+(-18)$

Objective 2   Add positive and negative numbers.

Find the sum.

435. $8 + (-5)$

436. $-9 + 8$

437. $10 + (-10)$

438. $-\dfrac{4}{5} + \dfrac{2}{5}$

439. $\dfrac{11}{12} + \left(-\dfrac{3}{4}\right)$

440. $-\dfrac{5}{7} + \dfrac{3}{5}$

441. $3\dfrac{5}{85} + \left(-1\dfrac{1}{4}\right)$

442. $13.5 + (-6.9)$

443. $-10.475 + 6.325$

444. $56 + (-73)$

445. $-42 + 63$

446. $-75 + 46$

Objective 3   Use the definition of subtraction.

Find the difference.

447. $10 - 20$

448. $29 - 36$

449. $-7 - 5$

450. $-9 - 5$

451. $7 - (-4)$

452. $1 - (-3)$

453. $-6 - (-10)$

454. $-8.6 - (-2.4)$

455. $4.5 - (-2.8)$

456. $\dfrac{1}{2} - \left(-\dfrac{1}{10}\right)$

457. $-\dfrac{3}{10} - \left(-\dfrac{4}{15}\right)$

458. $3\dfrac{3}{4} - \left(-2\dfrac{1}{8}\right)$

Work the problem.

459. $7 - [6 + (-2)]$

460. $-3 - [6 + (-3)]$

461. $[6-(-12)]-21$

462. $[5-(-2)]-8$

463. $[3+(-9)]-(-2)$

464. $3-[-8+(11-14)]$

465. $\left[\dfrac{1}{3}-\left(-\dfrac{1}{5}\right)\right]-\left(\dfrac{4}{15}\right)$

466. $\dfrac{2}{9}-\left[\dfrac{5}{6}-\left(-\dfrac{2}{3}\right)\right]$

467. $-3+[(-6-7)-(-1+3)]$

468. $-7+[(-12+10)-(-3+8)]$

Objective 4   Use the order of operations with real numbers.

Find the sum.

469. $10+[7+(-17)]$

470. $-4+[5+(-19)]$

471. $-16+[-14+(-6)]$

472. $-8+[7+(-18+13)]$

473. $[15+(-8)]+[7+(-11)]$

474. $-7.6+[5.2+(-11.4)]$

475. $\dfrac{3}{8}+\left[-\dfrac{2}{3}+\left(-\dfrac{5}{12}\right)\right]$

476. $-\dfrac{5}{9}+\left[\dfrac{1}{6}+\left(-\dfrac{2}{3}\right)\right]$

477. $\left[\dfrac{9}{10}+\left(-\dfrac{3}{5}\right)+\dfrac{1}{2}\right]$

478. $\left[2\dfrac{1}{2}+\left(-3\dfrac{1}{4}\right)\right]+1\dfrac{1}{8}$

Objective 5   Interpret words and phrases involving addition and subtraction.

Write a numerical expression for the phrase, and then simplify the expression.

479. The sum of -8 and 15

480. 20 more than -15

481. -3 increased by 11

482. The sum of -8 and 3 and 5

483. 10 added to the sum of -2 and -3

484. 6 more than -2, increased by 8

485. 15 increased by the sum of -20 and 2

486. The sum of -8 and -9 and -11

487. The sum of -5 and -8, increased by 12

488. -10 added to the sum of 20 and -8

Write a numerical expression for the phrase, and then simplify the expression.

489. The difference between -9 and 4

490. The difference between 10 and -5

491. The difference between -6 and -8

492. 4 less than the difference between -9 and -3

493. 8 less than -4

494. 9 less than -2

495. -4 decreased by 6 less than -2

496. -3 decreased by 1 less than -4

497. -6 subtracted from the sum of 2 and -3

498. -10 subtracted from the sum of -4 and -2

Solve the problem by writing a sum of real numbers and adding. No variables are needed.

499. A football team gained 2 yards from the scrimmage on the first play, lost 11 yards on the second play, and gained 21 yards on the third play. How many yards did the team gain or lose altogether?

500. Mr. Goodman has $302 in his checking account. He writes two checks, one for $38 and the other for $75. Finally, he deposits $55 in the account. How much does he now have in his account?

501. A mountain climber starts to climb at an altitude of 4325 feet. He climbs so that he gains 208 feet in altitude. Then he finds that, because of an obstruction, he must descend 25 feet. Then he climbs 58 feet up. What is his final altitude?

502. Virgil owes $78 to a credit card company. He makes a purchase of $45 with his card, and then pays $100 to the company. How much does he still owe?

503. The temperature at dawn in Chicago was 24° F. During the day the temperature increased 14° F. Then it fell 8° by sunset. What was the temperature at sunset?

504. A hot-air balloon rises 200 feet above the ground. It drops 50 feet, then drops another 25 feet, and finally rises 135 feet. How high is the balloon now?

505. A football team lost 10 yards on the first play, lost another 3 yards on the second play, and then gained 8 yards on the third play. How many yards did the team gain or lose altogether?

506. Roberto has $42 in his checking account. If he writes a check for $16, how much does he have left?

507. Marie has $51 in her checking account. Then she deposits $25. If she writes a check for $55, how much money is left in her account?

Solve the problem by writing a difference between real numbers and subtracting. No variables are needed.

508. A chemist runs an experiment at $-97.2°$ C. She then lowers the temperature by $4.9°$ C. What is the new temperature for the experiment.?

509. The highest point in the state of Washington, Mt. Rainier, has an elevation of 14,410 feet. The highest point in Oregon, Mt. Hood, has an elevation of 11,235 feet. How much taller is Mt. Rainier than Mt. Hood?

510. David has a checking account balance of $239.32. He overdraws his account by writing a check for $302.58. Write his new balance as a negative number.

511. At 1:00 A.M., the temperature on the top of Mt. Washington in New Hampshire was $-16°$ F. At 11:00 A.M., the temperature was $25°$ F. What was the rise in temperature?

512. The highest point in Israel, Mt. Meron, has an elevation of 1208 meters. The lowest point, the Dead Sea, is 396 meters below sea level. Using zero as sea level, find the difference between the two elevations.

513. Marlene owes $342.58 on her VISA account. She makes additional purchases which total $173.76. Express her new balance as a negative number.

514. Mr. Whitney has a bank balance of $728. Mr. Anderson has a debt of $427. Find the difference between these amounts.

515. The record high temperature for a midwestern city is $108°$ F, while the record low temperature is $-27°$ F. Find the difference between these temperatures.

516. A yogurt shop showed a profit of $3624 in July and a loss of $893 in January. Find the difference between these amounts.

517. In the card game of hearts, it is possible to have a negative score. (In this game, the player with the lowest score wins.) At the end of one game, Bill had 46 points and Larry had -27 points. By how many points did Larry win the game?

Objective 6   Use signed numbers to interpret data.

Use the chart below to answer the questions.

Year	Price ($ Per Peck)	Production (Millions Of Pecks Of Pickled Peppers)
1984	3.15	3200
1985	3.00	4000
1986	2.95	4750
1987	3.25	3500
1988	3.50	3100
1989	3.70	2800
1990	3.65	2900
1991	3.80	3050
1992	3.90	2600
1993	4.20	1750

518. What signed number represents the change in production from 1988 to 1989?

519. What signed number represents the change in production from 1990 to 1991?

520. What signed number represents the change in price from 1984 to 1985?

521. What signed number represents the change in price from 1992 to 1993?

Mixed Exercises

Find the sum.

522. $-3+(-2)$

523. $-2+(-5)$

524. $-8+(-5)$

525. $-10+(-15)$

526. $-3+(-3)$

527. $8+(-7)$

528. $17+(-14)$

529. $13+(-15)$

530. $-8.5+10.1$

531. $-\dfrac{3}{10}+\dfrac{1}{6}$

532. $-\dfrac{1}{15}+\dfrac{11}{12}$

533. $-2.5+1.5$

534. $8+\left[-7+(-3)\right]$

535. $\left[5+(-3)\right]+12$

536. $\left[6+(-3)\right]+\left[4+(-7)\right]$

537. $\left[-5+(-3)\right]+\left[4+(-2)\right]$

538. $\left[-9+(-4)\right]+\left[-6+(-5)\right]$

539. $\left[-2+(-5)\right]+\left[10+(-1)\right]+\left[15+(-8)\right]$

540. $\left[-8+(-2)\right]+\left[-7+(-3)\right]+\left[-9+(-1)\right]$

Write a numerical expression for each phrase, and then simplify the expression.

541. The sum of 5 and –8, increased by 6

542. –8 added to the sum of 4 and –13

543. The sum of –3 and –9 and –12

544. 20 more than the sum of –2 and –8

Solve the problem by writing a sum of real numbers and adding. No variables are needed.

545. Maya puts a casserole of leftovers that has a temperature of 72° F into the refrigerator. The casserole has cooled 30° when Maya removes it the next day. When she reheats the casserole, its temperature increases by 155°. What is the temperature of the casserole when she removes it from the oven?

546. Samantha owes $107.28 on her credit card. She makes purchases of $52.04 and $37.49. Then she makes a payment of $127.00. How much does she still owe?

Work the problem.

547. $15 - 25$

548. $-8 - (-12)$

549. $12 - (24)$

550. $-100 - (-100)$

551. $4.5 - (-7.8)$

552. $-6.2 - 8.9$

553. $6 - [-9 - (-2)]$

554. $[3 - (-12)] - 8$

555. $\left(\dfrac{1}{2} - \dfrac{1}{3}\right) - \dfrac{5}{6}$

556. $\left[\dfrac{5}{8} - \left(-\dfrac{1}{16}\right)\right] - \left(-\dfrac{3}{8}\right)$

557. $(5 - 9) - 3$

558. $(13 - 5) - (-2)$

559. $(15 - 20) - (-10)$

560. $(-4.4 - 8.6) - 1.3$

561. $(-9 - 4) - (-8 - 6)$

562. $(-1 - 3) - (-4 - 7)$

563. $-2 + [(-5 - 8) - (-6 + 2)]$

564. $-4 + [(-13 + 2) - (-2 - 10)]$

Write a numerical expression for the phrase, and then simplify the expression.

565. The difference between 6 and –10

566. The sum of –5 and 12, decreased by 6

567. 7 less than the difference between 10 and –6

568. The difference between –7 and –5, decreased by –2

Solve the problem by writing a difference between real numbers and subtracting. No variables are needed.

569. At 10:00 P.M., a plant worker found that a dial reading was 6.872. At 11:00 P.M., he found the reading to be –4.059. By how much had the reading declined?

570. Priya owes $82.75 on her credit card bill. She makes an additional purchase of $37.28 and a payment of $95.00. Express her new balance as a negative number.

Writing/Conceptual Exercises

Choose the phrase that completes the statement correctly.

571. The sum of two negative numbers is _____.

        (a) always a negative number
        (b) always a positive number
        (c) sometimes positive and sometimes negative
        (d) sometimes zero

572. The sum of a number and its opposite is _____.

        (a) always a positive number
        (b) always a negative number
        (c) always zero
        (d) always 1

573. The sum of a positive number and a negative number is _____.

        (a) always a positive number
        (b) always a negative number
        (c) sometimes a positive number and sometimes a negative number
        (d) sometimes a positive number, sometimes a negative number, and sometimes zero

574. The sum of a positive number and a negative number, where the positive number has the larger absolute value, is _____.

        (a) always a positive number
        (b) always a negative number
        (c) sometimes a positive number and sometimes a negative number
        (d) sometimes a positive number, sometimes a negative number, and sometimes a zero

Suppose that x represents a negative number, y represents a positive number, and z represents a positive or a negative number.

575. $x - y$

576. $y - x - z$

577. $|x| + |y| + |z|$

578. $x - |z|$

579. $|x + y| + |z|$

580. $|y + z| - x$

Section 1.6 Multiplication and Division of Real Numbers

Objective 1 Find the product of a positive number and a negative number.

Find the product.

581. $3(-4)$

582. $(-10)(12)$

583. $(-12)(12)$

584. $13(-11)$

585. $(-80)(8)$

586. $9(-5)$

587. $\left(\frac{1}{5}\right)\left(-\frac{2}{3}\right)$

588. $\left(-\frac{3}{8}\right)\left(\frac{14}{9}\right)$

589. $\left(-\frac{2}{7}\right)\left(\frac{21}{26}\right)$

590. $3(-2.5)$

591. $(-5)(4.4)$

592. $(-3.2)(4.1)$

Objective 2 Find the product of two negative numbers.

Find the product.

593. $(-3)(-5)$

594. $(-8)(-9)$

595. $(-13)(-13)$

596. $(-10)(-100)$

597. $(-25)(-25)$

598. $(-3)(-10)$

599. $\left(-\frac{2}{7}\right)\left(-\frac{14}{5}\right)$

600. $\left(-\frac{1}{12}\right)\left(-\frac{6}{7}\right)$

601. $\left(-\frac{3}{10}\right)\left(-\frac{5}{9}\right)$

602. $(-1.3)(-2.1)$

603. $(-5.1)(-.2)$

604. $(-.9)(-3.7)$

Objective 3 Identify factors of integers.

Find all the integer factors of the given number.

605. 24

606. 12

607. 40

608. 13

609. 7

610. 10

611. 36

612. 39

613. 23   614. 50   615. 43   616. 30

Objective 4   Use the reciprocal of a number to apply the definition of division.

Use the definition of division to find each quotient.

617. $\dfrac{15}{3}$

618. $\dfrac{-16}{8}$

619. $\dfrac{26}{-13}$

620. $\dfrac{36}{-4}$

621. $\dfrac{-100}{-5}$

622. $\dfrac{0}{-8}$

623. $\dfrac{-72}{24}$

624. $\dfrac{9}{0}$

625. $\dfrac{3}{4} \div \dfrac{5}{6}$

626. $-\dfrac{8}{9} \div \dfrac{2}{3}$

627. $\dfrac{12}{3} \div \left(-\dfrac{16}{2}\right)$

628. $-\dfrac{3}{2} \div \left(-\dfrac{3}{4}\right)$

Objective 5   Use the order of operations when multiplying and dividing signed numbers.

Perform the indicated operations.

629. $12 - 8 \cdot 4$

630. $7 \cdot 3 - 2 \cdot 13$

631. $7(-8) - (-2)(-4)$

632. $(-3)(5) - (-8)(2)$

633. $-3(8) - 4(-7)$

634. $-6(7 - 4)$

635. $-4[(-2)(6) - 7]$

636. $-6[-4 - (-2)(-7)]$

637. $(3 - 8)(-2) - 10$

638. $8 - (-6)(4 - 7)$

639. $-20 - (-1)(-7 - 11)$

640. $40 - (-2)[8 - 3(-1)]$

Find the quotient.

641. $\dfrac{50}{-5}$

642. $\dfrac{-28}{4}$

643. $\dfrac{99}{-11}$

644. $\dfrac{-120}{-15}$

645. $\dfrac{10}{0}$

646. $\dfrac{0}{-8}$

647. $-\dfrac{5}{8} \div \dfrac{3}{4}$

648. $-\dfrac{2}{3} \div \left(-\dfrac{8}{9}\right)$

649. $-\dfrac{27}{35} \div \left(-\dfrac{3}{5}\right)$

650. $\dfrac{14}{33} \div \left(-\dfrac{7}{11}\right)$

651. $(-12.3) \div (-3)$

652. $(-5.5) \div 2.2$

Objective 6   Evaluate expressions involving variables.

Evaluate the expression if $x = -2$, $y = 3$, and $a = 4$.

653. $-x + y - 3a$

654. $-3x + 2y - (a - x)$

655. $(3x - y)(-a)$

656. $-x^2 + 3y$

657. $(2y + 3)(a) + |y|$

658. $(x - 2)(y - 3)$

659. $(x - y) - (a - 2y)$

660. $(-2y + 5a) - (3x + y)$

661. $2(x - 3)^2 + 3y^2$

662. $(x + 6)^3 - y^3$

663. $\left(\dfrac{1}{2}x + \dfrac{2}{3}y\right)\left(-\dfrac{1}{4}a\right)$

664. $\left(\dfrac{2}{3}x - \dfrac{1}{6}y\right)\left(\dfrac{1}{2}a\right)$

Objective 7   Interpret words and phrases involving multiplication and division.

Write a numerical expression for the phrase and simplify.

665. The product of 4 and –3, added to 5

666. The product of -8 and 3, added to –4

667. The product of 11 and –4, subtracted from 5

668. The difference between –8 and the product of –7 and 4

669. Twice the sum of 14 and –6, added to –2

670. Three-tenths of the difference between 50 and –10, subtracted from 100

671. 40% of the sum of 6 and –2

672. 85% of the difference between 32 and –8

673. –24 subtracted from two-thirds of the sum of 16 and –10

674. –7 added to seven-eighths of the difference between –2 and 6

675. The quotient of –72 and –17

676. The quotient of 90 and the sum of 35 and –5

677. The sum of –11 and the quotient of 49 and –7

678. The difference between 8 and the quotient of –21 and 3

679. The difference between –12 and the quotient of –18 and –9

680. The product of 30 and –8, divided by the difference between 5 and –10

681. The product of –8 and 6, divided by the sum of –3 and 15

682. The quotient of the sum of 11 and –7 and the difference between –8 and –6

683. The quotient of the difference between 30 and –6 and the product of –3 and 4

684. The sum of the quotient of –16 and 4 and the quotient of 25 and –5

Objective 8   Translate simple sentences into equations.

Write the statement in symbols and find the solution by guessing or trial and error. Use x as the variable. All solutions come from the list of integers between –6 and 6, inclusive.

685. 9 times a number is –45.

686. The quotient of a number and –2 is –3.

687. The sum of a number and –4 is 1.

688. The difference between a number and –10 is 12.

689. Two-thirds of a number is –4.

690. The product of a number and –1 is 4.

691. When a number is divided by –3, the result is 1.

692. 8 less than a number is –5.

693. –5 times a number is 10.

694. The product of a number and 7 is the number.

Mixed Exercises

Perform the indicated operations.

695. $6(-12)$

696. $(-10)(-18)$

697. $-11(-3-4)$

698. $-7[-3-5(-2)]$

699. $(4-2)(-3-1)$

700. $(-8-3)(-2)-4$

701. $(-3-6)(-5)-10$

702. $|-3(-7)|-|-8|$

Evaluate the expression if $x = -2$, $y = 3$, and $a = 4$.

703. $-x^2 + 2a$

704. $(-5+x)(-a)-|x|$

705. $(5-x)(2-a)$

706. $(2x+3y)(2a-x)$

707. $3y^2 - 4x^2 + a^2$

708. $(y-7)^2 + 2x^3$

Find all the integer factors of the given number.

709. 15   710. 35   711. 18   712. 79

713. 6   714. 17   715. 25   716. 49

Write a numerical expression for the phrase and simplify.

717. 30% of the difference of 18 and 12

718. The product of –7 and 6 subtracted from –15

719. Twice the sum of –8 and 6, added to –9

720. The product of –6 and 30 minus the product of 8 and –10

Perform the indicated operations.

721. $\dfrac{-7}{-7}$

722. $\dfrac{-96}{6}$

723. $\dfrac{0}{-12}$

724. $\dfrac{12}{0}$

725. $\dfrac{1}{2} \div \left(-\dfrac{1}{2}\right)$

726. $\left(-\dfrac{3}{4}\right) \div \left(-\dfrac{5}{8}\right)$

727. $\dfrac{-3(5)-(-6)}{4+(-1)}$

728. $\dfrac{6(-10+3)}{15(-2)-3(-9)}$

729. $\dfrac{8^2-12}{5(-5)+3(4)}$

730. $\dfrac{7(-3)-(-6)(-8)}{-3(6)-(5)(1)}$

731. $\dfrac{-8[6-(-3+1)]}{-3[2-(-4)]-9(-2)}$

732. $\dfrac{-8[9-(-9+2)]}{14(-3)+(-9-8)(-5+3)}$

Write a numerical expression for the phrase and simplify the expression.

733. The quotient of 100 and the sum of –16 and –9

734. The product of –24 and 5, divided by the difference between 12 and –3

735. The quotient of the sum of –10 and –20 and the difference between 3 and –3

736. The sum of the quotient of –30 and –5 and the quotient of 100 and –25

737. The quotient of 90 and the sum of 35 and –5

738. The quotient of the difference between –3 and 5 and the sum of –5 and 3

Write the statement in symbols and find the solution. Use x as the variable. All solutions come from the list of integers between –6 and 6, inclusive.

739. One-half of a number is –3.

740. The quotient of a number and –2 is 3.

741. Fifteen divided by a number is –3.

742. The sum of a number and –3, divided by 2, is –4.

Writing/Conceptual Exercises

Decide whether the statement is *true* or *false*.

743. The product of three negative numbers is a positive number.

744. The product of two negative numbers and a positive number is a negative number.

745. When a negative number is cubed, the result is a negative number.

746. When the sum of a positive number and a negative number is multiplied by zero, the result is a negative number.

747. When a negative number is raised to the sixth power, the result is a negative number.

748. If x is replaced by a positive number and y is replaced by a negative number, the value of the expression $5x - 6y$ will be a positive number.

749. Which one of the following expressions is undefined?

(a) $\dfrac{10-10}{10}$ (b) $\dfrac{10-10}{10+10}$ (c) $\dfrac{10+10}{10-10}$ (d) $\dfrac{10+10}{10+10}$

750. Which one of the following expressions is equal to zero?

(a) $\dfrac{15+15}{15-15}$ (b) $\dfrac{15+15}{15}$ (c) $\dfrac{15}{15+15}$ (d) $\dfrac{15-15}{15+15}$

Assume that x is positive, y is negative, and z is negative. Tell whether the value of the given expression is positive or negative.

751. $\dfrac{x}{y \cdot z}$

752. $\dfrac{x \cdot y^2}{z}$

753. $\dfrac{x}{y} + \dfrac{x}{z}$

754. $\dfrac{x - z}{y^2}$

755. $\dfrac{y(z - x)}{x^2}$

756. $\dfrac{x^2 \cdot y}{z^2}$

Section 1.7   Properties of Real Numbers

Objective 1   Use the commutative properties.

Use a commutative property to complete the statement.

757. $y + 9 = \underline{\hspace{1cm}} + y$

758. $5(3) = \underline{\hspace{1cm}}(5)$

759. $-3\left(\dfrac{1}{4}\right) = \left(\dfrac{1}{4}\right)\underline{\hspace{1cm}}$

760. $3m = \underline{\hspace{1cm}}(3)$

761. $-7(p + 9) = \underline{\hspace{1cm}}(-7)$

762. $10\left(\dfrac{1}{4} \cdot \dfrac{1}{3}\right) = \underline{\hspace{1cm}}(10)$

763. $(pq)(-5) = (-5)\underline{\hspace{1cm}}$

764. $-3 + 18 = 18 + \underline{\hspace{1cm}}$

765. $3 + [11 + (-9)] = \underline{\hspace{1cm}} + 3$

766. $-3(4 + z) = \underline{\hspace{1cm}}(-3)$

Objective 2   Use the associative properties.

Use an associative property to complete the statement.

767. $m(np) = \underline{\hspace{1cm}}(p)$

768. $(3 \cdot 5)(-6) = \underline{\hspace{1cm}}[5(-6)]$

769. $[-3 + (-5)] + y = \underline{\hspace{1cm}} + (-5 + y)$

770. $(2m)(-3) = (2)\underline{\hspace{1cm}}$

771. $3(ab) = \underline{\hspace{1cm}} \cdot b$

772. $[2 \cdot (-3)] \cdot 5 = (2)\underline{\hspace{1cm}}$

773. $(-12m)(-n) = (-12)\underline{\hspace{1cm}}$

774. $(-r)[(-p)(-q)] = \underline{\hspace{1cm}}(-q)$

775. $[t + (-z)] + m = t + \underline{\hspace{1cm}}$

776. $3r + (6s + 14t) = \underline{\hspace{1cm}} + 14t$

Objective 3 Use the identity properties.

Complete the statement so that it is an example of an identity property.

777. $-2 + \underline{\hspace{1cm}} = -2$

778. $\underline{\hspace{1cm}} \cdot 0 = 0$

779. $-2 \cdot \underline{\hspace{1cm}} = -2$

780. $\underline{\hspace{1cm}} + |-8| = |-8|$

781. $18 + \underline{\hspace{1cm}} = 18$

782. $-9.87 \cdot \underline{\hspace{1cm}} = -9.87$

783. $\underline{\hspace{1cm}} \cdot (-7) = -7$

784. $2.18 + \underline{\hspace{1cm}} = 2.18$

785. $\underline{\hspace{1cm}} + \left(-\dfrac{3}{4}\right) = -\dfrac{3}{4}$

786. $\underline{\hspace{1cm}} \cdot \left(\dfrac{11}{7}\right) = \dfrac{11}{7}$

Objective 4 Use the inverse properties.

Complete the statement so that it is an example of an inverse property.

787. $\underline{\hspace{1cm}} + 8 = 0$

788. $.25 \cdot \underline{\hspace{1cm}} = 1$

789. $\underline{\hspace{1cm}} \cdot \left(\dfrac{3}{2}\right) = 1$

790. $.25 + \underline{\hspace{1cm}} = 0$

791. $-16 + \underline{\hspace{1cm}} = 0$

792. $72 + \underline{\hspace{1cm}} = 0$

793. $-\dfrac{2}{5} \cdot \underline{\hspace{1cm}} = 1$

794. $\dfrac{3}{25} \cdot \underline{\hspace{1cm}} = 1$

795. $-18 + \underline{\hspace{1cm}} = 0$

796. $\underline{\hspace{1cm}} \cdot (-8) = 1$

Objective 5 Use the distributive property.

Use the distributive property to rewrite the expression. In Exercises 805 and 806, simplify the results.

797. $6y + 4y$

798. $10r - r$

799. $a(z + 1)$

800. $9 \cdot x + 9 \cdot y$

801. $8(a - b)$

802. $3c - 3d$

803. $n(2a - 7b + 6c)$

804. $4(2m - 3n)$

805. $-2(5y - 7z)$

806. $-(-2k - 5)$

807. $6x + 6m$

808. $5(7x) + 5(8z)$

Mixed Exercises

Use a commutative property to complete the statement.

809. $x + 3 = 3 + \underline{\quad}$

810. $y + 9 = \underline{\quad} + y$

811. $2 + 7 = \underline{\quad} + 2$

812. $5(3) = \underline{\quad}(5)$

813. $7(18) = (18)\underline{\quad}$

814. $-3\left(\dfrac{1}{4}\right) = \left(\dfrac{1}{4}\right)\underline{\quad}$

Use an associative property to complete the statement.

815. $a(5 \cdot 3) = (a \cdot 5)\underline{\quad}$

816. $m(np) = \underline{\quad}(p)$

817. $(pq)r = \underline{\quad}(qr)$

818. $(3 \cdot 5)(-6) = \underline{\quad}[5(-6)]$

819. $(2 + 3) + 7 = 2 + \underline{\quad}$

820. $[-3 + (-5)] + y = \underline{\quad} + (-5 + y)$

Complete the statement so that it is an example of either an identity property or an inverse property. Identify which property is used.

821. $-3 + \underline{\quad} = 0$

822. $5 + \underline{\quad} = 5$

823. $\dfrac{1}{2} + \underline{\quad} = \dfrac{1}{2}$

824. $\underline{\quad} + \dfrac{1}{3} = 0$

825. $1 \cdot \underline{\quad} = 1$

826. $\dfrac{2}{5} \cdot \underline{\quad} = \dfrac{2}{5}$

827. $\dfrac{2}{3} \cdot \underline{\quad} = 1$

828. $-\dfrac{3}{5} \cdot \underline{\quad} = 1$

829. $\underline{} \cdot 1 = -\dfrac{7}{9}$

830. $\underline{} \cdot \dfrac{1}{4} = 1$

831. $-\dfrac{14}{5} \cdot \underline{} = 1$

832. $\dfrac{4}{5} + \underline{} = \dfrac{4}{5}$

Use the distributive property to rewrite the expression.

833. $2(m+3)$

834. $a(z+1)$

835. $-3(a+b)$

836. $9 \cdot x + 9 \cdot y$

837. $4m + 4n$

838. $8(a-b)$

Writing/Conceptual Exercises

839. Give an example which shows that subtraction is not commutative.

840. Give an example which shows that division is not commutative.

841. On an algebra quiz, Ruth was asked to state which property is used to justify the following statement:

$$5 + (9+12) = 5 + (12+9)$$

Ruth's answer was "the associative property". When her teacher marked this question wrong, Ruth said, "I thought it was the associative property because there are three numbers." How should Ruth's teacher respond?

842. Explain in your own words why 0 has an additive inverse but no multiplicative inverse.

Section 1.8   Simplifying Expression

Objective 1   Simplify expressions.

Simplify the expression.

843. $14 + 3y - 6$

844. $6(5r - 8s)$

845. $10+5(4y+6)$

846. $-1+7(2n-3)$

847. $-(9-3b)-8$

848. $11-(d-5)+(2-5)$

849. $13-6y+(5-7)$

850. $-6+s=(13-26)$

851. $-3(-3m+5)+10$

852. $4(-6p-1)+(2-5)$

Objective 2 Identify terms and numerical coefficients.

Give the numerical coefficient of the term.

853. $8x$

854. $-9y$

855. $-6r^2$

856. $3s^4$

857. $.7a^2b$

858. $-12xy$

859. 25

860. z^5

861. $-r^2s^2$

862. $\frac{1}{3}km^2$

863. $-\frac{5}{9}v^6w^4$

864. $\frac{8t}{7}$

Objective 3 Identify like terms.

Identify the group of terms as *like* or *unlike*.

865. $5x, 2x$

866. $4y, -6y$

867. $-7q^2, 7q^2$

868. $-8m, -8m^2$

869. $9w, 3w, -w$

870. $4k, -7k, 3$

871. $t, -3t$

872. $6p^2, 6s^2$

873. $4x, -10x, -9x^2$

874. $18, 18y$

875. $8, -4, 16$

876. $4xy, -6xy^2$

Objective 4 Combine like terms.

Simplify the expression by combining like terms.

877. $12x-5x$

878. $6r+7-8r+5$

879. $2.3r + 6.9 + 2.8 + 3.6r$ 880. $2a^2 - 4a^3 - 5a^2 + 6a^3$

881. $\dfrac{1}{3} + \dfrac{3}{4}y - \dfrac{5}{6} - \dfrac{2}{3}y$ 882. $\dfrac{7}{10}r + \dfrac{3}{10}s - \dfrac{2}{5}r - \dfrac{4}{5}s$

883. $7(3x + 8)$ 884. $-5(2y - 3)$

885. $6r - (2r + 5)$ 886. $8(-3 + t) - 9t$

887. $9(5q + 1) - (q + 10)$ 888. $-6(a + 1) + 8(2a - 1)$

889. $-5(s + 6) + 4(2s + 7)$ 890. $7(4t + 1) - (t + 6)$

891. $2.5(3y + 1) - 4.5(2y - 3)$ 892. $-.8(7t - 5) + .6(9t - 7)$

Objective 5   Simplify expressions from word phrases.

Write the phrase as a mathematical expression and simplify by combining like terms. Use x as the variable.

893. Nine times a number, added to five times the number

894. Seven times a number, subtracted from four times the number

895. The sum of eight times a number and 13, added to three times the number

896. Six times the sum of 14 and three times a number, added to nine times the number

897. The sum of four times a number and 7, subtracted from twice the number

898. The difference between seven times a number and 14, added to four times the sum of the number and 2

899. The sum of eleven times a number and 4, subtracted from the difference between 8 and nine times the number

900. Twelve times the difference between 6 and twice a number, subtracted from 14

901. Twice the difference between three times a number and seven times the number, added to three times the sum of the number and 8

902. Four times the difference between twice a number and –10, subtracted from three times the sum of –7 and five times the number

Mixed Exercises

Simplify the expression.

903. $4k + 3 - 2$

904. $15 + 3m - 8$

905. $4(3p + 2q)$

906. $5(2a - 5b)$

907. $9 + 3(4x - 2)$

908. $7 + 2(3y - 7)$

909. $6 - 5(2n - 3)$

910. $-10 - 4(2p - 5)$

Give the numerical coefficient of the term.

911. $4y$

912. $3x$

913. $17z$

914. $-12p^4$

915. $-2k^5$

916. $3a^4b^2$

917. $5m^{12}n^4$

918. -9

Identify the group of terms as *like* or *unlike*.

919. $5m, -12m$

920. $4a, -3a$

921. $13z^3, -13z^3$

922. $15m^3, 15m$

923. $-3n^{10}, -3n^9$

924. $13y, 12y, -5y$

Simplify the expression by combining like terms.

925. $4m + 7m$

926. $3y + 5y + 2y$

927. $3r - 2r$

928. $5x - 4x$

929. $5z + 8z - 18z$ 930. $8y + 8y^2$

931. $4a^3 - 4a^2$ 932. $5p + 3(4 + 2p)$

933. $7q - 2 - 4(1 + q)$ 934. $-(2 + 5r) + 4r$

935. $-(4 - 3k) + 2k$ 936. $4(2 - 3m) - (6 - 4m)$

937. $5(4 - 2n) - (3 - 5n)$ 938. $-2z + 3 + 4z - 17 + 20$

Write the phrase as a mathematical expression and simplify by combining like terms. Use x as the variable.

939. Ten times a number subtracted from seven times the number

940. The sum of eleven times a number and twenty, added to five times the number

941. Three times the sum of two and four times a number, added to six times the number

942. Ten times the difference between a number and three, subtracted from twelve

943. Five times the difference between three times a number and eight, subtracted from seventeen times the number

Writing/Conceptual Exercises

Write each expression in words.

944. $3x + 8x$ 945. $3(4 - 1) - 4$ 946. $\dfrac{x + 5}{7}$

947. $5 - 2(x + 1)$ 948. $2(x + 7) + 4x$

CHAPTER 2   SOLVING EQUATIONS AND INEQUALITIES

Section 2.1   The Addition and Multiplication Properties of Equality

Objective 1   Identify linear equations

Tell whether the equation is a linear equation.

1. $6x - 7 = 0$

2. $x^2 + 2x + 3 = 0$

3. $3x = 8$

4. $3x^3 = 2x^2$

5. $\dfrac{1}{x} - \dfrac{1}{2} = 0$

6. $6x + 2 = 4x - 5$

7. $3x - 4x^2 = 5$

8. $9x - 7x = 9 - 7$

9. $2x - \dfrac{4}{x} = 5$

10. $3 + 3x = 3(1 - x)$

Objective 2   Use the addition property of equality.

Solve the equation and check your solution.

11. $y - 5 = 16$

12. $r + 9 = 3$

13. $4x = 3x + 5$

14. $3y = 2y - 3$

15. $p - \dfrac{2}{3} = \dfrac{1}{6}$

16. $y + \dfrac{3}{4} = \dfrac{5}{8}$

17. $\dfrac{2}{3}t - 5 = \dfrac{5}{3}t$

18. $\dfrac{9}{8}p - \dfrac{1}{2} = \dfrac{1}{8}p$

19. $x - 4.5 = -3.2$

20. $9.5y - 2.4 = 10.5y$

21. $7x + 2x - 8x + 5 = 10$

22. $2(b + 4) - (3 + b) = 8$

23. $11t + 2 + 4t - 3t = 5t - 8 + 6t$

24. $3(2 - 5n) + 4(7n - 1) = 4(3n + 5)$

25. $9(2z - 3) - 4(5 + 3z) = 5(4 + z)$

26. $-5(8 - 2k) + 4(7 - k) = 7(8 + k) - 3$

Objective 3   Use the multiplication property of equality.

Solve the equation and check your solution.

27. $6x = 24$

28. $-7w = 42$

29. $-12a = -48$

30. $\dfrac{m}{6} = 6$

31. $\dfrac{b}{-7} = 21$

32. $\dfrac{3}{5}r = -27$

33. $-\dfrac{5}{8}q = 40$

34. $\dfrac{3}{4}x = \dfrac{1}{4}$

35. $\dfrac{6}{7}y = \dfrac{2}{3}$

36. $9.8q = -34.3$

37. $-2.7v = -17.28$

38. $5.6z = -21.28$

Objective 4   Simplify equations, and then use properties of equality.

Solve the equation.

39. $5m + 3m = 16$

40. $7p - 18p = 22$

41. $4r - 9r = -15$

42. $8r - 3r + 7r = -84$

43. $-11z + 12z + 15z = -80$

44. $2p + 5p - 6p = 14$

45. $3(2y - 1) = 12y + 21$

46. $-2(r + 7) = 5(r + 7)$

47. $3m + 5 - 6m + 2 = 4m - 14$

48. $6x - x + 7x + 1 = 13x - 4 - 5x - 3$

49. $4(2 + 3x) - (1 - 4x) = -25$

50. $4(2 - y) - 3(2y + 1) = 45$

Mixed Exercises

Solve the equations and check your solution.

51. $z - 8 = -12$

52. $7z + 2 = 6z - 7$

53. $12y = -72$

54. $\dfrac{x}{8} = -11$

55. $t + \dfrac{2}{5} = -\dfrac{3}{10}$

56. $\dfrac{2}{7}r + \dfrac{1}{3} = -\dfrac{5}{7}r + \dfrac{5}{6}$

57. $\dfrac{4}{5}z = 8$

58. $\dfrac{3}{8}z = -\dfrac{5}{4}$

59. $3w - 8w = 20$

60. $6q - 10q = -24$

61. $3.5b = 24.85$

62. $1.9k = 11.02$

63. $8s - 3s + 5s = 90$

64. $-11h - 6h + 14h = -21$

65. $\dfrac{7}{8}n + 5 = \dfrac{15}{8}n$

66. $-2 + \dfrac{4}{5}p = \dfrac{9}{5}p$

67. $7(r - 2) - 4 = 2r - r$

68. $3(p - 8) = -2(p - 2) + 7$

69. $-t = -9.5$

70. $-3.5q = 33.25$

71. $\dfrac{7}{2}k + 1 = \dfrac{9}{2}k - 5$

72. $\dfrac{3}{4} + \dfrac{5}{6}t = \dfrac{2}{3} + \dfrac{11}{6}t$

73. $4(q - 9) = -2q + 7$

74. $-2(s + 6) = 8 - (s + 7)$

Writing/Conceptual Exercises

Decide whether the given item is an expression that can be simplified or an equation that can be solved. If it is an expression, simplify it. If it is an equation, solve it.

75. $-2z + 6 + 11z - 13$

76. $6j + 15 = -8j + 1$

77. $7(r - 2) + 8(r + 3)$

78. Which of the following are not linear equations?

    (a) $6y + 5 = 2y - 1$ (b) $y^2 - 2y = 1$

    (c) $y^3 = 27$ (d) $7(y - 1) = 5$

79. Which one of the following equations does not require the use of the multiplication property of equality?

(a) $\dfrac{2}{3}r = 8$ (b) $\dfrac{r}{-8} = -9$

(c) $5r = 7 - 2r$ (d) $9r = 6 + 8r$

80. Write an equation that requires the use of the multiplication property of equality, where both sides must be multiplied by $-\dfrac{3}{5}$, and where the solution is a negative integer.

Section 2.2   More on Solving Linear Equations

Objective 1   Learn and use the four steps for solving a linear equation.

Solve the equations and check your solution.

81. $7t + 6 = 11t - 6$ 82. $4w + 5 = 6w - 1$

83. $5j + 1 = 10j - 29$ 84. $4 + x = -(x + 2)$

85. $7y - (2y - 9) = 39$ 86. $5(z - 2) - (3z - 1) = 3z - 6$

87. $6 - (2 - b) = 18 + 4b$ 88. $3(x + 7) = 3 - 2(x - 4)$

89. $3 - (1 - y) = 4 + 5y$ 90. $-(v + 7) = 3 + v$

91. $4w - 5w + 3(w - 6) = -4(w + 3)$ 92. $3a - 6a + 4(a - 3) = -2(a + 3)$

Objective 2   Solve equations with fractions or decimals as coefficients.

Solve the equation by first clearing of fractions or decimals.

93. $\dfrac{1}{5}(z + 5) = \dfrac{1}{4}(z + 7)$ 94. $\dfrac{1}{4}x + \dfrac{2}{3}x = \dfrac{1}{6}$

95. $\dfrac{1}{3}y - \dfrac{1}{4}y = -\dfrac{1}{12}y + \dfrac{1}{2}$ 96. $\dfrac{1}{3}(2m - 1) + \dfrac{3}{4}m = \dfrac{5}{6}$

97. $\dfrac{5}{6}(r - 4) - \dfrac{2}{9}(r + 4) = \dfrac{7}{18}$ 98. $\dfrac{1}{5}s + \dfrac{2}{3}s = s + \dfrac{1}{15}$

99. $.30(50)+.25(200)=.20x$

100. $.35(20)+.45y=.125(200)$

101. $.88x+.16(10-x)=.40x$

102. $.76t+.80(11-t)=.45(20)$

103. $.24x-.38(x+2)=-.34(x+4)$

104. $.45a-.35(20-a)=.02(50)$

Objective 3   Recognize equations with no solutions or infinitely many solutions.

Solve the equation.

105. $3(a-4)=3a-12$

106. $5x+15=5(x-6)$

107. $6y-3(y+5)=3(y-4)$

108. $2x-4(x+1)=-2(x+2)$

109. $8k+17=2(k+1)+3(2k+5)$

110. $9(2p-5)-3(6p+1)=-48$

111. $6(6t+1)=9(4t-3)+11$

112. $5(3j-6)+15=15(j-1)$

113. $3b-7b-16+b=2b+16-5b$

114. $4w-7+11-5w=10w-3+12-11w$

Objective 4   Write expressions for two related unknown qualities.

Write the answer to the problem as an algebraic expression.

115. Two numbers have a sum of 17. One of the numbers is y. Find an expression for the other number.

116. The product of two numbers is 18. One of the numbers is z. Find an expression for the other number.

117. Rachel is t years old. How old will she be in 15 years?

118. Roberto is r years old. How old was he 10 years ago?

119. Denise has x nickels. Find the value of the nickels in cents.

120. A bank teller has d dollars, all of it in twenty-dollar bills. How many twenty-dollar bills does he have?

121. Tickets for a play cost $5 for adults and $3 for students. Find the cost in dollars for x adult tickets and y student tickets.

122. Find the cost in cents of r 25¢ stamps and s 20¢ stamps.

123. Marc has n nickels, d dimes, and q quarters. How much money does he have in cents?

124. Together, Jenna and Ilana have 18 sweaters. If Jenna has t sweaters, how many sweaters does Ilana have?

Mixed Exercises

Solve the equation and check your solution.

125. $3 - 6(j - 3) = 4(2j + 1) + 3$

126. $3(t + 4) = 6 - 2(t - 3)$

127. $6f - 8f + 4(f - 6) = -5(f + 2)$

128. $4r - 8r + 3(r - 5) = -2(r + 9)$

129. $\frac{1}{8}(t - 5) + \frac{1}{4}(t + 2) = t - 3$

130. $\frac{3}{8}x - \left(x - \frac{3}{4}\right) = \frac{5}{8}(x + 3)$

131. $.07(10{,}000) + .02x = .03(10{,}000 + x)$

132. $.01r + .325(1000) = .05(1000 + r)$

133. $7p + 9p - 5 = 43$

134. $-8z + 2z - 7 + 3z = 8$

135. $5y - 7y + 6y - 9 = 2y + 3$

136. $11 + 3(b + 1) = 16 + 5b$

137. $3(c + 5) = -2c + 5 + 2(c - 10)$

138. $-(5 + m) - 3(m + 4) = 6 - 3$

139. $-(4x + 2) - (-3x - 5) = 3$

140. $3(p - 2) + 4p = 8 + p + 1 - p$

141. $8.9r - 2.3 = 9.9r + 1.0$

142. $2(2y - 7) = 4y + 10$

143. $5(3r - 6) = 4(2r + 7) + 7r$

144. $6(x - 9) = 3(2x - 18)$

145. $4(2.2s + 5) = 7.8s - 15$

146. $-5(7a - 3) + 2(9a - 7) = -4(4a - 1)$

147. $4(6y - 5) + 10 = 2(12y + 5) + 5y$

148. $5(4z - 8) + 25 = 2(10z - 6) - 3$

149. $-\frac{1}{6}(a - 2) + \frac{1}{3}(a + 7) = 2a - 1$

150. $-\frac{2}{5}k - \left(k - \frac{1}{2}\right) = \frac{1}{4}(k + 3)$

Write the answer to the problem as an algebraic expression.

151. Gena is t years old. How old will she be in 12 years?

152. The sum of two numbers is 163. One of the numbers is y. Write an expression for the other number.

153. A collection of coins consists of n nickels and d dimes. How much money is in the collection in cents?

154. Adult tickets for a play cost $10 while children's tickets cost $7.50. Find the cost in dollars for x adult tickets and y children's tickets.

Writing/Conceptual Exercises

155. Which one of the following equations has no solutions?

    (a) $2(x+3)=3x+6$ (b) $7(x-2)=7x-14$

    (c) $3(x-1)+5=3x+4$ (d) $x-5=5-x$

156. Which one of the following equations has all real numbers as solutions?

    (a) $3(x-5)=3x-5$ (b) $2x-7=2x+7$

    (c) $x-8=8-x$ (d) $6(2x-1)=3(4x-2)$

Section 2.3   An Introduction to Applications of Linear Equations

Objective 2   Solve problems involving unknown numbers.

Solve the problem.

157. The product of 8 and the sum of a number and 3 equals 40. Find the number.

158. A number added to twice a number is 24. Find the number.

159. Three times the quotient of a number and 2 equals 18. Find the number.

160. Eight times the difference between a number and 8 equals the product of the number and –4. Find the number.

161. When the difference between a number and 2 is multiplied by –7, the result is one more than –10 times the number. Find the number.

162. If three times a number is decreased by 5, the result is 16. Find the number.

163. When the sum of a number and 6 is multiplied by 3, the result is 30. Find the number.

164. When twice a number is added to 15, the result is 9 more than the number. Find the number.

165. When 6 is subtracted from a number, the result is −3 times the number. Find the number.

166. If seven times a number is added to 3, the result is two less than eight times the number. Find the number.

Objective 3 Solve problems involving sums of quantities.

Solve the problem.

167. Dave has a board 33 inches long. He wishes to cut it into two pieces so that one piece is 3 inches longer than the other. How long is the shorter piece?

168. Jackson and Brown were opposing candidates in the school board election. Jackson received 87 more votes than Brown, with 439 votes cast. How many votes did Brown receive?

169. On a psychology test, the highest grade was 38 points more than the lowest grade. The sum of the two grades was 142. Find the lowest grade.

170. In a physical fitness test, Brian did 30 more pushups than Vincent. The total number of pushups for both men is 160. How many pushups did Vincent do?

171. Walt Thompson runs a ski train. One day he noticed that the train contained 13 more women than men (including himself). If there were a total of 165 people on the train, how many of them were men?

172. A piece of rope is 79 centimeters long. It is cut into three pieces. The longest piece is 3 times as long as the shortest piece, and the middle-sized piece is 14 centimeters longer than the shortest piece. Find the lengths of the three pieces.

173. On a history test, the highest grade was 57 points higher than the lowest grade. The sum of the two grades is 95. Find the highest and lowest grades.

174. Fred bought a cassette player and a calculator for a total of $208. If the cassette player cost $140 more than the calculator, how much did the calculator cost?

175. Craig purchased two house plants for a total of $58. If the grape ivy cost $16 more than the Boston fern, how much did the Boston fern cost?

176. Jan paid $37 more for telephone service in December than in November. She paid a total of $143 for telephone service during the two months. How much did she pay in November?

Objective 4   Solve problems involving supplementary and complementary angles.

Solve the problem.

177. Find the measure of an angle if the measure of the angle is 5 times the measure of its supplement.

178. Find the measure of an angle if the measure of the angle is 4 times the measure of its complement.

179. Find the measure of an angle whose supplement is 20° more than twice its complement.

180. Find the measure of an angle whose complement is 30° less than half its supplement.

181. Find the measure of an angle such that the sum of the measures of its complement and its supplement is 138°.

182. Find the measure of an angle such that the sum of the measures of its supplement and its complement is 114°.

183. Find the measure of an angle such that the difference between the measure of its supplement and twice the measure of its complement is 43°.

184. Find the measure of an angle such that the difference between the measure of the angle and the measure of its complement is 16°.

185. Find the measure of an angle such that the difference between the measure of the angle and the measure of its supplement is 96°.

186. Find the measure of an angle such that the difference between the measure of its supplement and four times the measure of its complement is 63°.

Objective 5   Solve problems involving consecutive integers.

Solve the problem.

187. Find two consecutive integers whose sum is –313.

188. Find two consecutive even integers such that three times the first subtracted from four times the second is 26.

189. Find two consecutive integers such that twice the first is 26 more than the second.

190. Find two consecutive integers such that when the smaller is subtracted from one and one-half times the larger, the result is 45.

191. Find three consecutive integers such that the sum of the three is 61 more than the largest.

192. Find two consecutive integers such that four times the larger is 334 more than the smaller.

193. Find three consecutive odd integers such that the sum of the smallest and the largest is eleven less than three times the middle integer.

194. Find two consecutive integers such that their sum is 31 less than three times the larger integer.

195. Find three consecutive integers such that when twice the second is subtracted from three times the third, the result is 84.

196. Find three consecutive even integers such that the sum of the three is 174 less than five times the first.

Mixed Exercises

Solve the problem.

197. If 9 is added to 3 times a number, the result is 6. Find the number.

198. If 6 is subtracted from a number and the difference is doubled, the result is 10. Find the number.

199. If –4 is multiplied by the difference between 9 and a number, the result is 24. Find the number.

200. Find two consecutive even integers whose sum is 294.

201. Find two consecutive even integers such that the smaller added to three times the larger is equal to 150.

202. Steve is making punch for a party. The recipe requires twice as much orange juice as cranberry juice and 8 times as much ginger ale as cranberry juice. If he plans to make 176 ounces of punch, how much of each ingredient should he use?

203. Carolyn, Lori, and Becky swim at a public pool each day for exercise. One day Carolyn swam twice as many laps as Lori, and Becky swam three more laps than Lori. If the women swam 23 laps altogether, how many laps did each one swim?

204. Seven times the sum of a number and 5 is equal to the difference between the number and 7. Find the number.

205. Find three consecutive odd integers whose sum is 351.

206. When the sum of a number and 12 is multiplied by –5, the result is –20. Find the number.

207. Find three consecutive integers such that the sum of the first two is 29 more than the third.

208. Two pages that face each other in this book have 657 as the sum of their page numbers. What are the page numbers?

209. Margo and Gladis sold a total of 348 boxes of Girl Scout cookies. Margo sold 52 more boxes than Gladis. How many boxes did each sell?

210. The measure of the supplement of an angle is 6° less than three times the measure of its complement. Find the measure of the angle.

211. Find two consecutive even integers whose sum is 178.

212. Damon and Nolanda jog a total of 4 miles each day. Damon jogs .4 mile more than Nolanda. How many miles does each jog?

213. Find the measure of an angle whose complement has a measure 6° more than 6 times the measure of the angle.

214. Find three consecutive even integers whose sum is 20 more than twice the second.

215. Desi had a reading assignment in two books totaling 160 pages. The number of pages in the second book was 25 more than half of the number of pages in the first book. How many pages did she have to read in each book?

216. The measure of the supplement of an angle is four times the measure of its complement. Find the measure of the angle.

217. Find two consecutive even integers such that their sum is 125 more then half of the first integer.

218. Helene has a collection of dimes and nickels. There are 354 coins in the collection. There are 18 more dimes than nickels. Find the number of each type of coin.

219. Find the measure of an angle whose supplement has a measure which is 42° less than twice the measure of the angle.

220. Find the measure of an angle whose supplement has a measure equal to 10 times the measure of its complement.

221. Gita baked a total of 200 cookies. Some were chocolate chip, and some were sugar cookies. The number of sugar cookies was 20 less than the number of chocolate chip cookies. How many of each type did she bake?

222. The measure of the supplement of an angle is one-fifth the measure of the angle. Find the measure of an angle.

223. Before leaving on a trip, Abby had to edit 500 pages of a manuscript. She did this in two days. On the first day, she edited 220 pages less than twice the number of pages she edited on the second day. How many page did she edit each day?

224. The measure of the supplement of an angle is 12° less than 5 times the measure of the angle. Find the measure of the angle.

225. George spent $60 on a shirt and pair of jeans. The price of the shirt was 2/3 the price of the jeans. Find the price of each.

226. The measure of the complement of an angle is 15° more than one-half the measure of the angle. Find the measure of the angle.

Section 2.4   Formulas and Applications from Geometry

Objective 1   Solve a formula for one variable given the values of the other variables.

In the following exercises a formula is given, along with the values of all but one of the variables in the formula. Find the value of the variable that is not given.

227. $d = rt$; $r = 55$, $t = 5$

228. $P = 2L + 2W$; $P = 52$, $W = 4$

229. $A = \dfrac{1}{2}bh$; $b = 8$, $h = 2.5$

230. $v = \dfrac{1}{3}Bh$; $B = 27$, $V = 63$

231. $C = 2\pi r$; $C = 56.52$, $\pi = 3.14$

232. $A = \pi r^2$; $r = 7$, $\pi = 3.14$

233. $I = prt$; $I = 990$, $r = .06$, $t = 3$

234. $C = \dfrac{5}{9}(F - 32)$; $F = 212$

235. $F = \dfrac{9}{5}C + 32$; $C = 25$

236. $A = \frac{1}{2}(b + B)h$; $b = 10$, $B = 12$, $A = 132$

237. $A = \frac{1}{2}(b + B)h$; $b = 7$, $B = 13$, $A = 140$

238. $V = \frac{1}{3}\pi r^2 h$; $r = 3$, $h = 4$, $\pi = 3.14$

Objective 2   Use a formula to solve a geometric application.

Use a formula to write an equation for the application; then solve it.

239. Find the length of a rectangular garden if its perimeter is 76 feet and its width is 15 feet.

240. Find the height of a triangular banner whose area is 36 square inches and base is 9 inches.

241. Tom has 30 feet of binding for a rectangular rug that he is weaving. If the rug is 7 feet wide, how long can he make the rug if he wishes to use all the binding on the perimeter of the rug?

242. The circumference of a circular pool is 20.41 meters. Find the diameter of the pool. (Use 3.14 as an approximation for π.)

243. The circumference of a circular garden is 628 feet. Find the area of the garden. (Use 3.14 as an approximation for π.)

244. A water tank is a right circular cylinder. The tank has a radius of 4 meters and a volume of 502.4 cubic meters. Find the height of the tank. (Use 3.14 as an approximation for π.)

245. A tent has the shape of a right pyramid. The volume is 200 cubic feet and the height is 12 feet. Find the area of the floor of the tent.

246. A spherical balloon has a radius of 15 centimeters. Find the amount of air required to fill the balloon. (Use 3.14 as an approximation for π.)

247. A playground has the shape of a trapezoid. The parallel sides measure 82.3 feet and 114.5 feet. The height of the trapezoid is 45.2 feet. Find the area of the playground.

248. Find the height of an ice cream cone if the diameter is 8 centimeters and the volume is 461.58 cubic centimeters. (Use 3.14 as an approximation for π. Round answer to the nearest hundredth.)

Objective 3   Solve problems about angle measures.

Find the measure of the marked angles.

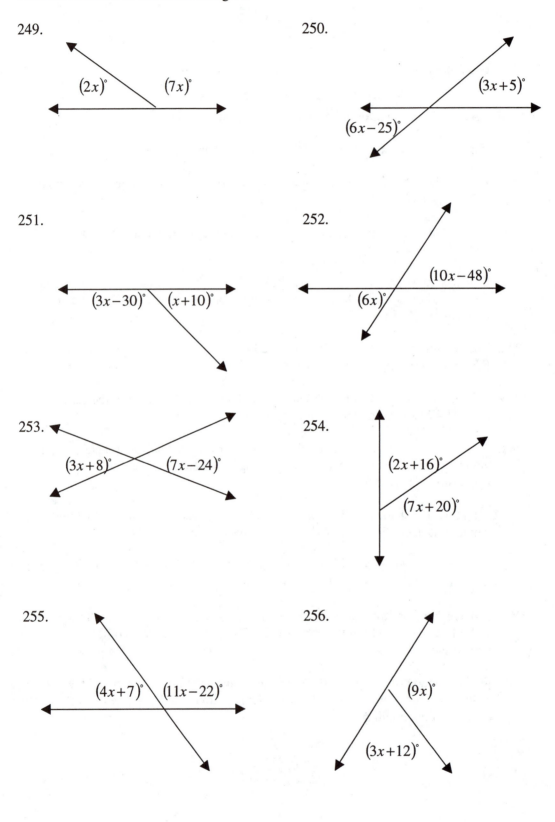

249.

$(2x)°$ $(7x)°$

250.

$(3x+5)°$

$(6x-25)°$

251.

$(3x-30)°$ $(x+10)°$

252.

$(10x-48)°$

$(6x)°$

253.

$(3x+8)°$ $(7x-24)°$

254.

$(2x+16)°$

$(7x+20)°$

255.

$(4x+7)°$ $(11x-22)°$

256.

$(9x)°$

$(3x+12)°$

257.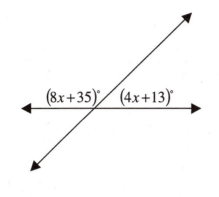

258.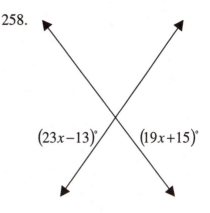

Objective 4   Solve a formula for a specified variable.

Solve the formula for the specified variable.

259. $V = LWH$ *for L*

260. $A = p + prt$ *for t*

261. $S = \dfrac{a}{1-r}$ *for r*

262. $V = \dfrac{1}{3} Bh$ *for B*

263. $S = 2\pi rh + 2\pi r^2$ *for h*

264. $a_n = a_1 + (n-1)d$ *for d*

265. $g = \dfrac{mv^2}{2k}$ *for k*

266. $V = \pi r^2 h$ *for h*

267. $A = \dfrac{1}{2}(b + B)h$ *for B*

268. $S_n = \dfrac{n}{2}(a_1 + a_n)$ *for n*

269. $C = \dfrac{5}{9}(F - 32)$ *for F*

270. $d = gt^2 + vt$ *for v*

271. $V = \dfrac{1}{3}\pi r^2 h$ *for r^2*

272. $A + B + C = 180°$ *for C*

Mixed Exercises

In the following exercises a formula is given, along with the values of all but one of the variables in the formula. Find the value of the variable that is not given.

273. $P = 4s;$ $s = 12$

274. $P = 2L + 2W;$ $L = 8, W = 3$

275. $P = 2L + 2W$; $L = 12, W = 4$

276. $d = rt$; $r = 55, t = 3$

277. $P = 4s$; $P = 36$

278. $A = \dfrac{1}{2}bh$; $b = 10, h = 5$

279. $A = \dfrac{1}{2}bh$; $b = 8, h = 2.5$

280. $V = \dfrac{1}{3}Bh$; $B = 20, h = 12$

281. $A = \dfrac{1}{2}bh$; $b = 14, A = 56$

282. $C = 2\pi r$; $C = 28.26, \pi = 3.14$

283. $V = \dfrac{1}{3}Bh$; $V = 4, B = 4.8$

284. $A = \pi r^2$; $r = 5, \pi = 3.14$

285. $C = \dfrac{5}{9}(F - 32)$; $F = 212$

286. $I = prt$; $I = 200, p = 500, r = .2$

Solve the formula for the specified variable. Then find the value of the specified variable, using the given values of the other variables.

287. $P = a + b + c$ *for c*;
$P = 28, a = 12, b = 3$

288. $V = LWH$ *for W*;
$V = 144, L = 8, H = 3$

289. $V = \dfrac{1}{3}\pi r^2 h$ *for h*;
$V = 36, r = 2, \pi = 3.14$
(Round answer to the nearest hundredth.)

290. $C = 2\pi r$ *for r*;
$C = 9.42, \pi = 3.14$

291. $S = \dfrac{a}{1 - r}$ *for a*;
$S = 54, r = .2$

292. $P = A - Art$ *for r*;
$P = 850, A = 1000, t = 2.5$

Solve the problem.

293. The height of a triangular frame is 4 centimeters and the base is 9 centimeters. Find the area of the frame.

294. A handyman has 1000 meters of fencing material to enclose a rectangular field. The width of the field will be 200 meters. Find the length.

295. The perimeter of a triangle is 52 centimeters. One of the sides is 12 centimeters long, and the other two sides are of equal length. Find the length of each equal side.

296. The area of a rectangular swimming pool is 4500 square feet.   The length is 90 feet. Find the width.

297. The area of a triangular lot is 150 square meters. If the height is 20 meters, find the base of the lot.

298. A soup can has a height of 11 centimeters and a diameter of 6 centimeters. How much soup will the can hold? (Use 3.14 as an approximation for π.)

299. A pair of vertical angles have measures $(5x - 20)^\circ$ and $(3x + 14)^\circ$. Find the measures of these two angles.

300. Two angles with measures $(3x + 42)^\circ$ and $(10x - 31)^\circ$ form a straight angle. Find the measures of these two angles.

301. A pair of vertical angles have measures $(5x + 2)^\circ$ and $(3x + 34)^\circ$. Find the measures of these two angles.

302. Two angles with measures $(11x)^\circ$ and $(6x + 10)^\circ$ form a straight angle. Find the measures of these two angles.

Writing/Conceptual Exercises

Decide whether perimeter or area would be used to solve a problem concerning the measure of the quantity.

303. Finding the cost of spraying insecticide around the edges of a room

304. Finding the cost of tiling a kitchen floor

305. Estimating the amount of paint needed to paint a room

306. Estimating the amount of caulk needed to caulk around the edges of a bathtub

Section 2.5   Ratios and Proportions

Objective 1   Write ratios.

Write a ratio for the word phrase as a fraction in lowest terms.

307. 5 women to 3 women

308. 14 marbles to 8 marbles

309. 5 days to 2 weeks

310. 4 gallons to 2 pints

311. 3 dollars to 18 quarters

312. 5 hours to 2 days

313. 100 inches to 5 yards

314. 35 nickels to 3 dollars

315. 7 days to 9 hours

316. 40 hours to 3 weeks

317. 12 ounces to 1 pound

318. 45¢ to $9

319. 40 centimeters to 2 meters

320. 75 years to 2 centuries

Objective 2   Decide whether proportions are true.

Decide whether the proportion is *true* or *false*.

321. $\dfrac{4}{7} = \dfrac{28}{49}$

322. $\dfrac{6}{8} = \dfrac{12}{16}$

323. $\dfrac{7}{10} = \dfrac{10}{30}$

324. $\dfrac{16}{40} = \dfrac{12}{30}$

325. $\dfrac{110}{18} = \dfrac{55}{9}$

326. $\dfrac{9}{10} = \dfrac{45}{40}$

327. $\dfrac{12}{18} = \dfrac{8}{12}$

328. $\dfrac{18}{20} = \dfrac{9}{10}$

329. $\dfrac{420}{300} = \dfrac{21}{15}$

330. $\dfrac{2}{3} = \dfrac{40}{60}$

331. $\dfrac{10}{15} = \dfrac{16}{24}$

332. $\dfrac{9}{12} = \dfrac{12}{16}$

Objective 3   Solve proportions.

Solve the equation.

333. $\dfrac{25}{4} = \dfrac{125}{x}$

334. $\dfrac{m}{32} = \dfrac{7}{24}$

335. $\dfrac{x}{15} = \dfrac{4}{5}$

336. $\dfrac{z}{80} = \dfrac{25}{125}$

337. $\dfrac{5}{m} = \dfrac{8}{5}$

338. $\dfrac{m}{5} = \dfrac{m-3}{3}$

339. $\dfrac{3k-1}{k} = \dfrac{6}{7}$

340. $\dfrac{x+8}{x-9} = \dfrac{1}{3}$

341. $\dfrac{6y-5}{y} = \dfrac{11}{5}$

342. $\dfrac{4}{z-1} = \dfrac{3}{z+7}$

343. $\dfrac{3x+5}{x-2} = \dfrac{1}{4}$

344. $\dfrac{s+6}{s+11} = \dfrac{9}{14}$

Objective 4   Solve applied problems using proportions.

Solve the problem by setting up and solving a proportion.

345. Michael can type 4 pages of his term paper in 30 minutes. How long will it take him to type the paper if it has 14 pages?

346. If 4 small pizzas cost $15, find the cost of 7 small pizzas.

347. On a road map, 4 inches represents 50 miles. How many inches would represent 125 miles?

348. A certain lawn mower uses 5 tanks of gas to cut 18 acres of lawn. How many acres could be cut using 8 tanks of gas?

349. If 3 ounces of medicine must be mixed with 5 ounces of water, how many ounces of medicine must be mixed with 15 ounces of water?

350. A certain lawn mower uses 4 tanks of gas to cut 15 acres of lawn. How many tanks of gas are needed to cut 30 acres of lawn?

351. One bowl of a certain cereal has 286 milligrams of sodium. How many bowls would one need to eat to obtain 429 milligrams of sodium?

352. If two pounds of fertilizer will cover 50 square feet of garden, how many pounds would be needed for 125 square feet?

353. A garden service charges $30 to install 50 square feet of sod. Find the charge to install 225 square feet.

354. The charge to move a load of freight 700 miles is $100. Find the charge to move the freight 1750 miles.

A supermarket was surveyed and the following prices were charged for items in various sizes. Find the best buy (based on price per unit) for the following particular items.

355. Rice
    1-pound box: $1.29
    2-pound box: $2.31
    3-pound box: $3.32
    5-pound box: $4.44

356. Tomato catsup
    14-ounce size: $ .93
    32-ounce size: $1.92
    44-ounce size: $2.59
    64-ounce size: $3.45

356. Tomato sauce
     8-ounce can: $ .39
     15-ounce can: $ .73
     29-ounce can: $1.33

357. Corn Oil
     18-ounce bottle: $1.27
     32-ounce bottle: $2.40
     48-ounce bottle: $3.19
     64-ounce bottle: $4.43

359. Applesauce
     8-ounce jar: $ .59
     16-ounce jar: $ .96
     24-ounce jar: $1.31
     48-ounce jar: $1.99

360. Freeze-dried coffee
     2-ounce size: $2.21
     4-ounce size: $3.78
     7-ounce size: $4.81
     10-ounce size: $6.14

361. Breakfast cereal
     10-ounce box: $2.91
     15-ounce box: $3.09
     20-ounce box: $3.79

362. Chilled orange juice
     32-ounce carton: $1.55
     64-ounce carton: $2.46
     96-ounce bottle: $3.89

363. Chlorine bleach
     16-ounce size: $ .43
     32-ounce size: $ .66
     64-ounce size: $1.05
     128-ounce size: $1.43

364. Liquid dishwashing detergent
     12-ounce size: $1.09
     22-ounce size: $1.79
     32-ounce size: $2.39
     42-ounce size: $3.07

Objective 5   Solve direct variation problems

365. If one dozen doughnuts cost $4.50, find the cost of 18 doughnuts.

366. If 4 six-packs of cola cost $5.16, find the cost of 9 six-packs of cola.

367. If 7 gallons of gasoline cost $7.91, find the cost of 11 gallons of gasoline.

368. If 20 trash bags cost $3.10, find the cost of 36 trash bags.

369. If a 24-count case of green beans costs $6.96, find the cost of 18 cans of green beans.

370. If a 36-pack box of baseball cards costs $39.24, find the cost of 14 packs of baseball cards.

371. If 5 tickets to an amusement park cost $187.50, find the cost of 13 tickets.

372. If a 4.9 GB hard-drive can hold 147 files, find the number of files that an 8.1 GB hard-drive can hold.

373. If 3.5 lbs. of coffee cost $9.80, find the cost of 7.5 lbs. of coffee.

374. If 16 reams of paper cost \$46.24, find the cost of 21 reams of paper.

Mixed Exercises

Write a ratio for each word phrase as a fraction in lowest terms.

375. 4 dollars to 7 dimes

376. 90 minutes to 2 days

377. 30 minutes to 8 hours

378. 60 inches to 5 yards

379. 7 quarters to 8 dollars

380. 5 pints to 2 quarts

Decide whether the proportion is *true* or *false*.

381. $\dfrac{5}{9} = \dfrac{15}{28}$

382. $\dfrac{7}{12} = \dfrac{1}{2}$

383. $\dfrac{56}{60} = \dfrac{14}{20}$

384. $\dfrac{99}{44} = \dfrac{54}{24}$

385. $\dfrac{42}{36} = \dfrac{24}{28}$

386. $\dfrac{11}{28} = \dfrac{176}{448}$

387. $\dfrac{25}{33} = \dfrac{425}{594}$

388. $\dfrac{13}{37} = \dfrac{455}{1295}$

389. $\dfrac{25}{49} = \dfrac{5}{7}$

Solve the equation.

390. $\dfrac{k}{24} = \dfrac{12}{144}$

391. $\dfrac{32}{s} = \dfrac{8}{14}$

392. $\dfrac{8}{r} = \dfrac{9}{7}$

393. $\dfrac{t-4}{t+4} = \dfrac{1}{3}$

394. $\dfrac{q+3}{q+8} = \dfrac{7}{12}$

395. $\dfrac{p}{32} = \dfrac{p+5}{48}$

396. $\dfrac{3}{r} = \dfrac{1}{r+1}$

397. $\dfrac{4}{3} = \dfrac{10}{r}$

398. $\dfrac{y}{6} = \dfrac{14}{12}$

Solve the problem by setting up and solving a proportion.

399. The distance between two cities on a road map is 11 inches. Actually, the cities are 308 miles apart. The distance between a second pair of cities on the map is 15 inches. How far apart are these cities?

400. A candy bar contains 200 calories. How many bars would you need to eat to get 700 calories?

401. If 9 pairs of jeans cost $135, find the cost for 5 pairs.

402. If twelve yards of material is needed for 5 dresses, how much material is needed for 12 dresses?

Find the best buy (based on unit price) for each of the following items.

403. Peaches
    12-ounce can: $1.02
    18-ounce can: $1.50
    24-ounce can: $1.98
    48-ounce can: $3.99

404. Grape juice
    20-ounce bottle: $2.20
    24-ounce bottle: $2.70
    32-ounce bottle: $3.49
    48-ounce bottle: $5.25

Section 2.6   More about Problem Solving

Objective 1   Use percent in problems involving rates.

Solve the problem.

405. How much pure alcohol is in 50 liters of a 45% alcohol solution?

406. How much pure alcohol is in 25 liters of a 30% alcohol solution?

407. How much pure antifreeze is in 72 liters of a 22% antifreeze solution?

408. If $10,000 is invested for one year at 8% simple interest, how much interest is earned?

409. How much interest is earned if $15,000 is invested at 7% simple interest for one year?

410. How much interest is earned if $7800 is invested at 12% simple interest for one year?

411. What is the monetary value of 63 nickels?

412. What is the monetary value of 95 dimes?

413. What is the monetary value of 47 quarters?

414. What is the monetary value of 39 half dollars?

Objective 2   Solve problems involving mixtures.

Solve the problem.

415. A pharmacist has 2 liters of a solution containing 30% alcohol. If he wants to have a solution containing 44% alcohol, how much pure alcohol must he add?

416. A car radiator contains 4 gallons of a coolant which is a mixture of antifreeze and water. If the coolant in the radiator is 30% antifreeze, how much coolant must be replaced with 80% antifreeze to have a 50% solution?

417. How many liters of a 25% salt solution must be added to 20 liters of 12% solution to make a solution that is 20% salt?

418. How many pounds of a certain alloy containing 35% silver must be melted with another alloy containing 65% silver to obtain 50 pounds of an alloy containing 50% silver?

419. A cereal manufacturer has 2500 pounds of a cereal mixture containing 25% corn, 40% sugar, 25% oat flour, and a 10% combination of wheat-starch and salt. If he wants to increase the percentage of corn content in the cereal mixture to 40%, how many pounds of corn should he add?

420. A chemist has two acid solutions. One is a 60% solution and the other a 30% solution. How many liters of each should she mix to obtain 10 liters of 51% acid solution?

421. A grocer has 60 pounds of candy worth $1.50 per pound. He wishes to upgrade this candy to $3.50 per pound by mixing it with candy worth $5.50 per pound. How many pounds of $5.50 candy should he use?

422. How many liters of water must be added to 2 liters of pure alcohol to obtain a 10% solution?

423. How many gallons of 50% chlorine (to the nearest thousandth) must be added to 20,000 gallons of swimming pool water to obtain a (4/100)% solution of chlorine?

424. A merchant has 25 pounds of cashews worth $4.30 per pound. He wishes to mix the cashews with peanuts worth $1.80 per pound to have a nut mixture he can sell for $2.80 per pound. How many pounds of peanuts should he use?

Objective 3   Solve problems involving simple interest.

Solve the problem. Assume that simple interest is being paid.

425. August Zarcone has an annual interest income of $3390 from two investments. He has $10,000 more invested at 8% than he has invested at 6%. Find the amount invested at each rate.

426. Felicia Whitcomb has some money invested at 5%, and $5000 more than this amount invested at 9%. Her total annual interest income is $1430. Find the amount invested at each rate.

427. Georgia Levy has 3 times as much money invested in 8% bonds as she has in stocks paying 9½ %. How much does she have invested in each if her yearly income from the investments is $5695?

428. A businessman wishes to invest $20,000 in two different funds which yield annual profits of 12½ % and 14%, respectively. Find how much he should invest in each in order to make an annual profit of $2710.

429. A total of $2000 is invested for one year, part at 7½ % and the remainder at 8½ %. If $156 interest is earned, how much is invested at 7½ %?

430. Michael Morris has $1000 more invested at 9% than he has invested at 11%. If the annual income for the two investments is $1290, find how much he has invested at each rate.

431. Ariel Mejia receives an inheritance, part of which she invests at 9%. She invests the same amount plus $3500 at 10%. If she earns $1490 in annual interest from the two investments, find the amount she inherited.

432. Diane Pietrzak earns a $17,000 bonus from her company. She invests part of the money at 9% and the balance at 11%. If the annual interest for the two investments is $1670, find the amount invested at each rate.

433. Adam Costello received an inheritance of $13,500. He wishes to divide the amount between investments at 4% and 7% to receive an average return of 6% on the two investments. How much should he invest at each rate?

434. Vincent Nganga has $37,600 invested in bonds paying 9%. How much additional money should he invest at 6% to have an average return of 8% on the two investments?

Objective 4   Solve problems involving denominations of money.

Solve the problem.

435. A collection of coins consisting of nickels and dimes has a value of $5.80. Find the number of nickels and dimes in the collection if there are 22 more dimes than nickels.

436. A woman has $2.05 in change consisting of three more dimes than quarters. Find the number of dimes and quarters.

437. A cashier has $645 in ten-dollar and five-dollar bills. There are 90 bills in all. How many of each bill does the cashier have?

438. Total receipts from the sale of 300 tickets to a school musical were $1130. If student tickets cost $3 each and adult tickets $5 each, how many student tickets were sold?

439. A stamp collector buys some 20¢ stamps and some 35¢ stamps, paying $9.35 for them. She buys one more than twice as many 35¢ stamps than 20¢ stamps. Find the number of 35¢ stamps she buys.

440. At a dinner theater, certain tickets sold for $13.95 and others sold for $10.45. If a total of 449 tickets were sold for $5374.55, find the number of each kind sold.

441. Meredith has $10.35 in nickels, dimes, and quarters. If she has six more dimes than nickels and twice as many quarters as nickels, how many of each kind of coin does she have?

442. Aaron has $9.70 in nickels, dimes, and quarters. If he has four more dimes than nickels and three times as many quarters as dimes, how many of each kind of coin does he have?

443. The admission to a soccer game is $4.50 for adults and $2.25 for children. The receipts total $974.25 for 278 paid admissions. Find the number of adults and children who paid admission to the game.

444. Tickets for a performance of *The Nutcracker* ballet cost $14 for adults and $12 for students and senior citizens. If total receipts from the sale of 650 tickets were $8390, how many adult tickets were sold?

Objective 5   Solve problems involving distance, rate, and time.

Solve the problem, using $d = rt$, $r = \dfrac{d}{t}$, or $t = \dfrac{d}{r}$, as necessary.

445. A driver averaged 48 miles per hour and took 6 hours to travel from Chicago to St. Louis. What is the distance between Chicago and St. Louis?

446. A small plane flew from San Francisco to Los Angeles, averaging 194 miles per hour. The trip took 2 hours. What is the distance between San Francisco and Los Angeles?

447. An Amtrak train traveled from Seattle to San Francisco, averaging 54 miles per hour. The distance between the two cities is 810 miles. How long did the trip take?

448. A driver averaged 49 miles per hour and took 4 1/2 hours to travel from New York City to Boston. Find the distance between the two cities.

449. A plane flew from Denver to Chicago, a distance of 1020 miles, in 2 1/2 hours. Find the plane's average speed for the trip.

450. A driver averaged 45 miles per hour on a trip from Boston to Philadelphia, a distance of 315 miles. How long did the trip take?

451. A plane flew from Dallas to Atlanta, a distance of 822 miles, in two hours. Find the plane's average speed.

452. An Amtrak train traveled from New York to Chicago, a distance of 810 miles. If the trip took 13 1/2 hours, find the train's average speed.

453. A runner ran the Boston Marathon, a distance of 42.2 kilometers, at an average speed of 16.88 kilometers per hour. Find the runner's time for the race.

454. In the 1988 Olympics, the 400-meter women's relay was won by the United States team of Brown, Echols, Griffith-Joyner, and Ashford in a time of 42 seconds. Find the relay team's average speed in meters per second. (Round to the nearest hundredth.)

Solve the problem.

455. A freight train travels 90 miles in the same time that a passenger train travels 150 miles. If the passenger train goes 30 miles per hour faster than the freight train, find the rate of each.

456. Ron and Doug leave the same point at the same time traveling in cars going in opposite directions. Ron travels at 40 miles per hour and Doug travels at 60 miles per hour. In how many hours will they be 350 miles apart?

457. Mitsuko walks to her friend's house at the rate of 3 miles per hour and then runs home at the rate of 6 miles per hour. How far does Mitsuko live from her friend if the round trip takes 30 minutes?

458. Kevin flies a distance of 1575 miles in the same amount of time Latasha drives 420 miles. If Kevin flies at a rate 165 miles per hour faster than Latasha drives, find the rate at which Kevin and Latasha are traveling.

459. Rodney and Evan are 72 miles apart. Both start riding their bicycles at the same time and travel toward each other. If Rodney bikes at 13 miles per hour and Evan bikes at 11 miles per hour, in how many hours will they meet?

460. A boat goes 3 miles upstream in the same time it takes the boat to go 5 miles downstream. If the rate of the current is 2 miles per hour, what is the speed of the boat in still water?

461. Elly and Sam are jogging. Elly runs from point A to point B in one hour at 2 miles per hour faster than Sam does. If Sam takes 1/2 hour more time than Elly to go the same distance, find the distance between points A and B.

462. Sandy and Jerome return home from college for spring vacation. Sandy, who travels by bus at 50 miles per hour, leaves one hour earlier than Jerome, who travels by car at 10 miles per hour faster than Sandy. They arrive home at the same time. How far is their home from their college?

463. Two planes leave Denver at the same time. One plane flies northwest toward Seattle at 350 miles per hour. The other plane flies southeast toward Atlanta at 420 miles per hour. In how many hours will they be 1925 miles apart?

464. At 42 miles per hour, a trip takes 1/2 hour longer than it does at 60 miles per hour. How long does the trip take at 42 miles per hour?

Mixed Exercises

Solve the problem.

465. How much pure alcohol is in 40 liters of a 95% solution?

466. How much pure antifreeze is in 50 liters of a 40% solution?

467. How much interest is earned if $8000 is invested at 6% simple interest for one year?

468. How much interest is earned if $15,000 is invested at 8% simple interest for one year?

469. What is the monetary value of 18 quarters?

470. What is the monetary value of 37 nickels?

471. How many liters of pure alcohol must be added to 20 liters of a 30% solution to obtain a 35% solution?

472. How many liters of pure water must be added to 15 liters of a 40% salt solution to obtain a 30% salt solution?

473. How many liters of a 20% alcohol solution must be added to 28 liters of a 42% alcohol solution to obtain a 30% alcohol solution?

474. How many pounds of nuts worth $1.50 per pound must be mixed with 20 pounds of nuts worth $2.50 per pound to obtain a mixture worth $1.75 per pound?

475. How many gallons of 85 octane gasoline must be mixed with 200 gallons of 92 octane gasoline to obtain a mixture of 90 octane gasoline?

476. How many pounds of candy worth $1.30 per pound must be mixed with 30 pounds of candy worth $1.60 per pound to obtain a mixture worth $1.40 per pound?

477. How much money should be invested at 8% simple interest to obtain an annual return of $3200?

478. If $10,000 is invested part at 7% and part at 9% and if the total annual return is $840, how much was invested at 7%?

479. If $5000 is invested at 6%, how much should be invested at 8% for a total annual return of $700?

480. If $14,000 is invested at 6%, and part at 7% and if the total annual return is $920, how much was invested at 7%?

481. If $8000 is invested at 9%, how much additional money should be invested at 9.5% for an annual yield of $910?

482. A collection of coins consists of dimes and nickels. The number of dimes is two more than twice the number of nickels. The value of the collection is $2.70. How many dimes are in the collection?

483. A collection of coins consists of quarters and nickels. The number of quarters is four times the number of nickels. The value of the collection is $5.25. How many quarters are in the collection?

484. A collection of coins consists of nickels, dimes, and quarters. The number of nickels is one less than the number of quarters. The number of dimes is three more than the number of nickels. The value of the collection is $3.35. How many nickels are in the collection?

485. A collection of coins consists of quarters and dimes. The total value of the collection is $6.00. The number of dimes is four more than the number of quarters. How many dimes are in the collection?

486. Tickets to a play cost $14 and $16. The number of $14 tickets sold was 50 more than the number of $16 tickets sold. The total receipts were $2200. How many $14 tickets were sold?

487. In a charity raffle, 140 tickets were sold, some at $5.50 and some at $7.50. If the total receipts were $930.00, how many $7.50 tickets were sold?

488. Tickets to a game sell for $4.00 and $6.00. The number of $4.00 tickets sold was one-third the number of $6.00 tickets sold. The total receipts were $4400. How many $4.00 tickets were sold?

489. A distance of 240 miles was traveled in 5 hours. Find the average speed.

490. A distance of 540 miles was traveled in 9 hours. Find the average speed.

491. A distance of 320 miles was traveled at an average speed of 60 miles per hour. How long did the trip take?

492. A distance of 600 miles was traveled at an average speed of 50 miles per hour. How long did the trip take?

493. A distance of 880 miles was traveled at an average speed of 55 miles per hour. How long did the trip take?

494. Find the distance traveled at an average speed of 45 miles per hour for 8 hours.

495. Find the distance traveled at an average speed of 60 miles per hour for 5 1/3 hours.

496. Find the distance traveled at an average speed of 55 miles per hour for 12 hours.

497. One car travels for 8 hours at an average speed of 55 miles per hour. A second car travels the same distance in 11 hours. Find the average speed of the second car.

498. On a 380-mile trip, the first 4 hours were traveled at an average speed of 50 miles per hour. The last part of the trip was traveled at an average speed of 60 miles per hour. How many hours did the last part of the trip take?

499. A boat traveled upstream for 14 hours. The return trip downstream took 6 hours. The average speed of the current is 2 miles per hour. Find the average speed of the boat still in the water.

500. Hyman and Sol start jogging in opposite directions. At the end of 1 1/2 hours they are 15 miles apart. Find Hyman's average speed if his average speed is 2 miles per hour more than Sol's average speed.

501. Justina and Desiree live 550 miles apart. They start driving toward each other at the same time and meet in 5 hours. If Justina's average speed is 10 miles per hour more than Desiree's average speed, find Justina's average speed.

Writing/Conceptual Exercises

502. Suppose that water is to be added to a 42% antifreeze solution in a car radiator. Which one of the following concentrations could not be obtained?

(a) 35%   (b) 28%   (c) 45%   (d) 40%

503. Suppose that a 22% alcohol solution is mixed with a 36% alcohol solution. Which one of the following concentrations could *not* be obtained?

(a) 28%   (b) 20%   (c) 34%   (d) 29%

Section 2.7   The Addition and Multiplication Properties of Inequality

Objective 1   Graph intervals on a number line.

Graph the inequality on a number line.

504. $x \geq 3$

505. $r < 4$

506. $t \leq 1$

507. $-5 \leq r$

508. $0 \geq z$

509. $7 < a$

510. $-5 \leq y < -1$

511. $4 < z \leq 6$

512. $0 < a \leq 5$

513. $-3 \leq y < 0$

514. $8 \le k \le 10$

515. $-6 < m < -2$

Objective 2 Use the addition property of inequality.

Solve the inequality and graph the solutions.

516. $y - 2 \ge 5$

517. $j + 7 \le 12$

518. $b + 3 \le 5$

519. $t + 5 > 6$

520. $z - 7 < -5$

521. $y - 4 > -9$

522. $3x \le 2x + 3$

523. $5r + 1 > 6r$

524. $4k \ge 5k - 2$

525. $7t + 2 < 6r$

526. $1 + m < 2m - 4$

527. $-2 + 6b \ge 5b - 1$

528. $6 + 7h > 8h + 7$

529. $3 + 9p \le 8p + 3$

Objective 3 Use the multiplication property of inequality.

Solve the inequality and graph the solutions.

530. $2x \le 10$

531. $6y > 18$

532. $4k \ge -16$

533. $5r > -20$

534. $-2s > 4$

535. $-7q \ge 35$

536. $-6k \le 0$

537. $5a > 0$

538. $-3w > -15$

539. $-5t \le -35$

540. $-\dfrac{2}{3}z > 4$

541. $-04t \le .2$

Objective 4 Solve linear inequalities.

Solve the inequality.

542. $12p + 5 \ge 13p$

543. $3(y - 5) + 2 > 2(y - 4)$

544. $-6(m+2)+3 \le -7(m-5)$

545. $2-3k \le -4k$

546. $5(2z+3)-2(z-4) > 3(2x+2)+z$

547. $2x - \dfrac{3}{4} \ge x + \dfrac{1}{3}$

Solve the inequality and graph the solutions.

548. $5m - 8 \ge 3m$

549. $8p + 12 < 10p$

550. $4r - 3 + 2r > 3r + 9$

551. $5p - 5 - p > 7p - 2$

552. $5(y+3) - 5y > 3(y+1) + 4$

553. $3(z+1) \le 5(2z-4) + 2$

554. $4 - \dfrac{1}{3} \le 6 + \dfrac{2}{3}y$

555. $3 - \dfrac{1}{4}z \le 2 + \dfrac{3}{8}z$

Objective 5   Solve applied problems by using inequalities.

Solve the problem by writing and solving an inequality.

556. If twice a number is added to 3, the result is more than 15. Find all numbers that satisfy this condition.

557. Find every number such that half the sum of that number and 14 is less than or equal to 5.

558. Sheila has grades of 78 and 86 on her first two chemistry quizzes. What must she score on her third quiz to have an average of at least 85 on the three quizzes?

559. Marcel has two more papers to write this term in his composition class. If he has scored 82, 74, 90, and 68 on papers so far, how much must he average on the two remaining papers to have an average of at least 80 for the six papers?

560. Dreyfus has a budget of $150 for gifts this year. So far he has bought gifts costing $27.50, $38.98, and $16.98. If he has three more gifts to buy, find the average amount he can spend on each gift and still stay within his budget.

561. Miranda is paid $.07 per newspaper to deliver papers for the Record-Courier. How many papers must she deliver in an hour to make at least $3.50 per hour?

562. Emil teaches dance classes in the evenings in a studio for which he pays $300 per month rent. If rent is his only expense and he charges each student $20 per month, how many students must he teach to make a profit of at least $600 per month?

563. An interior decorator has 45 yards of a certain fabric which she intends to use for making draperies and slipcovers. If she makes three slipcovers requiring five yards each, find the number of pairs of draperies she can make if each drapery requires three yards.

564. Two sides of a triangle are equal in length, with the third side 8 feet longer than one of the equal sides. The perimeter of the triangle cannot be more than 38 feet. Find the longest possible value for the length of the equal sides.

565. The perimeter of a triangle must be no more than 45 centimeters. One side of the triangle is 18 centimeters and a second side is 12 centimeters. Find the largest possible length for the third side.

Objective 6   Solve three-part inequalities.

Solve the inequality and graph the solutions.

566. $4 \leq x - 3 \leq 7$

567. $2 \leq y + 5 \leq 11$

568. $-1 < a + 6 < 3$

569. $-10 < b - 2 < 0$

570. $1 < 2r - 3 \leq 5$

571. $5 \leq 3k + 2 < 11$

572. $-2 < 5t + 3 \leq 10$ 573. $-4 < 8b - 7 < 9$

574. $-5 < 2x - 1 \leq 9$ 575. $-10 \leq 4t - 2 < 6$

576. $-10 < 3x + 14 \leq 17$ 577. $-5 \leq 3x - 8 < 6$

Mixed Exercises

Graph the inequality on a number line.

578. $y \leq -4$ 579. $y \geq -2$

580. $-5 > z$ 581. $x \geq 0$

582. $-2 \leq x \leq -1$ 583. $-4 \leq x < 4$

584. $0 < x < 2$ 585. $-1 < x < 3$

Solve the inequality and graph the solutions.

586. $-1 + 17b \leq 16b - 4$ 587. $5r + 1 > 6r$

588. $7p + 4 > 3p - 8$ 589. $42 - 6 \geq 9w - 15$

590. $6t - 4 < 4t + 7$ 591. $16 - 5x > 13 - 2x$

592. $4(z - 3) + 3(2z + 5) < -7$ 593. $6(x - 3) + 4 \leq 3(x - 2) + 2x$

594. $7 - 2x \le 5 - 3x$

595. $-9t + 4(t - 2) > t - (6 + 5t) + 9$

596. $\frac{3}{4}(r - 5) > \frac{1}{6}(r + 2)$

597. $\frac{1}{2}(x - 5) \le \frac{3}{10}(x - 3)$

598. $-1 \le \frac{1}{2}y + 4 \le 1$

599. $-2 < \frac{1}{3}z - 3 < 0$

600. $0 < \frac{1}{3}r + 1 \le 1$

601. $-2 \le \frac{1}{4}w - 2 \le -1$

602. $-4 < 2(x - 1) < 8$

603. $-3 \le 3(j + 2) \le 9$

604. $0 < 2(t + 3) < 5$

605. $-2 < 4(b - 1) < 0$

Solve the problem by writing and solving an inequality.

606. If 4 is subtracted from twice a number, the result is less than 8. Find all numbers that satisfy this condition.

607. If the sum of a number and 5 is multiplied by 6, the result is greater than 72. Find all such numbers.

608. If twice the sum of a number and 5 is subtracted from three times a number, the result is more than −7. Find all such numbers.

609. The perimeter of a triangle must be no less than 109 centimeters. One side of the triangle is 43 centimeters and a second side is 29 centimeters. Find the smallest possible length for the third side.

609. The perimeter of a triangle must be no less than 109 centimeters. One side of the triangle is 43 centimeters and a second side is 29 centimeters. Find the smallest possible length for the third side.

610. Naomi earned $250 in January, $320 in February, and $165 in March as a student aide in the college library. How much must she earn in April in January through April?

611. The Hastings School P.T.A. wishes to build a playground with a rectangular fence. If 136 meters of fencing are available and the width of the playground must be 25 meters. Find all possible lengths of the playground.

Writing/Conceptual Exercises

612. Why is it wrong to write $-2 < x < -3$?

613. Why is it wrong to write $0 < x < -7$?

614. If $-2x < -6$, why is it wrong to write $x < 3$?

615. In solving an inequality, when must we reverse the direction of the inequality symbol?

CHAPTER 3   LINEAR EQUATIONS IN TWO VARIABLES

Section 3.1   Linear Equations in Two Variables

Objective 1   Interpret graphs.

The pie chart below shows how the Patel family used its income in 1995. The total family income was $58,000. Use this chart to answer the questions in Exercises 1-3.

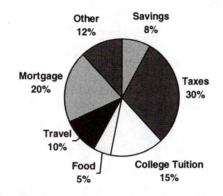

1. What percentage of the family's income was spent on the combination of the following expenses: mortgage payments, college tuition, and travel?

2. How much money was spent on college tuition?

3. How much more money was spent on mortgage payments than went to savings?

Use the bar graph below to answer to questions in Exercises 4-6.

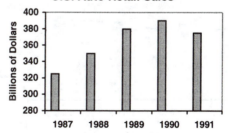

4.   In which year were auto retail sales the highest?

5. Between what two years did auto retail sales decrease?

6. Between what two years did auto retail sales increase by the largest amount?

The line graph below shows the rate of inflation during recent years in the United States. Use this graph to answer the questions in Exercises 7-9.

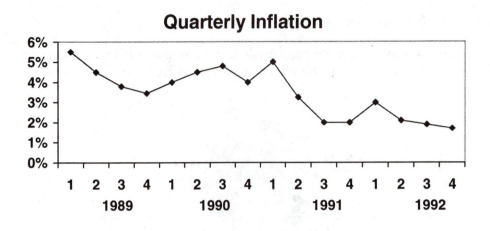

7. What is the highest inflation rate shown on the graph? When did it occur?

8. What is the lowest inflation rate shown on the graph? When did it occur?

9. During which quarter did the greatest increase in inflation rate occur?

The line graph below shows crime rates per one million people in the areas in and around certain American cities. Use this graph to answer the questions in Exercises 10-12.

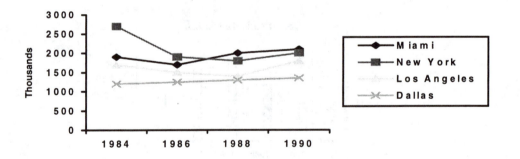

10. Which city showed a decline in its crime rate from 1984 to 1990?

11. Which city had the greatest crime rate in 1990?

12. Which city had about the same crime rate throughout the period 1984-1990?

Objective 2 Write a solution as an ordered pair.

Write the solution as an ordered pair.

13. $x = 3$ and $y = 2$

14. $x = -1$ and $y = 6$

15. $a = -5$ and $b = 0$

16. $a = 2$ and $b = 3$

17. $p = \dfrac{1}{3}$ and $q = -2$

18. $p = -4$ and $q = -7$

19. $r = 8$ and $s = 5$

20. $s = \dfrac{1}{3}$ and $r = 0$

21. $m = .2$ and $n = .3$

22. $n = -20$ and $m = 4$

Objective 3 Decide whether a given ordered pair is a solution of a given equation.

Decide whether the given ordered pair is a solution of the given equation.

23. $x + y = 6$; $(2, 4)$

24. $2a - b = 6$; $(2, -2)$

25. $4m - 3n = 6$; $(1, 2)$

26. $5a - 3b = 1$; $\left(0, \dfrac{1}{3}\right)$

27. $x = -4y$; $(0, 0)$

28. $m = 3n$; $(3, 9)$

29. $y + 3 = 0$; $(0, -3)$

30. $x = -7$; $(-7, 3)$

31. $x - 6 = 2$; $(8, 3)$

32. $x = 1 - 2y$; $\left(0, -\dfrac{1}{2}\right)$

Objective 4   Complete ordered pairs for a given equation.

Complete the ordered pairs beneath the equation.

33. $y = 3x - 2$

(a) $(4, \quad)$
(b) $(0, \quad)$
(c) $(\quad , 4)$
(d) $(\quad , -8)$
(e) $(\quad , 10)$

34. $y = -4x + 6$

(a) $(1, \quad)$
(b) $(-3, \quad)$
(c) $(\quad , 6)$
(d) $(\quad , 26)$
(e) $(\quad , -6)$

35. $y = 4 - 2x$

(a) $(-2, \quad)$
(b) $(4, \quad)$
(c) $(\quad , 0)$
(d) $(-1, \quad)$
(e) $(\quad , -8)$

36. $5x + 2y = 3$

(a) $(1, \quad)$
(b) $(-7, \quad)$
(c) $(\quad , -6)$
(d) $(0, \quad)$

(e) $(\quad , 0)$

37. $x = -2$

(a) $(\quad , -2)$
(b) $(\quad , 0)$
(c) $(\quad , 12)$
(d) $(\quad , 5)$

(e) $\left(\quad , -\dfrac{2}{3} \right)$

38. $y = 4$

(a) $(3, \quad)$
(b) $(0, \quad)$
(c) $(7, \quad)$
(d) $(-4, \quad)$

(e) $(.75, \quad)$

Complete the table of values.

39. $2x + y = 4$

x	2		1
y		4	

40. $3x = y$

x	2		
y		-9	0

41. $3x + 2y = 6$

x	0		4
y		0	

42. $y = 2x - 1$

x	0	3	
y			7

43. $4m - n = 8$

m	0		-1
n		0	

44. $3s - 4t = 12$

s	0		-8
t		0	

45. $x = -5$

x			
y	0	4	-5

46. $y = 7$

x	-5	0	7
y			

47. $x - 8 = 0$

x			
y	-8	4	8

48. $y + 6 = 0$

x	-6	0	6
y			

Objective 5   Plot ordered pairs.

Plot the ordered pair on a coordinate system.

49. $(7, 1)$

50. $(-2, 4)$

51. $(5, 6)$

52. $(-2, -7)$

53. $(4, -2)$

54. $(-3, 4)$

55. $(1, 0)$

56. $(0, -4)$

57. $(-5, 0)$

58. $(0, 2)$

59. $(0, 0)$

60. $(2, 5)$

Without plotting the given point, name the quadrant in which it lies or the axis on which it lies.

61. $(1, 2)$

62. $(5, -3)$

63. $(-7, 2)$

64. $(-10, -1)$

65. $\left(-\dfrac{1}{2}, \dfrac{4}{5}\right)$

66. $(0, -7)$

67. $(5, 0)$

68. $(0, 0)$

Mixed Exercises

Decide whether the given ordered pair is a solution of the given equation.

69. $3x + y = 8$; $(8, 0)$

70. $2x + y = 5$; $(2, 1)$

71. $y = 5x$; $(5, 1)$

72. $3z + y = 8$; $(8, 0)$

73. $x = 2y$; $(4, 2)$

74. $x + 4 = 0$; $(0, -4)$

Complete the given ordered pairs for the equation.

75. $y = -2x + 5$ $(6,\ )\left(-\dfrac{1}{2},\ \right)(\ ,0)$

76. $3x - 4y = 12$ $(0,\ )(\ ,0)(2,\ )$

77. $-x + y = 7$ $(3,\ )(\ ,5)(2.5,\ )$

78. $y - 4 = -1$ $(0,\ )(-1,\ )(100,\ )$

Complete the table of values.

79. $-2x + y = 6$

x	3		-4
y		0	

80. $2r - 5s = 10$

r	0		-3
s		0	

81. $x + 3 = -2$

x			
y	-5	0	1

82. $5u + 7v = 35$

u		0	4
v	0		

Without plotting the given point, name the quadrant in which it lies or the axis on which it lies.

83. $\left(-2, \dfrac{3}{4}\right)$ 84. $(5, 7)$ 85. $(-8.5, 0)$

86. $\left(-\dfrac{2}{3}, -4\right)$ 87. $(3, -6)$ 88. $(0, 17)$

The line graph below shows industrial production of three countries (1985=100).   Use this graph to answer the questions in Exercises 89-92.

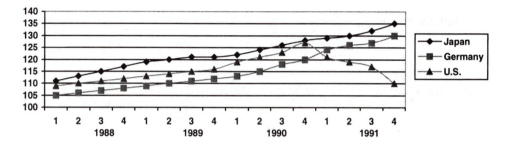

89. Which country had the smallest change in industrial production over the entire period?

90. Which country showed a decline in industrial production during a certain period?

91. Which country showed the greatest increase over production during a certain period?

92. Which country showed the most steady rate of growth in industrial production over the entire period?

Writing/Conceptual Exercises

93. Do $(-3, 5)$ and $(5, -3)$ represent the same ordered pair? Explain.

94. Do the ordered pairs $(-2, 0)$ and $(0, -2)$ correspond to the same point in the plane? Explain.

95. The ordered pair $(4, -1)$ is a solution of the equation $2x + 3y =$ _____.

96. The ordered pair $(-1, \underline{\quad})$ is a solution of the equation $y = -x$.

97. The ordered pair $(\underline{\quad}, -3)$ is a solution of the equation $x = 4$.

98. The ordered pair $(5, \underline{\quad})$ is a solution of the equation $y = -2$.

99. The point whose graph has coordinates $(0, -6)$ lies along the $\underline{\quad}$-axis.

100. The point whose graph has coordinates $(-7, 0)$ lies along the $\underline{\quad}$-axis.

Section 3.2   Graphing Linear Equations in Two Variables

Objective 1   Graph linear equations.

Complete the ordered pairs using the given equation. Then graph the equation by plotting the points and drawing a line through them.

101. $x + y = 3$
    $(0, \quad)$
    $(\quad, 0)$
    $(2, \quad)$

102. $x + y = -2$
    $(0, \quad)$
    $(\quad, 0)$
    $(3, \quad)$

103. $x - y = 4$
    $(0, \quad)$
    $(\quad, 0)$
    $(-2, \quad)$

104. $x - y = -1$
    $(0, \quad)$
    $(\quad, 0)$
    $(4, \quad)$

105. $y = x + 7$
    $(0, \quad)$
    $(\quad, 0)$
    $(-4, \quad)$

106. $y + 4 = x$
    $(0, \quad)$
    $(\quad, 0)$
    $(\quad, -2)$

107. $y = 3x - 2$
       $(0, \quad)$
       $(\quad, 0)$
       $(2, \quad)$

108. $y = 4x + 8$
       $(0, \quad)$
       $(\quad, 0)$
       $(-4, \quad)$

109. $y + 3 = 0$
       $(0, \quad)$
       $(4, \quad)$
       $(-3, \quad)$

110. $x - 4 = 0$
       $(\quad, 0)$
       $(\quad, -2)$
       $(\quad, 3)$

111. $x = -2y + 2$
       $(0, \quad)$
       $(\quad, 0)$
       $(-2, \quad)$

112. $2x + 3y = 6$
       $(0, \quad)$
       $(\quad, 0)$
       $(-3, \quad)$

Objective 2   Find intercepts.

Find the intercepts for the graph of the equation.

113. $5x + 2y = 10$

114. $3x - 2y = 12$

115. $2x + y = 0$

116. $4x - 7y = -8$

117. $5x - 3y = 10$

118. $4x + 3y = 6$

119. $3x - 2y = 8$

120. $7x + 3y = 14$

121. $2x - 9y = -9$

122. $3x + 5y = 9$

Find the intercepts for the equation. Then graph the equation.

123. $2x + y = 4$ 124. $4x - y = 4$

125. $3x + 4y = -12$ 126. $x + 2y = 4$

127. $6x + 5y = 15$ 128. $2x - 3y = 6$

129. $5x - 2y = -10$ 130. $3x - y = 5$

131. $x + 2y = -3$ 132. $5x + 6y = -30$

Objective 3   Graph linear equations of the form $Ax + By = 0$.

Graph the equation.

133. $3x - y = 0$ 134. $x + 5y = 0$

135. $x + y = 0$ 136. $4x - y = 0$

137. $2x + y = 0$ 138. $3x + y = 0$

139. $2x-3y=0$ 140. $5x+2y=0$

141. $3x+4y=0$ 142. $-4x+5y=0$

Objective 4   Graph linear equations of the form $y=k$ **or** $x=k$.

Graph the equation.

143. $x=3$ 144. $y=2$

145. $y=-5$ 146. $x=-3$

147. $x-1=0$ 148. $y-4=0$

149. $x+4=0$ 150. $y+1=0$

151. $x=0$ 152. $y=0$

Objective 5   Define a function.

Decide whether each set of ordered pairs defines a function.

153. $\{(0,1),(1,0),(2,0),(0,2),(3,0),(0,3)\}$

154. $\{(1,1),(2,2),(3,3),(4,2),(5,0),(-6,3)\}$

155. The ordered pairs that satisfy $2x - y = -5$.

156. The ordered pairs that satisfy $y = -9$.

157. The ordered pairs that satisfy $x = 2$.

158. The ordered pairs that satisfy $5x + 2 = -3$.

159. The ordered pairs that satisfy $x = y$.

Mixed Exercises

Complete the ordered pairs for the equation. Then graph the equation by plotting the points and drawing a line through them.

160. $y - 3 = x$
    (0,)
    (, 0)
    (, 4)

161. $x = 2y + 1$
    (0,)
    (, 0)
    (, -2)

162. $x = 3y - 6$
    (0,)
    (, 0)
    (, 4)

163. $2x = y - 4$
    (0,)
    (, 0)
    (-4,)

164. $x = 4y - 1$
    (3,)
    (, 0)
    (, -1)

165. $4x + 5y = 5$
    (0,)
    (, 0)
    (5,)

Find the intercepts for the graph of the equation.

166. $3x - y = -3$

167. $6x + 4y = -24$

168. $4x - 5y = 10$

169. $-4x + 7y = 8$

170. $3x - 2y = 6$

171. $4x - y = 8$

172. $x + 3y = 0$

173. $6x - 7y = 3$

Find the intercepts for the graph of the equation. Then graph the equation.

174. $3x + 2y = 12$

175. $4x - 7y = -8$

176. $4x - 3y = 0$

177. $5x + 3y = 0$

178. $4x + 3y = 6$

179. $3x - 2y = 8$

180. $3x - 2y = 0$

181. $3x - 5y = 0$

182. $y + 5 = 0$

183. $x - 6 = 0$

Writing/Conceptual Exercises

184.   The graph of the equation $2x - 4y = 0$ goes through the _____.

185.   The graph of the equation $x - 4 = 0$ is a _____ line.

186.   The graph of the equation $y + 1 = 0$ is a _____ line.

187.   The graph of the equation $x = 0$ is the ___-axis.

188.   The graph of the equation $y = 0$ is the ___-axis.

189.   If a line has no y-intercept, it must be a _____ line.

Section 3.3   The Slope of a Line

Objective 1   Find the slope of a line given two points.

Find the slope of the line.

190. Through $(5, 3)$ and $(7, 4)$ 191. Through $(6, 2)$ and $(9, 1)$

192. Through $(-3, 1)$ and $(5, 4)$ 193. Through $(5, -6)$ and $(2, 3)$

194. Through $(2, -2)$ and $(-6, 1)$

195. Through $(3, 4)$ and $(-7, 4)$

196. Through $(0, -2)$ and $(7, 0)$

197. Through $(-5, -1)$ and $(4, -7)$

198. Through $(-3, -4)$ and $(-2, -5)$

199. Through $(0, 0)$ and $(6, -8)$

200. Through $(-8, 6)$ and $(-8, -1)$

201. Through $(3, -4)$ and $(-3, 4)$

Objective 2   Find the slope from the equation of a line.

Find the slope of the line.

202. $y = -\dfrac{8}{7}x + 12$

203. $3y = x + 7$

204. $5y = 2x - 5$

205. $2x + 4y = 7$

206. $8y - 2x = 11$

207. $3x + 5y = 0$

208. $y = -5$

209. $x = 0$

210. $9x = 6y$

Objective 3　Use the slope to determine whether two lines are parallel, perpendicular, or neither.

For the pair of equations, give the slopes of the lines, and then determine whether the two lines are *parallel, perpendicular,* or *neither.*

211.　$y = 5x - 2$
　　　$y = 5x + 11$

212.　$y = 2x + 7$
　　　$y = 8 - \dfrac{1}{2}x$

213.　$x + y = 7$
　　　$x - y = -3$

214.　$3x + y = 7$
　　　$x - 3y = 2$

215.　$5x + y = 8$
　　　$x + 5y = -3$

216.　$9x + 3y = 7$
　　　$x - 3y = 4$

217.　$4x + 5y = 6$
　　　$5x + 4y = 11$

218.　$8x + 2y = 5$
　　　$4x = 1 - y$

219.　$y - 6 = 0$
　　　$y + 4 = 0$

220.　$y = 7$
　　　$x = 2$

221.　$3x - y = 4$
　　　$3x - y = 2$

222.　$6x + 5y = 28$
　　　$6x - 5y = 7$

Mixed Exercises

Find the slope of the line.

223. Through $(6, 6)$ and $(3, 8)$

224. Through $(4, 11)$ and $(-6, 1)$

225. Through $(3, -2)$ and $(7, 1)$

226. Through $(-4, 5)$ and $(1, 3)$

227. Through $(0, 4)$ and $(0, -4)$

228. Through $(5, -2)$ and $(-5, -2)$

229. Through $(0, 0)$ and $(4, -2)$

230. Through $\left(3, -\dfrac{1}{2}\right)$ and $(5, 4)$

231. $y = \dfrac{1}{3}x - 6$

232. $y = \dfrac{4}{5}x + 7$

233. $6x - 3y = 8$

234. $x = -7$

235. $x = 3y$

236. $3x + 3y = 13$

237. $y + 6 = 0$

238. $3x + 4y = 24$

For the pair of equations, give the slopes of the lines, and then determine whether the two lines are *parallel, perpendicular,* or *neither.*

239. $4x + 7y = 1$
     $7x - 4y = 2$

240. $8x - 9y = 2$
     $9x + 8y = 11$

241. $x = 29$
       $y = 4$

242. $y = 5$
       $y + 2 = 8$

243. $8x - 3y = 12$
       $-8x + 3y = 1$

244. $7x - 11y = 9$
       $7x + 11y = 2$

245. $3x - 2y = 4$
       $-9x + 6y = 10$

246. $3x - 2y = 0$
       $2x + 3y = 0$

In Exercises 247-251, two views of the same line are shown in the accompanying calculator screen, along with coordinates of two points displayed at the bottom. Find the slope of the line.

247.

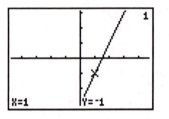 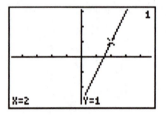

248.

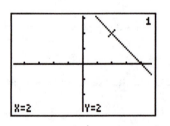 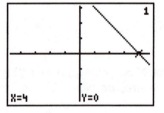

249.

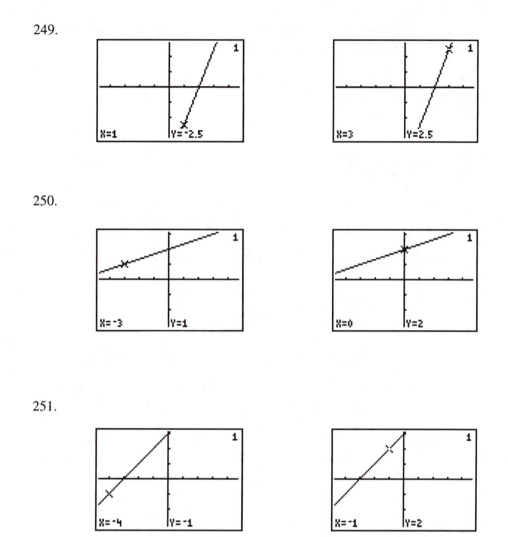

250.

251.

Writing/Conceptual Exercises

Tell whether each statement is *true* or *false*. If the statement is false, show how it can be changed to a correct statement.

252. A line with negative slope rises from left to right.

253. Horizontal lines have undefined slope.

254. If two lines are perpendicular, their slopes are reciprocals.

For the line shown, decide whether (a) the slope is *positive, negative,* or *zero,* and (b) the *y*-value of the *y*-intercept is *positive, negative,* or *zero.*

255.

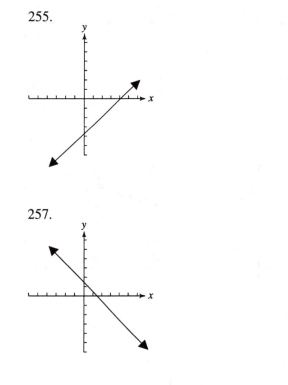

256.

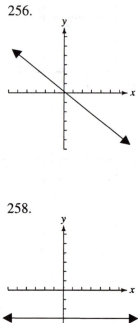

257.

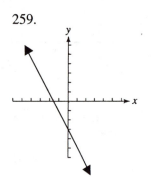

258.

259.

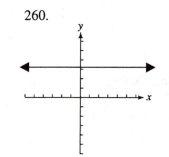

260.

CHAPTER 4   POLYNOMIALS AND EXPONENTS

Section 4.1   Addition and Subtraction of Polynomials; Graphing Simple Polynomials

Objective 1   Identify terms and coefficients.

For the polynomial, determine the number of terms and name the coefficients of the terms.

1. $12r$

2. $-6y^2$

3. $a^5 - a^3$

4. $-2b^3 + b^2 - b$

5. $8x^3 + 7x^2 - 5x + 4$

6. $3 - 7q$

7. $\frac{1}{2}x + y$

8. $-\frac{4}{5}y^2 + \frac{2}{3}y$

9. $.75x - .25y$

10. $.10a^2 - .20b^2$

Objective 2   Add like terms.

In the polynomial, add like terms whenever possible. Write the result in descending powers of the variable.

11. $-6s^3 + 9s^3$

12. $2t^6 + \left(-7t^6\right)$

13. $7x^2 - 6x^3 + 3x^2 - 5x^3$

14. $-5y^3 - y^4 + 2y^3 + y^2$

15. $.3m^5 - .9m^5$

16. $3.2r^2 - 5.8r - 6.5r - 7.9r^2$

17. $8c^3 - 11c^2 - c^2 + 11 + 3c^2 - 5c - 5 + 2c^3$

18. $14y^4 - 3y^2 + 2 - 7y^3 + 6y^4 - 2y^2 - y$

19. $8a^2 - 3a^3 + 5a^2 - 7a^2$

20. $-14x^3 - 2x^2 + 6x - x^3$

21. $-\frac{1}{2}r^3 + \frac{1}{3}r + \frac{1}{4}r^3 - \frac{1}{3}r$

22. $\frac{4}{5}m^3 - \frac{2}{3}m^2 + \frac{1}{10}m^3 - \frac{1}{3}m^2$

Objective 3   Know the vocabulary for polynomials.

Choose one or more of the following descriptions for each expression: (a) polynomial, (b) polynomial written in descending order, (c) not a polynomial.

23. $-6w^3 + 7w^2 + 2w - 11$

24. $x^5 + x^9$

25. $4y^3 - 3y + \dfrac{2}{y}$

26. $3a^5 - 2a^3 - 6a^{-1} + 15$

27. j^{-1}

28. $k^6 - k^4 + 3k - 2$

Simplify the polynomial, if possible, and write the resulting polynomial in descending powers of the variable. Then give the degree of this polynomial, and tell whether it is a *monomial*, a *binomial*, a *trinomial*, or *none of these*.

29. $5y^2 + 2y - 7$

30. $3m^2 + 4m - m^5$

31. $z^4 + 3z^3 - 2z^3$

32. $p^3 - p^5 + 2p^5$

33. $n^8 - n^7 + 2n^8$

34. $x^2 - 4x^5 + 5x^2$

35. $\dfrac{7}{8}x^2 - \dfrac{3}{4}x - \dfrac{1}{4}x^2 - \dfrac{1}{2}x$

36. $\dfrac{5}{6}y^2 - \dfrac{1}{3}y + \dfrac{1}{6} - \dfrac{2}{3}y^2 + \dfrac{1}{3} - \dfrac{5}{6}y$

Objective 4   Evaluate polynomials.

Find the value of the polynomial when (a) $x = 3$ and (b) $x = -2$.

37. $5x - 11$

38. $7x + 5$

39. $3x^2 - 4x + 1$

40. $x^2 - 3x + 8$

41. $-x^2 - 4x + 7$

42. $-x^2 + 5x - 9$

43. $2x^3 + 4x^2 - 7x + 3$

44. $x^3 - 4x^2 - 8x - 5$

45. $5x^4 - 2x^2 + 8x - 1$

46. $3x^4 + x^3 - 6x + 11$

Objective 5 Add and subtract polynomials.

Add.

47.
$$3x^2 + 6x + 2$$
$$\underline{5x^2 - 2x + 3}$$

48.
$$8y^3 - 2y^2 + 5$$
$$\underline{2y^3 - 3y^2 + 11}$$

49.
$$6m^3 + 2m^2 - m + 1$$
$$\underline{-4m^3 - 11m^2 - 2m + 4}$$

50.
$$w^4 + 3m^2 - 2m + 6$$
$$\underline{4w^4 - 2m^2 + 8m + 4}$$

51.
$$5x^5 + 2x^2 - 3x - 1$$
$$\underline{-6x^5 \qquad + x + 1}$$

52.
$$4y^4 - 3y^3 + 6y + 9$$
$$\underline{3y^3 - 5y + 1}$$

53.
$$7p^4 \qquad + 2p^2 - 3p + 11$$
$$\underline{5p^4 + 3p^3 - 8p^2 - 3p + 4}$$

54.
$$8z^5 - 2z^4 + 12z^3 - 5z^2 + z$$
$$\underline{-8z^5 - 4z^4 - 9z^3 + z^2 + z}$$

55. $(8x^3 + 2x^2 - 6x + 1) + (3x^3 - 4x^2 - 7)$

56. $(9x^3 - 2x^2 - 6x + 1) + (5x^3 + 4x^2 - 3)$

57. $(6y^4 + 3y^3 + 2y - 11) + (2y^3 + 6y + 5)$

58. $(4z^7 - 6z^5 - 3z^4 + 1) + (z^6 + 3z^5 + 2z^3 + 10)$

59. $(x^5 + 2x^3 - x^2) + (4x^5 - 3x^4 - 2x^3 + 8x^2 - x)$

60. $(m^4 - 3m^3 + 2m^2 + m + 6) + (m^3 - 3m^2 - m + 4)$

Subtract.

61.
$$8x^2 + 2x$$
$$\underline{5x^2 + x}$$

62.
$$10m^3 - 2m$$
$$\underline{6m^3 + 3m}$$

63.
$$4y^2 + 3y - 8$$
$$\underline{7y^2 - 2y + 5}$$

64.
$$9k^2 - 3k - 18$$
$$\underline{-11k^2 - 5k + 1}$$

65.
$$\begin{array}{r} 6n^3 + 3n \\ 5n^3 - n - 2 \\ \hline \end{array}$$

66.
$$\begin{array}{r} 4x^2 - 2x - 1 \\ -3x^2 + x - 8 \\ \hline \end{array}$$

67.
$$\begin{array}{r} 7y^4 - 3y^3 - 2y^2 - y \\ 11y^4 + 2y^3 - 2y^2 \quad -1 \\ \hline \end{array}$$

68.
$$\begin{array}{r} z^5 - z^4 + 3z^3 \\ 4z^4 \quad -3z^2 + 8z \\ \hline \end{array}$$

69. $\left(7x^3 + 4x^2 - 10x + 7\right) - \left(5x^3 - 6x + 4\right)$

70. $\left(9x^3 - x^2 + 7x + 10\right) - \left(6x^3 - 4x + 7\right)$

71. $\left(5m^3 - 3m + 7\right) - \left(-5m - 8\right)$

72. $\left(14z^4 - z^2 + 3z\right) - \left(3z^3 - z^2 - 2z - 1\right)$

73. $\left(8a^5 - 3a^2 - a + 11\right) - \left(-a^5 + 2a^3 - 5a^2 + a - 4\right)$

74. $\left(p^4 - 9p^3 - 6p^2 + 8p - 1\right) - \left(-2p^4 - 3p^2 + 1\right)$

Objective 6   Graph equations defined by polynomials with degree 2.

Select several values for x; then find the corresponding y-values, and graph.

75. $y = x^2 - 1$

76. $y = x^2 + 2$

77. $y = 2x^2 - 3$

78. $y = 2 - x^2$

79. $y = -x^2 - 4$

80. $y = 3x^2 - 2$

Mixed Exercises

Perform the indicated operations. Write each resulting polynomial in descending powers of the variable. Then give the degree of this polynomial, and tell whether it is a *monomial*, a *binomial*, a *trinomial*, or *none of these.*

81. $\left(8x^3 + 2x^2 - 6x + 1\right) + \left(3x^3 - 4x^2 - 7\right)$

82. $\left(6y^2 + 2y - 3\right) - \left(2y^2 - 2y + 1\right)$

83. $\left(6y^4 + 3y^3 + 2y - 11\right) + \left(2y^3 + 6y + 5\right)$

84. $\left(x^5 + 2x^3 - x^2\right) + \left(4x^5 - 3x^4 - 2x^3 + 8x^2 - x\right)$

85. $\left(p^4 + 9p^3 - 6p^2 + 8p - 1\right) - \left(-2p^4 - 3p^2 + 1\right)$

86. $10 - \left(4 - 2y - 11y^2\right)$

87. $\left(3a^2 - 2a + 5\right) + \left(-2a^2 + 2a - 1\right) + \left(a^2 + 3\right)$

88. $\left(5x^3 + x - 3\right) - \left(2x^3 - x - 5\right) + \left(x^3 - 2x - 2\right)$

For the following polynomials, state the coefficient of the second term. Then evaluate the polynomial when $x = 4$ and $x = -3$.

89. $2x^2 + 3x - 1$

90. $\left(3x^2 - 2x + 4\right)$

91. $x^2 - x + 5$

92. $x^3 + x - 7$

93. $\dfrac{1}{2}x^2 - 2x + 3$

94. $-\dfrac{1}{3}x^2 + 4x - 2$

95. $-3x^3 - 2x^2 + 4x + 3$

96. $-x^2 + 7$

Add or subtract as indicated.

97. $\left(4x^2 y + 5xy + 3y^2\right) + \left(-2x^2 y - 7xy + 5y^2\right)$

98. $\left(3c^2 - 5cd + 6d^2\right) - \left(-2d^2 + 3cd - c^2\right)$

99. $\left(3rs + 4r - 5s\right) - \left(6s + 2r - 5sr\right)$

100. $\left(7ab + 2bc - 4ac\right) + \left(5ca - 2cb - 9ba\right)$

101. $\left(2x^2 y + 5xy - 6xy^2\right) + \left(7xy + 8xy^2\right) - \left(3x^2 y + 4xy\right)$

102. $\left(.01ab + .03a^2 - .05b^2\right) - \left(-.08a^2 + .02b^2 + .01ab\right)$

Add.

103. $\begin{array}{r} 4x^3y - 2x^2y^2 + 5xy^3 \\ -3x^3y + 2x^2y^2 - 6xy^3 \\ \hline \end{array}$

104. $\begin{array}{r} 12r^2t + 3rt - 8rt^2 \\ -4r^2t - 5rt - 6rt^2 \\ \hline \end{array}$

Subtract.

105. $\begin{array}{r} 13a^3b - 6a^2b^2 + 5ab^3 + 7 \\ 6a^3b \qquad\quad - 7ab^3 - 2 \\ \hline \end{array}$

106. $\begin{array}{r} -6rs + 7rt - 5st \\ 2rs - 9rt - \quad st \\ \hline \end{array}$

Writing/Conceptual Exercises

107. Explain why the degree of the term 4^5 is not 5. What is its degree?

108. Can the sum of two polynomials in x, both of degree 4, be of degree 3? If so, give an example.

109. Can the sum of two polynomials in w, both of degree 3, be of degree 1? If so, give an example.

110. Is it possible to add two trinomials and obtain a sum which is a binomial? If so, give an example.

Section 4.2   The Product Rule and Power Rules for Exponents

Objective 1   Identify bases and exponents.

Write the expression in exponential form and evaluate.

111. $2 \cdot 2 \cdot 2 \cdot 2 \cdot 2$

112. $10 \cdot 10 \cdot 10 \cdot 10$

113. $(-1)(-1)(-1)(-1)(-1)(-1)$

114. $12 \cdot 12 \cdot 12$

115. $\left(\dfrac{1}{3}\right)\left(\dfrac{1}{3}\right)\left(\dfrac{1}{3}\right)\left(\dfrac{1}{3}\right)$

116. $\left(-\dfrac{2}{5}\right)\left(-\dfrac{2}{5}\right)\left(-\dfrac{2}{5}\right)$

Write the expression in exponential form.

117. $r \cdot r \cdot r \cdot r \cdot r \cdot r$

118. $(-2y)(-2y)(-2y)$

119. $(ab)(ab)(ab)(ab)$

120. $(.5st)(.5st)(.5st)$

Evaluate the exponential expression. Name the base and the exponent.

121. $(-4)^3$

122. -3^4

123. 2^6

124. $(-7)^2$

125. -8^2

126. $(-9)^3$

Objective 2   Use the product rule for exponents.

Use the product rule to simplify the expression, if possible. Write the answer in exponential form.

127. $4^5 \cdot 4^3$

128. $(-3)^3 \cdot (-3)^3$

129. $2^5 \cdot 2 \cdot 2^6$

130. $2^3 \cdot 3^2$

131. $\left(\dfrac{1}{2}\right)^4 \cdot \left(\dfrac{1}{2}\right)^3$

132. $\left(\dfrac{2}{5}\right)^6 \cdot \left(\dfrac{2}{5}\right)$

133. $(-5)^{10} \cdot (-5)^5$

134. $10^2 + 10^3$

Multiply.

135. $4p^4 \cdot 6p^2$

136. $3k \cdot 13k^4$

137. $5a \cdot 12a^6$

138. $12b \cdot \left(-12b^{14}\right)$

139. $\left(-3m^3\right) \cdot \left(-7m^5\right)$

140. $\left(-4x^3\right)\left(9x^{12}\right)$

In the following exercises, add the given terms. Then start over and multiply them.

141. $3y^2, 5y^2$

142. $x^4, -5x^4$

143. $5a^3, 6a^3, -a^3$

144. $t^2, 6t^2, -7t^2$

Objective 3   Use the rule $\left(a^m\right)^n = a^{mn}$.

Simplify the expression. Write the answer in exponential form.

145. $\left(3^4\right)^2$

146. $\left(7^3\right)^5$

147. $\left(9^2\right)^6$

148. $\left(12^4\right)^7$

149. $\left(8^5\right)^4$

150. $\left(11^5\right)^9$

151. $\left[(-3)^3\right]^9$

152. $\left[(-2)^7\right]^3$

153. $-\left(13^3\right)^8$

154. $\left(14^5\right)^3$

155. $\left(16^{11}\right)^{10}$

156. $-\left(21^5\right)^7$

Objective 4   Use the rule $(ab)^m = a^m b^m$.

Simplify the expression.

157. $(3b)^2$

158. $\left(yz^4\right)^2$

159. $\left(p^2 q^3\right)^5$

160. $\left(r^5 s^2\right)^3$

161. $5\left(ab^3\right)^5$

162. $\left(2w^2 z^7\right)^3$

163. $\left(5r^7 t^2\right)^3$

164. $(-3xy)^3$

165. $\left(-2r^2 s\right)^4$

166. $2\left(3c^2 d^3\right)^4$

167. $3\left(xy^5\right)^3$

168. $\left(4c^3 d^4\right)^3$

Objective 5   Use the rule $\left(\dfrac{a}{b}\right)^m = \dfrac{a^m}{b^m}$.

Simplify the expression. Assume all variables represent nonzero real numbers.

169. $\left(\dfrac{2}{3}\right)^3$

170. $\left(\dfrac{7}{8}\right)^2$

171. $\left(\dfrac{x}{y}\right)^5$

172. $\left(\dfrac{w}{5}\right)^4$

173. $\left(\dfrac{z}{10}\right)^3$

174. $\left(\dfrac{2x}{3}\right)^2$

175. $\left(\dfrac{xy}{z}\right)^4$

176. $\left(\dfrac{x^3 y}{z}\right)^4$

177. $\left(-\dfrac{4x}{5}\right)^2$

178. $\left(\dfrac{2a^2}{b^3}\right)^4$

179. $\left(\dfrac{r}{s}\right)^3$

180. $\left(\dfrac{1}{g}\right)^3$

Mixed Exercises

Evaluate the exponential expression. Name the base and the exponent.

181. $(-2)^6$

182. -2^6

183. $(-1)^7$

184. -1^7

185. -5^3

186. $(-5)^3$

Simplify the expression. Write the answer in exponential form.

187. $\left(8^5\right)^4$

188. $\left(x^2y\right)^3$

189. $\left(3a^2b^3\right)^5$

190. $\left(\dfrac{3x}{5y}\right)^6 (y \neq 0)$

191. $\left(7x^8y\right)^5$

192. $(4wt)^5(4wt)^2$

193. $\left(-5a^2\right)\left(-5a^3\right)^4$

194. $\left(5a^2b^3c\right)^3\left(ab^3c^2\right)^4$

195. $\left(z^3\right)^5\left(z^2\right)^6$

196. $\left(\dfrac{6}{7}\right)^2 (6y)^3$

197. $\left(\dfrac{9wx^3}{y^2}\right)^4 (y \neq 0)$

198. $\left(\dfrac{r^2st}{2n}\right)^2 (n \neq 0)$

199. $(4x)^2(4x)^3$

200. $(-6q)^3(-6q)^6$

201. $\left(\dfrac{4}{7}\right)^3 \cdot 4^2$

202. $\left(\dfrac{6b^2}{11}\right)^5$

203. $\left(\dfrac{3}{4}\right)^3 (5x)^4$

204. $\left(\dfrac{1}{3}\right)^4 \left(4ab^2\right)^3$

205. $\left(z^4\right)^3\left(z^2\right)^5$

206. $\left(5x^2y^3\right)^3\left(5xy^4\right)^2$

207. $\left(\dfrac{7a^2b^3}{2}\right)^5$

208. $\left(\dfrac{km^2p^3}{3n^4}\right) (n \neq 0)$

209. $(-2pq)^3(-2pq)^4$

210. $\left(3x^2yz^3\right)^2\left(x^4y\right)^3$

Writing/Conceptual Exercises

211. Explain how $(-2)^6$ and -2^6 are different.

212. Explain how $(2y)^5$ and $2y^5$ are different.

213. Explain why the product rule for exponents does not apply to the expression $2^5 \cdot 5^2$. Then evaluate the expression by finding the individual powers and multiplying the results.

214. An algebra student applied the product rule for exponents in the following way:

$$\left(3x^4\right)\left(5x^6\right)=15x^{24}.$$

Explain to the student why this is not correct and show how to obtain the correct answer.

215. After listening to your explanation, the student in Exercise #214 said "Now I understand. Let me try another one." This time the student applied the product rule in the following way:

$$\left(8x^3\right)\left(9x^6\right)=17x^9.$$

Explain to the student why this is also incorrect, and then show how to obtain the correct answer.

216. Explain why $\left(3a^3b^4\right)^2$ is not equivalent to $(3\cdot2)a^6b^8$ or $6a^6b^8$.

217. Is $\left(3^4\right)^5$ equivalent to $\left(3^5\right)^4$? Explain your answer.

218. On a math test covering the material in this chapter, a student was asked to simplify the following expressions:

1. $\left(\dfrac{3a^3b^4}{c^5}\right)^2 \quad (c \neq 0)$

2. $\left(3a^3b^4c^5\right)^2.$

She wondered why the first expression was written with the restriction "$c \neq 0$" while the second wasn't, even though the expressions contain the same numbers and variables. Explain the reason for this.

Section 4.3   Multiplication of Polynomials

Objective 1   Multiply a monomial and a polynomial.

Find the product.

219. $\left(4y^3\right)\left(7y^2\right)$

220. $\left(-9y^3\right)\left(-8y^5\right)$

221. $3z\left(5z^2+7\right)$

222. $-4p^3\left(8p^2-7p-5\right)$

223. $2k\left(k^5+4k^4-2k^3+6k^2-9\right)$

224. $-3y^2\left(4y^3+2y^2-8y+11\right)$

225. $2k\left(3+2k+5k^4\right)$

226. $7m\left(3-5m^2+2m^3\right)$

227. $-4a\left(3-2a+7a^5\right)$

228. $-3r^4\left(2r^2-7r+5\right)$

229. $8mn\left(4m^2-5mn+6n^2\right)$

230. $-2r^2s\left(10r^2s^2-3rs+5rs^2\right)$

Objective 2   Multiply two polynomials.

Find the product.

231. $(x+3)(x+5)$

232. $(x-6)(x-2)$

233. $(2p+3)(p+4)$

234. $(8+3a)(7-2a)$

235. $(5n+1)(2n-7)$

236. $(2x-5)(3x-1)$

237. $(9r-7)(3r+5)$

238. $(x+2)\left(x^2-2x+4\right)$

239. $(y-5)\left(y^2+5y+25\right)$

240. $(x+1)\left(2x^3-3x^2+4\right)$

241. $(2y-1)\left(3y^3-2y^2+y-7\right)$

242. $\left(5m^2+1\right)\left(2m^3-3m^2-5m\right)$

243. $\left(2z^2-3\right)\left(z^4-z^3+z^2-z+1\right)$

244. $\left(5x^2+3x+1\right)\left(2x^2-3x+5\right)$

Objective 3   Multiply binomials by the FOIL method.

Use the FOIL method to find the product.

245. $(x+3)(x-4)$

246. $(y+6)(y+2)$

247. $(r-8)(r+7)$ 248. $(z-5)(z-11)$

249. $(2k-1)(2k+3)$ 250. $(4m+5)(m-8)$

251. $(8y+5)(2y+1)$ 252. $(3x+2y)(2x-y)$

253. $(3m+n)(4m-7n)$ 254. $(2p+5q)(3p-2q)$

255. $(2+5m)(4-3m)$ 256. $(11k-10)(11k+10)$

257. $(1-3x)(3+2x)$ 258. $(3+8a)(1+6a)$

259. $(5-4y)(1+2y)$ 260. $(7p+2)(3p-1)$

261. $(5r-3s)(3r+2s)$ 262. $(12m+5n)(-3m+2n)$

263. $(3x^2+2)(2x^2-5)$ 264. $(5v^2+w^2)(v^2-3w^2)$

Mixed Exercises

Find the product.

265. $(5x^3)(4x^2)$ 266. $4m(3m+7)$

267. $(q-7)(q-6)$ 268. $(9p+4r)(6p-5r)$

269. $(x-7y)(x+7y)$ 270. $-11p^3(-1+6p-7p^3+4p^5)$

271. $(6x^2+19x+14)(2x^2-1)$ 272. $(3x^2-8x-6)(x-2)$

273. $(6x^3-15x^2+4x-3)(x^2+2x-1)$ 274. $(3x^3-2x^2+3x+1)(2x^2-x+2)$

275. $(x+1)(4x^3-2x+6)$ 276. $(3y-1)(4y^3+2y^2-y+5)$

277. $(2a^2+1)(4a^3-2a^2-a)$ 278. $(4b^2-2)(b^4-3b^3+b^2-b+1)$

279. $(2x^2-3x+2)(4x^3+x+6)$ 280. $(3y^3-y^2+2)(-3y^3+2y+1)$

281. $(2m^3+3m-4)(-2m^3-4m+1)$ 282. $(4x^2+x)(4x^2+5x-4)$

Section 4.4   Special Products

Objective 1   Square binomials.

Find the square by using the pattern for the square of a binomial.

283. $(z+6)^2$

284. $(t-10)^2$

285. $(2x+3)^2$

286. $(5y-1)^2$

287. $(2m-7)^2$

288. $(3p+8)^2$

289. $(3-x)^2$

290. $(5+6y)^2$

291. $(4p-3q)^2$

292. $(2m-7p)^2$

293. $\left(5z-\dfrac{1}{2}\right)^2$

294. $\left(2x-\dfrac{1}{4}y\right)^2$

295. $\left(3x+\dfrac{1}{3}y\right)^2$

296. $\left(7a-\dfrac{1}{2}b\right)^2$

Objective 2   Find the product of the sum and difference of two terms.

Find the product by using the pattern for the sum and difference of two terms.

297. $(z-7)(z+7)$

298. $(k-12)(k+12)$

299. $(4b-7)(4b+7)$

300. $(8k-3p)(8k+3p)$

301. $(4p+3q)(4p-3q)$

302. $(5+3x)(5-3x)$

303. $(9-2y)(9+2y)$

304. $\left(x+\dfrac{1}{2}\right)\left(x-\dfrac{1}{2}\right)$

305. $\left(y+\dfrac{2}{3}\right)\left(y-\dfrac{2}{3}\right)$

306. $\left(4m-\dfrac{1}{4}\right)\left(4m+\dfrac{1}{4}\right)$

307. $\left(7a+\dfrac{2}{3}b\right)\left(7a-\dfrac{2}{3}b\right)$

308. $\left(\dfrac{3}{4}s+\dfrac{2}{5}t\right)\left(\dfrac{3}{4}s-\dfrac{2}{5}t\right)$

309. $\left(y^2 + 3\right)\left(y^2 - 3\right)$

310. $\left(5m^2 - 2n^3\right)\left(5m^2 + 2n^3\right)$

Objective 3   Find higher powers of binomials.

Find the product.

311. $(x - 1)^3$

312. $(r + 2)^3$

313. $(y + 5)^3$

314. $(2x - 1)^3$

315. $(2x + 5)^3$

316. $(k + 1)^4$

317. $(t - 2)^4$

318. $(3x + 2y)^4$

319. $(z - 5)^4$

320. $(4s + 3t)^4$

321. $(2x - 3)^3$

322. $(j - 1)^4$

Mixed Exercises

Find the product.

323. $(4y - 3)^2$

324. $(6b - 11)(6b + 11)$

325. $(8b - 3)^2$

326. $(7t + 4u)(7t - 4u)$

327. $(9k + 2m)^2$

328. $(25 - 16w)(25 + 16w)$

329. $\left(3x + \dfrac{5}{4}\right)\left(3x - \dfrac{5}{4}\right)$

330. $\left(6y - \dfrac{1}{3}\right)^2$

331. $\left(5j + \dfrac{1}{2}k\right)^2$

332. $\left(\dfrac{4}{7}t + 9u\right)\left(\dfrac{4}{7}t - 9u\right)$

333. $\left(\dfrac{1}{2}x - \dfrac{1}{3}y\right)^2$

334. $\left(\dfrac{2}{3}x + \dfrac{3}{4}y\right)\left(\dfrac{2}{3}x - \dfrac{3}{4}y\right)$

335. $(2x - 3y)^3$

336. $(3a - 4b)^4$

337. $\left(\dfrac{1}{2}x + y\right)^4$

338. $\left(2x - \dfrac{1}{2}y\right)^3$

339. $(6x - 2y)(6x + 2y)$

340. $(x + 2y)^4$

341. $\left(5a - \dfrac{1}{5}b\right)^2$

342. $(2x - y)^4$

343. $\left(\dfrac{1}{2}x - \dfrac{1}{3}y\right)^3$

344. $\left(3x - \dfrac{1}{6}y\right)\left(3x + \dfrac{1}{6}y\right)$

Writing/Conceptual Exercises

345. Explain how the expressions $x^2 - y^2$ and $(x - y)^2$ differ.

346. A student is asked to find the product $(x - 3)^2$, and gives the answer $x^2 - 9$. Use a numerical example to explain why this is incorrect.

347. A student remembers that the square of a binomial is a trinomial. When asked to find $(x - 3y)^2$, he gives the answer $x^2 - 3xy + 9y^2$. Explain why this answer is incorrect.

348. Based on your experience in finding powers of binomials such as $(a + b)^2$, $(a + b)^3$, and $(a + b)^4$, how many terms would you expect to find in the simplified answer for $(a + b)^5$? Explain your answer. (Do not actually find the product.)

Section 4.5   Integer Exponents and the Quotient Rule

Objective 1   Use zero as an exponent.

Evaluate the expression.

349. 8^0

350. $(-6)^0$

351. -13^0

352. $\left(\dfrac{1}{5}\right)^0$

353. $5^0 + 6^0$

354. $7^0 + (-7)^0$

355. $\left(\dfrac{2}{3}\right)^0 - \left(\dfrac{1}{3}\right)^0$

356. $(-5)^0 + (-5)^0$

357. $-12^0 + (-12)^0$

358. $-25^0 - (-25)^0$

359. $-r^0 \ (r \neq 0)$

360. $\dfrac{0^8}{8^0}$

Objective 2   Use negative numbers as exponents.

Evaluate the expression.

361. 7^{-2}

362. 12^{-1}

363. $(-5)^{-2}$

364. $(-2)^{-5}$

365. $\left(\dfrac{3}{4}\right)^{-2}$

366. $\left(\dfrac{8}{9}\right)^{-1}$

367. $3^{-1} + 5^{-1}$

368. $10^{-2} + 5^{-2}$

369. $8^{-1} - 2^{-1}$

Simplify by using the definition of negative exponents. Write the expression with only positive exponents. Assume all variables represent nonzero real numbers.

370. r^{-4}

371. y^{-8}

372. $\dfrac{2}{r^{-4}}$

373. $\dfrac{r^{-5}}{6}$

374. $\dfrac{2x^{-3}}{3y^{-2}}$

375. $\dfrac{4^{-2}}{5^{-2}}$

Objective 3   Use the quotient rule for exponents.

Use the quotient rule to simplify the expression. Write answers with only positive exponents. Assume that all variables represent nonzero real numbers.

376. $\dfrac{4^7}{4^3}$

377. $\dfrac{5^{12}}{5^7}$

378. $\dfrac{(-2)^{10}}{(-2)^3}$

379. $\dfrac{(-4)^{-4}}{(-4)^3}$

380. $\dfrac{7^4 \cdot x^3}{7^5 \cdot x^8}$

381. $\dfrac{4k^7 m}{8k^3 m^5}$

382. $\dfrac{12x^3 y^5}{12^4 x^3 y^4}$

383. $\dfrac{3^{-2}}{3^{-4}}$

384. $\dfrac{12^{-3}}{12^{-2}}$

385. $\dfrac{x^4}{x^{-3}}$

386. $\dfrac{3^{-1} m^{-2} p^5}{3^4 m^{-1} p^{-2}}$

387. $\dfrac{8b^{-3} c^7}{8^{-4} b^{-7} c^{-3}}$

388. $\dfrac{1}{z^{-1}}$

389. $\dfrac{a^3 b^4}{a^{-3} b^{-4}}$

390. $\dfrac{5^{-3} x^{-2} y^0}{5^2 x^{-3} y^{-2}}$

Objective 4   Use combinations of rules.

Use a combination of the rules for exponents to simplify the expression. Write answers with only positive exponents. Assume that all variables represent nonzero real numbers.

391. $\dfrac{\left(7^2\right)^4}{7^6}$

392. $\dfrac{\left(9^3\right)^2}{9^5}$

393. $8^7 \cdot 8^{-1} \cdot 8^4$

394. $a^{-3} \cdot a^3 \cdot a^7$

395. $\left(2^{-3}\right)^4$

396. $\left(3^2 x^{-3} y\right)^2$

397. $\left(5w^2 y^2\right)^{-2}\left(4wy^{-3}\right)^2$

398. $\left(2p^{-3} q^2\right)^2\left(4p^4 q^{-1}\right)^{-1}$

399. $\dfrac{\left(2y\right)^{-5}}{\left(3y\right)^{-2}}$

400. $\left(9xy\right)^5 \left(9xy\right)^{-6}$

401. $\dfrac{\left(t^{-3}\right)^2 \left(t^5\right)^3}{\left(t^{-6}\right)^{-2}}$

402. $\dfrac{\left(q^{-2}\right)^{-4}\left(q^0\right)^2}{\left(q^5\right)^{-3}}$

403. $\left(\dfrac{x^4 y^2}{x^6 y^{-2}}\right)^{-2}$

404. $\left(\dfrac{k^4 t^2}{k^2 t^{-3}}\right)^{-2}$

Mixed Exercises

Evaluate the expression.

405. $10^0 + \left(-10\right)^0$

406. 5^{-2}

407. $-8^0 + \left(-8\right)^0$

408. $\left(3^{-4}\right)^0$

409. $\left(3^0\right)^{-4}$

410. $\left(4^0\right)^0$

411. $\left(\dfrac{2}{3}\right)^{-3}$

412. $\left(-\dfrac{3}{4}\right)^{-4}$

413. $6^{-1} + 6^{-2}$

Simplify the expression. Write answers with only positive exponents. Assume that all variables represent nonzero real numbers.

414. $\dfrac{7^{-8}}{7^{-9}}$

415. $\dfrac{a^{-8}}{a^{2}}$

416. $\dfrac{\left(2^{3}\right)^{2}}{2^{7}}$

417. $\dfrac{\left(5^{3}\right)^{-2}}{5^{-1}}$

418. $\left(\dfrac{3}{4}\right)^{-2} \cdot \left(\dfrac{4}{3}\right)$

419. $\dfrac{\left(5k^{-1}\right)^{2}}{5k^{3}}$

420. $\dfrac{\left(2xy^{-1}\right)^{3}}{2^{3}x^{-3}y}$

421. $\dfrac{x^{3} \cdot x^{-1}}{x^{5} \cdot x^{-4}}$

422. $\left(\dfrac{5x^{2}}{2x^{-3}}\right)^{-1}$

423. $\dfrac{\left(z^{3}\right)^{2}\left(z^{2}\right)^{4}z^{-8}}{\left(z^{4}\right)^{-3}}$

424. $\dfrac{\left(3y^{-1}z^{4}\right)^{-1}\left(3y^{-2}\right)}{\left(y^{3}z^{2}\right)^{-2}}$

425. $\dfrac{\left(2^{-1}m^{-1}n^{-1}p\right)^{-2}}{\left(p^{2}m^{-1}\right)^{-3}}$

Writing/Conceptual Exercises

Decide whether the expression is positive, negative, or zero.

426. $(-5)^{3}$

427. 5^{-3}

428. $(-6)^{2}$

429. -6^{-2}

430. $\left(\dfrac{1}{6}\right)^{-2}$

431. $\left(-\dfrac{1}{6}\right)^{-2}$

432. $8^{0} - 9^{0}$

433. $(-5)^{2} - 5^{2}$

434. 10^{-10}

435. -10^{-10}

436. If one side of a square measures $3x^5$ centimeters, what is the area of the square?

437. If one edge of a cube measures $2y^4$ feet, what is the volume of the cube?

438. On an algebra quiz, a student evaluated the expression $2^{-2}+2^{-3}$ as follows:

$$2^{-2}+2^{-3}=2^{-5}=\frac{1}{2^5}=\frac{1}{32}.$$

Explain why this is incorrect and show the correct solution.

Section 4.6   The Quotient of a Polynomial and a Monomial

Objective 1   Divide a polynomial by a monomial.

Divide the polynomial by $3m^2$.

439. $6m^3+9m^2$

440. $12m^4-9m^3+6m^2$

441. $15m^5-9m^3$

442. $30m^3-18m^2+3m$

443. $27m^3+18m^2-6m$

444. $4m^3-2m^2$

445. $-54m^3+30m^2+6m$

446. $3m^2-3$

447. $6m^2-3m+9$

448. $15m^5-5m^4+3m^2-2m+3$

Perform the division.

449. $\dfrac{6p^4+18p^7}{3p^3}$

450. $\dfrac{12x^6+18x^5+30x^3}{6x^2}$

451. $\left(8y^7-9y^2\right)\div(4y)$

452. $\left(9z^5+27z^3-3z+10\right)\div(3z)$

453. $\left(60x^4-20x^2\right)\div(2x)$

454. $\left(6m^5-4m^3+12m\right)\div(4m)$

455. $\left(m^2+3m-12\right)\div(2m)$

456. $\dfrac{20p^4-25p^3-5p}{5p^2}$

457. $\dfrac{50q^4 - 30q^2 + 20q}{10q^2}$

458. $\dfrac{8y^7 + 9y^6 - 11y - 12}{y^3}$

459. $\dfrac{12z^5 + 8z^4 - 6z^3 + 5z}{3z^3}$

460. $\dfrac{36x + 24x^4 + 3x^8}{4x}$

461. $\dfrac{6y^5 - 3y^4 + 9y^2 + 27}{-3y}$

462. $\dfrac{14y^2 - 14y + 70}{-7y^2}$

Objective 2   Divide a polynomial by a polynomial.

Perform the division.

463. $\dfrac{x^2 - x - 6}{x + 2}$

464. $\dfrac{y^2 - 2y - 24}{y - 6}$

465. $\dfrac{2x^2 + 9x - 35}{2x - 5}$

466. $\dfrac{p^2 + 8p - 20}{p - 2}$

467. $\left(x^2 + 4x + 4\right) \div \left(x + 2\right)$

468. $\left(r^2 - 2r - 20\right) \div \left(r - 5\right)$

469. $\left(2a^2 - 11a + 16\right) \div \left(2a + 3\right)$

470. $\left(9w^2 + 6w + 1\right) \div \left(3w + 1\right)$

471. $\dfrac{5w^2 - 22w + 4}{w - 4}$

472. $\dfrac{6b^2 + 37b + 7}{b + 7}$

473. $\dfrac{25m^2 - 40m + 16}{5m - 4}$

474. $\dfrac{y^2 - 1}{y + 1}$

475. $\dfrac{4x^2 - 9}{2x - 3}$

476. $\dfrac{12y^3 - 11y^2 + 9y + 18}{4y + 3}$

477. $\dfrac{2z^3 - 3z^2 + 8z + 6}{2z + 1}$

478. $\dfrac{6m^3 + 5m^2 - 34m + 13}{3m - 5}$

479. $\left(27p^4 - 36p^3 - 6p^2 + 26p - 24\right) \div \left(3p - 4\right)$

480. $\left(12x^3 - 17x^2 + 30x - 10\right) \div \left(3x^2 - 2x + 5\right)$

481. $\dfrac{6x^4 - 15x^3 + 14x^2 - 5x - 1}{3x^2 + 1}$

482. $\dfrac{4y^5 - 8y^4 - 3y^3 + 22y^2 - 15}{4y^2 - 3}$

483. $\dfrac{y^4 + 4y^3 - 3y^2 + 4y - 5}{y^2 + 1}$

484. $\dfrac{3x^4 + 2x^3 - 2x^2 - 2x - 2}{x^2 - 1}$

485. $\dfrac{y^3 - 1}{y - 1}$

486. $\dfrac{b^4 - 1}{b^2 + 1}$

487. $\dfrac{27x^5 - 3x^3 + 6x^2 - 2x}{3x - 1}$

488. $\dfrac{32x^5 - 243}{2x - 3}$

Writing/Conceptual Exercises

489. Suppose that a polynomial in the variable y has degree 6, and it is divided by a monomial in the variable y having degree 2. Describe the quotient in mathematical terms, giving the type of expression and the degree.

In her algebra class, Mara volunteered to work the following division problem on the blackboard:

$$\frac{9z^3 - 12z^2 + 3z}{3z}.$$

490. Mara's answer was $3z^2 - 4z$. She said that the $3z$ terms in the numerator and denominator would "cancel out.' Was she correct? Explain.

491. Marc, who is one of Mara's classmates, said that her answer could not possibly be correct because it only had 2 terms. Was he right?

492. Sophie, who is another one of Mara's classmates, told Mara she could have found her mistake if she had checked her answer. Show the result if Sophie checked Mara's answer and if she checked the correct answer.

493. Josh's algebra instructor put the following division problem on a quiz:

$$\left(x^5 - 1\right) \div (x - 1).$$

Josh rewrote the problem in the following way:

$$x - 1 \overline{\smash{)}x^5 - 1}.$$

What difficulty do you expect Josh will have when he tries to perform this division?

494. You are given the following division problem:

$$\left(-12x^3 + 10x^2 + x - 8\right) \div \left(4x - 2\right).$$

By looking at this problem, but without performing the division, determine which of the following would be the first term of the quotient.

(a) $-3x$ (b) $3x^2$ (c) $-3x^2$ (d) $-8x^2$

495. A student performs the division

$$\frac{3x^3 - 2x^2 + x - 5}{x - 4},$$

and obtains a remainder of $x + 13$. Without working the problem, explain why this remainder cannot be correct.

496. Two students give the following answers to a problem in which a polynomial is divided by the binomial $2x - 5$:

$$\text{Miriam:}\quad 3x + 5 + \frac{-4}{2x - 5}$$

$$\text{Ed:}\quad 3x + 5 - \frac{4}{2x - 5}.$$

Can both answers be correct? Explain.

Section 4.7   An Application of Exponents: Scientific Notation

Objective 1   Express numbers in scientific notation.

Write the number in scientific notation.

497. 675

498. 3295

499. 35,744

500. 609,906

501. 8.5

502. 13,205,000

503. .02

504. .176

505. .00005

506. .00736

507. −93,572

508. −.00048

Objective 2   Convert numbers in scientific notation to numbers without exponents.

Write the number without exponents.

509. 3.5×10^2

510. 6.37×10^4

511. -2.75×10^5

512. 7.045×10^0

513. 4×10^7

514. 1.5×10^8

515. 6.2×10^{-2}

516. 7.55×10^{-4}

517. 2.001×10^{-1}

518. 2.753×10^{-5}

519. -8.02×10^0

520. -9.99×10^{-3}

Objective 3   Use scientific notation in calculations.

Perform the indicated operations with the numbers in scientific notation, and then write the answer without exponents.

521. $\left(5 \times 10^3\right) \times \left(6 \times 10^2\right)$

522. $\left(2.3 \times 10^4\right) \times \left(1.1 \times 10^{-2}\right)$

523. $\left(2.3 \times 10^{-4}\right) \times \left(3.1 \times 10^{-2}\right)$

524. $\left(3 \times 10^6\right) \times \left(4 \times 10^{-2}\right) \times \left(2 \times 10^{-1}\right)$

525. $\dfrac{5.2 \times 10^4}{1.3 \times 10^{-2}}$

526. $\dfrac{7.2 \times 10^2}{5 \times 10^4}$

527. $\dfrac{8.5 \times 10^{-3}}{1.7 \times 10^{-7}}$

528. $\dfrac{4.6 \times 10^{-3}}{2.3 \times 10^{-1}}$

529. $\left(3 \times 10^4\right) \times \left(4 \times 10^2\right) \div \left(2 \times 10^3\right)$

530. $\left(6 \times 10^4\right) \times \left(3 \times 10^5\right) \div \left(9 \times 10^7\right)$

531. $\dfrac{\left(2.1 \times 10^{-3}\right) \times \left(4.8 \times 10^4\right)}{\left(1.6 \times 10^6\right) \times \left(7 \times 10^{-6}\right)}$

532. $\dfrac{\left(7.5 \times 10^6\right) \times \left(4.2 \times 10^{-5}\right)}{\left(6 \times 10^4\right) \times \left(2.5 \times 10^{-3}\right)}$

Mixed Exercises

If a number is written without exponents, rewrite it in scientific notation. If a number is written in scientific notation, rewrite it without exponents.

533. 4,905,000

534. 32

535. .1

536. .0075

537. −5.32

538. −.0000609

539. 3.42×10^8

540. 2.71×10^{-6}

541. -2.03×10^5

542. -9.24×10^{-7}

543. 12.36×10^8

544. -437.62×10^{-4}

545. 6×10^5

546. -7×10^{-5}

Perform the indicated operations with the numbers in scientific notation, and then write the answer without exponents.

547. $\left(1.6 \times 10^5\right) \times \left(3.4 \times 10^{-3}\right)$

548. $\left(2.5 \times 10^{-7}\right) \times \left(5.6 \times 10^4\right)$

549. $\dfrac{5 \times 10^{-4}}{5 \times 10^{-5}}$

550. $\dfrac{9 \times 10^3}{3 \times 10^{-2}}$

551. $\dfrac{6 \times 10^{-4}}{2 \times 10^{-7}}$

552. $\dfrac{6 \times 10^{-3}}{4 \times 10^{-7}}$

553. $\dfrac{\left(3\times10^4\right)\times\left(4\times10^2\right)}{2\times10^3}$

554. $\dfrac{\left(6\times10^4\right)\times\left(3\times10^5\right)}{\left(9\times10^7\right)}$

555. $\dfrac{\left(8\times10^2\right)\times\left(3\times10^{-5}\right)}{\left(6\times10^3\right)\times\left(4\times10^{-7}\right)}$

556. $\dfrac{\left(4\times10^{-5}\right)\times\left(3\times10^4\right)}{\left(6\times10^{-2}\right)\times\left(2\times10^{-1}\right)}$

557. $\dfrac{\left(3.8\times10^6\right)\times\left(7.2\times10^{-5}\right)}{\left(9\times10^2\right)\times\left(1.9\times10^{-4}\right)}$

558. $\dfrac{\left(12.5\times10^3\right)\times\left(4.9\times10^{-4}\right)}{\left(7\times10^{-8}\right)\times\left(2.5\times10^5\right)}$

Writing/Conceptual Exercises

Determine whether the given number is written in scientific notation, as defined in the textbook. If it is not, write it as such.

559. 3.804×10^3

560. 20

561. 3852

562. .0005

563. 92×10^3

564. 4×10^{-6}

565. $.3\times10^4$

566. $.075\times10^2$

567. Explain in your own words some reasons why you think scientists prefer to work with numbers that are written in scientific notation rather than those written without exponents.

568. Why do you think that 10 was chosen as the base for all numbers written in scientific notation?

CHAPTER 5   FACTORING AND APPLICATIONS

Section 5.1   The Greatest Common Factor; Factoring by Grouping

Objective 1   Find the greatest common factor of a list of terms.

Find the greatest common factor for the list of numbers.

1. $12, 18, 24$

2. $72, 36, 18$

3. $50, 25, 30$

4. $14, 28, 56$

5. $30, 20, 40$

6. $7, 18, 20$

7. $12, 18, 24, 36$

8. $28, 70, 126, 42$

9. $64, 96, 480$

10. $84, 28, 280$

Find the greatest common factor for the list of terms.

11. $30x, 20x^2, 15x$

12. $42y^3, 28y^2, 35y^3$

13. $12ab, 18a^2b, 26ab^2, 32a^2b^2$

14. $10m^3, 15m^5, 21m^7$

15. $6x^4, 9x^2, 12x^5$

16. y^4z^2, y^6z^8, z^9

17. $12k^2m^3n^5, 36km^2n^4, k^4m^8n^7$

18. $49w^3x^2y, w^4x^5y^7, 42w^2x^3y^5$

19. $45a^3y^4, 75a^3y^2, 60a^2y, 90a^3y^3$

20. $9xy^3, 72x^2y, 90xy^2, 108x^2y^2$

Objective 2   Factor out the greatest common factor.

Complete the factoring.

21. $56 = 8(\qquad)$

22. $9x^3 = 3x(\qquad)$

23. $-18y^8 = -6y^6(\qquad)$

24. $-75a^4y^2 = 25a^2y(\qquad)$

Factor out the greatest common factor.

25. $9r + 24t$

26. $57m + 38n$

27. $18q^2 - 45q$

28. $27xy + 18x + 36x^2$

29. $15w^2 + 90wx$

30. $24ab + 8a + 16ac$

31. $42tw + 28t + 70t^2$

32. $33rs + 22r + 55r^2$

33. $5a^4 + 25a^3 - 20a^2$

34. $11y^2 - 10$

35. $26x^8 + 13x^{12} - 78x^{10}$

36. $100b^4 + 15b^2$

37. $16x^2y + 24xy + 40xy^2$

38. $5(a+b) + x(a+b)$

39. $c(a-b) - d(a-b)$

40. $x^2(r-2s) + y^2(r-2s)$

Objective 3   Factor by grouping.

Factor by grouping.

41. $y^2 + 4y + 6y + 24$

42. $a^2 + 7a + 3a + 21$

43. $d^2 - 5d + 2d - 10$

44. $y^2 - 7y + 3y - 21$

45. $2m^2 + 6m + 2mn + 6n$

46. $18z^2 + 12zq - 3zq - 2q^2$

47. $1 - m + mn - n$

48. $4w + 12 - zw - 3z$

49. $7y - xy + 9x - 63$

50. $6a^4 + 3a^3b^2 - 2ab^3 - b^5$

51. $2a^3 - 3a^2b + 2ab^2 - 3b^3$

52. $8r^3 - 2rs - 4s^2r^2 + s^3$

53. $2x^4 + x^3y + 4xy^3 + 2y^4$

54. $20 - 4d - 5c + cd$

Mixed Exercises

Find the greatest common factor for the list.

55. $56, 28, 14$

56. $81, 18, 126$

57. $52y^4, 39y^3, 91y^4$

58. $51pq^3, 17p^2q^2, 34pq^2$

59. $35r^2s^5, 15r^3t^2, 21s^3t$

60. $-42u^2v, -84uv^2, -63uv^4$

Factor completely.

61. $21a^2 - 35a$

62. $8pq + 24q^2$

63. $5a^2b + 10ab + 15ab^2$

64. $2a(x-y) - 5b(x-y)$

65. $m^2 - 15m + 4m - 60$

66. $28 - 14q - 2q + q^2$

67. $10m^2 + 18m - 5m - 9$

68. $3r^3 - 4r^2s + 3s^2r - 4s^3$

69. $9m^2n - 27mn + 9$

70. $1 + p - q - pq$

71. $48x^{12} - 36x^{10} + 24x^7$

72. $20y^7m^2 - 36y^3m^4$

73. $13y^3 + 39y^2$

74. $45q^4p^4 - 36q^2p^3 + 81q^2p^4$

75. $8 - 12p - 6p^3 + 9p^4$

76. $12c^2 + 18c - 10c - 15$

77. $4mn^2 - 28mn$

78. $96a^5 + 48a^3 + 4a^2$

79. $28m^2 + 20mn - 7m - 5n$

80. $3x^3 + 3xy^2 + 4x^2y + 4y^3$

Writing/Conceptual Exercises

81. Give an example of four numbers whose greatest common factor is 7.

82. Give an example of three terms whose greatest common factor is $4x^2yz^3$.

Determine whether the expression is *in factored form* or *not in factored form*.

83. $(2r+s)(3r-s)$

84. $2(3r-s) + s(3r-s)$

85. $x^2(y^2+z^2) - 4(y^2+z^2)$

86. $(5-x^2)(9+y^2)$

Tell whether it is possible to factor the expression. If so, factor it. If not, explain why it cannot be factored.

87. $2x(3y-4)+5(3y+4)$

88. $r^2(2u+v)+s^2(2u+v)$

Section 5.2   Factoring Trinomials

Objective 1   Factor trinomials with a coefficient of 1 for the squared term.

List all pairs of integers with the given product. Then find the pair whose sum is given.

89. Product: 28; sum: 11

90. Product: 40; sum: -13

91. Product: -16; sum: 6

92. Product: -54; sum: -3

Complete the factoring.

93. $x^2+7x+10=(x+5)(\quad)$

94. $t^2-14t+45=(t-5)(\quad)$

95. $m^2+10m-24=(m-2)(\quad)$

96. $p^2-2p-15=(p+3)(\quad)$

Factor completely. If the polynomial cannot be factored, write *prime*.

97. $y^2+11y+30$

98. $x^2-12x+27$

99. b^2+b-2

100. $z^2-10z+25$

101. $x^2-12x-45$

102. $n^2-8n-33$

103. $r^2+15r+14$

104. $x^2-15xy+56y^2$

105. $x^2-4xy-21y^2$

106. $m^2-2mn-3n^2$

Objective 2   Factor such polynomials after factoring out the greatest common factor.

Factor completely.

107. $3a^2+27a+60$

108. $2p^3+8p^2-10p$

109. $3h^3k + 33h^2k - 36hk$

110. $8b^2 + 8b - 160$

111. $7a^2 - 42a + 56$

112. $2p^7 - 10p^6 - 28p^5$

113. $a^3b + a^2b^2 - 6ab^3$

114. $2x^2 - 24x + 64$

115. $10n^2 - 110n + 300$

116. $3h^2j - 18hj + 24j$

117. $10k^6 + 120k^5 + 320k^4$

118. $16x^7 - 48x^6 + 32x^5$

119. $6x^3y^2 + 30x^2y^3 + 36xy^4$

120. $4a^4b - 32a^3b^2 + 60a^2b^3$

Mixed Exercises

Factor completely. If the polynomial cannot be factored, write *prime*.

121. $a^2 - 11a + 28$

122. $w^2 - w + 1$

123. $x^3 - 8x^2 + 7x$

124. $5r^3 - 55r^2 + 150r$

125. $y^2 - 3y - 40$

126. $2m^3 - 8m^2 - 10m$

127. $3n^5 - 18n^4 + 15n^3$

128. $q^2 - 11q - 26$

129. $b^2 - 10bc + 25c^2$

130. $r^{10} - 4r^9q - 21r^8q^2$

131. $c^2 - 12cd + 35d^2$

132. $3m^2 - 6m + 3$

133. $2s^2t + 4st - 70t$

134. $2x^3 - 14x^2y + 20xy^2$

135. $2m^2 + 12m + 10$

136. $3n^2 + 12n + 9$

137. $x^2 - 2xy - 15y^2$

138. $p^2 + 12p + 32$

139. $4r^2 - 44r + 72$

140. $10z^2 + 20z - 350$

141. $5a^2 - 5ab - 10b^2$

142. $f^2 - 3fg - 10g^2$

143. $2s^2 - 40s + 198$

144. $x^3 - x^2y - 12xy^2$

145. $2r^3 - 6r^2 - 20r$ 146. $3a^3 + 21a^2 - 54a$

Writing/Conceptual Exercises

147. Which one of the following is the correct factored form of $z^2 - 7z - 18$?

    (a) $(z-9)(z-2)$ (b) $(z+9)(z+2)$
    (c) $(z+9)(z-2)$ (d) $(z-9)(z+2)$

148. Which one of the following is the correct factored form of $r^2 - 10r + 21$?

    (a) $(r-3)(r-7)$ (b) $(r+3)(r+7)$
    (c) $(r+3)(r-7)$ (d) $(r-3)(r+7)$

149. What would be your first step in factoring $3x^2 - 39x + 120$?

150. What would be your first step in factoring $4k^3 + 8k^2 - 48k$?

151. If you are asked to completely factor $4p^2 + 4p - 24$, would it be correct to give $(p-2)(4p+12)$ as your answer? Explain.

152. Explain how you can be sure that $s^2 - 4s + 7$ is a prime polynomial.

Section 5.3   More on Factoring Trinomials

Objective 1   Factor trinomials by grouping when the coefficient of the squared term is not 1.

Factor each trinomial by grouping.

153. $8b^2 + 6b + 1$ 154. $2x^2 + 13x + 21$

155. $15a^2 - a - 2$ 156. $2m^2 + 7m + 6$

157. $7b^2 + 12b - 4$ 158. $3m^2 + 11m + 6$

159. $3p^3 + 16p^2 + 5p$ 160. $8m^2 + 26mn + 6n^2$

161. $7a^2b + 10ab - 8b$ 162. $2s^2 + 19st - 10t^2$

163. $4y^2 - 2yz - 2z^2$ 164. $25a^2 + 25ab + 6b^2$

Objective 2   Factor trinomials using FOIL.

Complete the factoring.

165. $2a^2 + 7a - 4 = (2a - 1)($ $)$ 166. $16k^2 + 24k + 5 = (4k + 1)($ $)$

167. $6t^2 + 7t - 20 = (2t + 5)($ $)$ 168. $24y^2 - 23y + 5 = (3y - 1)($ $)$

Factor each trinomial by trial and error (using FOIL backwards).

169. $10x^2 + 9x + 2$ 170. $4z^2 - 4z - 3$

171. $2a^2 - 11a - 6$ 172. $15w^2 - 29w - 2$

173. $10q^2 + 11q + 3$ 174. $14m^2 - 33m - 5$

175. $3m^2 + 7m - 6$ 176. $15q^2 + 23q + 6$

177. $3a^2 - 5ab - 2b^2$ 178. $25w^2 + 20wz + 4z^2$

179. $8s^2 - 6sw - 9w^2$ 180. $6x^2 + 5xy - 6y^2$

181. $18x^2 - 27xy + 4y^2$ 182. $16y^2 + 8y - 15$

Mixed Exercises

Factor completely.

183. $2a^2 + 9a + 9$ 184. $6p^2 + 13p + 5$

185. $6y^2 + 5y - 1$ 186. $8z^2 + 14z + 5$

187. $3p^2 - 14p - 5$ 188. $12r^2 + 11r - 5$

189. $7x^2 - 27x - 4$ 190. $6x^2 + 8xy - 14y^2$

191. $2x^4 - 11x^3 + 12x^2$ 192. $12a^3 + 10a^2b - 12ab^2$

193. $27r^2 - 6rt - 16t^2$ 194. $2y^5z^2 - 5y^4z^3 - 3y^3z^4$

195. $72x^4y + 6x^3y^2 - 105x^2y^3$

196. $108x^2y^5z^2 - 6x^2y^4z^3 - 80x^2y^3z^4$

197. $288x^7 - 68x^6y - 288x^5y^2$

198. $90x^2 + 15xyz - 50y^2z^2$

199. $63x^2 + 59xy - 84y^2$

200. $12 - z - z^2$

201. $15 - 19x + 6x^2$

202. $-16x^3 - 28x^2 + 30x$

203. $6y^3 - 19y^2 - 7y$

204. $-x^2 - x + 20$

205. $-8a^2 - 2ab + b^2$

206. $-6p^2 + pq + 2q^2$

Writing/Conceptual Exercises

207. Which one of the following is the correct factored form of $6x^2 + x - 2$?

(a) $(3x - 2)(2x + 1)$ (b) $(3x + 2)(2x - 1)$
(c) $(3x + 1)(2x - 2)$ (d) $(3x - 1)(2x + 2)$

208. Which one of the following is the correct factored form of $6r^2 - rs - 5s^2$?

(a) $(3r + 5s)(2r - s)$ (b) $(3r - 5s)(2r + s)$
(c) $(6r + 5s)(r - s)$ (d) $(6r - s)(r + 5s)$

209. For the trinomial $6x^2 - 11x + 3$, 3 is not a common factor. Explain how we know that, $3x - 6$ then, cannot be a factor of this polynomial.

210. If $6y^2 - 83y + 100$ is written in factored form, what will be the signs of the last terms of the two binomial factors?

Section 5.4   Special Factoring Rules

Objective 1   Factor the difference of two squares.

Factor the binomial completely. If it cannot be factored, write *prime*.

211. $t^2 - 100$

212. $x^2 + 36$

213. $64 - a^2$

214. $y^2 - 225$

215. $4a^2 - 81$

216. $36j^2 - \dfrac{25}{49}$

217. $144b^2 - 49$

218. $121m^2 - 36n^2$

219. $x^4 - 81$

220. $z^4 - 144$

221. $a^4 - 10{,}000$

222. $256y^4 - 1$

223. $9x^2 + 1$

224. $m^4n^2 - m^2$

Objective 2   Factor a perfect square trinomial.

Factor each trinomial completely. It may be necessary to factor out the greatest common factor first.

225. $y^2 - 6y + 9$

226. $q^2 + 18q + 81$

227. $t^2 - 12t + 36$

228. $c^2 - 24c + 144$

229. $z^2 - z + \dfrac{1}{4}$

230. $9w^2 + 6w + 1$

231. $36q^2 - 60q + 25$

232. $49j^2 + 28j + 4$

233. $64p^4 + 48p^2q^2 + 9q^4$

234. $100p^2 - \dfrac{25}{2}pr + \dfrac{25}{64}r^2$

235. $r^2 - 2.6r + 1.69$

236. $-4x^2 + 8x - 4$

237. $-12a^2 - 60ab - 75b^2$

238. $18x^2 - 48xy + 32y^2$

Objective 3   Factor the difference of two cubes.

Factor completely.

239. $a^3 - 1$

240. $b^3 - 27$

241. $z^3 - 8$

242. $c^3 - 216$

243. $d^3 - 125$

244. $m^3 - 64$

245. $8y^3 - 1$

246. $125z^3 - 8$

247. $27r^3 - 1$

248. $343t^3 - 1$

249. $c^9 - d^6$

250. $t^6 - r^3$

251. $q^3 - 8p^3$

252. $125m^3 - 8p^3$

253. $8z^3 - 27y^3$

254. $64x^3 - 27y^3$

255. $8m^3 - \dfrac{1}{27}$

256. $p^3 - \dfrac{1}{8}$

257. $125c^3 - 216d^3$

258. $1000a^3 - 27b^3$

Objective 4   Factor the sum of two cubes.

Factor completely.

259. $x^3 + 1$

260. $y^3 + 27$

261. $z^3 + 8$

262. $m^3 + 64$

263. $n^3 + 216$

264. $p^3 + 125$

265. $27a^3 + 1$

266. $8b^3 + 1$

267. $125c^3 + 8$

268. $343d^3 + 27$

269. $t^6 + 1$

270. $s^6 + 125t^3$

271. $w^6 + 8z^3$

272. $64x^3 + 27y^3$

273. $27a^3 + 8b^3$

274. $125c^6 + 1$

275. $216m^3 + 125p^3$

276. $y^6 + z^9$

277. $z^3 + \dfrac{1}{8}$

278. $27t^3 + \dfrac{1}{64}$

Mixed Exercises

Factor completely. If the polynomial cannot be factored, write *prime*.

279. $a^2 - 16a + 64$

280. $100k^2 - m^2$

281. $9p^2 - 121$

282. $4f^2 + 4f + 1$

283. $16r^2 + 49$

284. $25h^2 + 20h + 4$

285. $8x^3 + 125$

286. $8y^3 + 27z^3$

287. $a^4 - 18a^2b + 81b^2$

288. $z^2 - 1.6z + .64$

289. $4x^4 - 32x$

290. $r^4 - 625$

291. $j^2 + \dfrac{2}{7}j + \dfrac{1}{49}$

292. $\dfrac{1}{8}x^3 + 27$

293. $x^6 - \dfrac{1}{125}$

294. $\dfrac{1}{9}x^2 + 4xy + 36y^2$

295. $x^6 - 64y^3$

296. $12m^2 + 75$

297. $x^9 - 1$

298. $64s^2 - 160s + 100$

299. $-2x^2 - 20x - 50$

300. $x^9 - y^8$

301. $125x^3 + 64y^6$

302. $12z^3 - 12$

303. $121t^4 - 144v^6$

304. $81p^4 - 16p^4$

Writing/Conceptual Exercises

Identify the monomial as *a perfect square, a perfect cube, both of these,* or *neither of these.*

305. $100r^8$

306. $27q^6$

307. $81r^3$

308. s^6t^{12}

309. $125p^5$

310. $64u^{18}v^{24}$

311. Give an example of a sum of two squares that *can* be factored. Explain how to find other examples of this type of binomial.

312. A student factors $x^2 + 25$ as $(x+5)(x+5)$ or $(x+5)^2$. Explain to the student why this is incorrect.

Section 5.5 Solving Quadratic Equations by Factoring

Objective 1 Solve quadratic equations by factoring.

Solve the equation and check the answers.

313. $(y+6)(2y-5)=0$

314. $(2c-1)(4c+3)=0$

315. $x^2 + 5x + 6 = 0$

316. $b^2 - 81 = 0$

317. $z^2 + 11z + 24 = 0$

318. $w^2 - 4w = 32$

319. $8r^2 = 72r$

320. $2y^2 + 15 = 11y$

321. $3y^2 + 7y = 6$

322. $4x^2 + 15x = 25$

323. $25x^2 + 4 = 20x$

324. $z^2 - 100 = 0$

325. $6p^2 + p = 15$

326. $15x^2 - 24x = 12$

327. $c(5c+17)=12$

328. $3x(x+3)=(x+2)^2 - 1$

Objective 2 Solve other equations by factoring.

Solve the equation.

329. $3x(x+6)(x+4)=0$

330. $x(2x^2 + 13x - 7)=0$

331. $z(4z^2 - 81)=0$

332. $z^3 - 36z = 0$

333. $b^3 = 121b$

334. $y^3 - 5y^2 + 4y = 0$

335. $2m^3 + m^2 - 6m = 0$

336. $(p-2)(p^2 - 49)=0$

337. $z^4 + 9z^3 - 10z^2 = 0$

338. $3z^3 + 11z^2 = 4z$

339. $(x+3)(x^2-7x+10)=0$

340. $(y^2-5y+6)(y^2-49)=0$

341. $x^3=-x^2+20x$

342. $(y-3)(2y^2+3y-2)=0$

343. $\left(x-\dfrac{1}{2}\right)(2x^2-x-15)=0$

344. $\left(y-\dfrac{2}{3}\right)\left(y^2-\dfrac{1}{4}\right)=0$

Mixed Exercises

Solve the equation.

345. $(x+8)(4x+9)=0$

346. $x(x+5)(x-7)=0$

347. $x^2+3x-10=0$

348. $x^2=5x$

349. $2x(x^2+2x-15)=0$

350. $3x^2=16x+12$

351. $4p^2-25=0$

352. $a^3=169a$

353. $(x-1)(x^2-25)=0$

354. $3y^2+8y=35$

355. $v^5-6v^4+8v^3=0$

356. $b(3b+11)=4$

357. $x^2=14x-48$

358. $(z-2)(z^3-z)=0$

359. $2x^2-7x=15$

360. $8x^2=0$

361. $x^2-5x=36$

362. $2(x+6)(x-5)=0$

363. $x^3=4x$

364. $4x^2-81=0$

365. $b^2(b+4)(b-9)=0$

366. $5x^2-65x+210=0$

367. $(x+6)(x-3)(x+5)=0$

368. $(x^2-4)(x^2-9)=0$

369. $x^2(x^2-81)(x^2-1)=0$

370. $8x\left(x+\dfrac{1}{2}\right)\left(x-\dfrac{1}{3}\right)=0$

Writing/Conceptual Exercises

Write a quadratic equation in standard form having the given solutions. Use x as the variable.

371. 3 and -7

372. 6 and $\dfrac{2}{3}$

373. $-\dfrac{1}{2}$ and $-\dfrac{3}{5}$

374. $\dfrac{5}{4}$ and $-\dfrac{3}{2}$

375. 0 and -5

376. 7 (only solution)

377. Explain why the solutions of the equation $(x+1)(x-1)=8$ are not found by solving the equations

$$x+1=8 \quad \text{and} \quad x-1=8.$$

378. What is wrong with the following solution?

$$6x^2 = 2x$$
$$3x = 1 \qquad \textit{Divide both sides by } x$$
$$x = \frac{1}{3}$$

Show the correct way to solve this equation.

Section 5.6   Applications of Quadratic Equations

Objective 1   Solve problems about geometric figures.

Solve the problem.

379. The length of a photograph is 3 centimeters more than the width. The area is 108 square centimeters. Find the length and width of the photograph.

380. The length of a rectangle is twice its width. If the width were increased by 3 and the length remained the same, the resulting rectangle would have an area of 140 square inches. Find the dimensions of the original rectangle.

381. The area of a rectangular rug is 88 square feet. Its width is 3 feet less than its length. Find the length and width of the rug.

382. Two rectangles with different dimensions have the same area. The length of the first rectangle is three times its width. The length of the second rectangle is 4 meters more than the width of the first rectangle, and its width is 2 meters more than the width of the first rectangle. Find the lengths and widths of the two rectangles.

383. Each side of one square is 1 meter less than twice the length of each side of a second square. If the difference between the areas of the two squares is 16 square meters, find the lengths of the sides of the two squares.

384. The area of a triangular sticker is 45 square centimeters. The base is 3 centimeters more than twice the height. Find the base and height of the sticker.

385. A rectangular bookmark is 4 centimeters longer than it is wide. Its area is numerically 4 less than its perimeter. Find the length and width of the bookmark.

386. A calculator is twice as long as it is wide. Find the length and width of the calculator in centimeters if its area is numerically 56 more than its perimeter.

387. The volume of a box is 432 cubic feet. If the width of the box is 6 feet and the length is 1 foot more than the height, find the height of the box.

388. A carpenter is building a box which will have a volume of 60 cubic meters. The height of the box will be 4 meters, and the length will be 2 meters more than the width. Find the width of the box.

Objective 2   Solve problems using the Pythagorean formula.

Solve the problem.

389. The hypotenuse of a right triangle is 8 inches longer than the shorter leg. The longer leg is 4 inches longer than the shorter leg. Find the length of the shorter leg.

390. The hypotenuse of a right triangle is 2 meters shorter than three times the shorter leg, and the longer leg is 2 meters longer than twice the shorter leg. Find the length of the hypotenuse.

391. A lot has the shape of a right triangle with one leg 3 meters longer than three times the length of the other leg. The hypotenuse is 3 meters less than four times the length of the shorter leg. Find the dimensions of the lot.

392. A train and a car leave a station at the same time, the train traveling due north and the car traveling due west. When they are 100 miles apart, the train has traveled 20 miles farther than the car. Find the distance each has traveled.

393. A ladder is leaning against the side of a building so that the distance from the bottom of the ladder to the building is 7 feet less than the distance up the building to the top of the ladder. If the ladder is 3 feet longer than twice the distance from the bottom of the ladder to the building. How long is the ladder?

394. Roxanne is standing directly beneath a kite attached to a string, which Ralph is holding, with his hand touching the ground. The height of the kite at the instant is 15 feet less than the distance between Roxanne and Ralph. The length of the kite string is 15 feet more than that distance. Find the length of the kite string.

395. A 15-foot ladder is leaning against a building. The distance from the bottom of the ladder to the building is 3 feet less than the distance from the top of the ladder to the ground. How far is the bottom of the ladder from the building?

396. A lot is in the shape of a right triangle. The shorter leg measures 60 meters. The hypotenuse measures 60 meters less than twice the longer leg. Find the dimensions of the lot.

397. Two cars left an intersection at the same time. One traveled south. The other traveled 14 miles farther, but to the east. How far apart were they when the distance between them was 2 miles more than the distance traveled east?

398. A ladder is leaning against a building. The distance from the bottom of the ladder to the building is 8 feet less than the length of the ladder. How high up the side of the building is the top of the ladder if that distance is 4 feet less than the length of the ladder?

Objective 3   Solve problems using quadratic models.

Use the quadratic model to answer the questions.

From the text, the equation $y = -.04x^2 + .93x + 21$ was developed to model fuel economy trends within the automobile industry starting in 1978. Suppose that an automotive engineer is revising the model to project fuel economy trends into the 21st century. She develops the following formula:

$$y = -.02x^2 + 1.19x + 27,$$

and determines that x is coded so that $x = 0$ represents 1999.

399. Calculate the expected miles per gallon in 2005. Round your answer to the nearest tenth.

400. Calculate the expected miles per gallon in 2049. Round your answer to the nearest tenth.

401. Calculate the expected miles per gallon in 2099. Round your answer to the nearest tenth.

402. Would you buy a car in 2099 based on the answer to #401? Why or why not?

Mixed Exercises

Solve the problem.

403. The length of a rectangular picture is 5 centimeters more than the width. The area is 66 square centimeters. Find the length and width of the picture.

404. The length of a rectangular card is twice its width. If the width were increased by 2 inches while the length remained the same, the resulting rectangle would have an area of 48 square inches. Find the dimensions of the original card.

405. The length of the shorter leg of a right triangle is tripled and 4 inches is added to the result, giving the length of the hypotenuse. The longer leg is 10 inches longer than twice the shorter leg. Find the length of the shorter leg of the triangle.

406. Thuy wishes to build a box to hold his tools. The box is to be 4 feet high, and the width of the box is to be 1 foot less than the length. The volume of the box will be 120 cubic feet. Find the length and width of the box.

407. The hypotenuse of a right triangle is 1 foot longer than twice the shorter leg. The longer leg is 1 foot shorter than twice the shorter leg. Find the length of the shorter leg of the triangle.

408. Judy and Sheri started walking from the same corner, with Judy walking west and Sheri walking north. When they were 15 kilometers apart, Sheri had walked 3 miles farther than Judy. Find the distance each of them had walked.

409. The length of a rectangular label is three times its width. If the length were decreased by 1 while the width stayed the same, the area of the new label would be 44 square centimeters. Find the length and width of the original label.

410. The length of the floor of a rectangular closet is 1 foot more than the width. The area of the floor is 56 square feet. Find the length and width of the floor.

411. A bicyclist heading east and a motorist traveling south left an intersection at the same time. When the motorcyclist had gone 17 miles farther than the bicyclist, the distance between them was 1 mile more than the distance traveled by the motorist. How far apart were they then?

412. The sides of one square have a length 2 meters more than the sides of another square. If the area of the larger square is subtracted from three times the area of the smaller square, the answer is 12 square meters. Find the lengths of the sides of each square.

Note to the instructor: The remaining exercises in this section cover types of applications that are included in the Section 5.6 Exercises but are not covered by the objectives or examples for this section in the textbook.

413. If a ball is thrown upward from ground level with an initial velocity of 80 feet per second, its height h (in feet) t seconds later is given by the equation

$$h = -16t^2 + 80t.$$

(a) After how many seconds is the height 100 feet?
(b) After how many seconds will the ball hit the ground?

414. An object is propelled upward from a height of 16 feet with an initial velocity of 48 feet per second, its height h (in feet) t seconds later is given by the equation

$$h = -16t^2 + 48t + 16.$$

(a) After how many seconds is the height 52 feet?
(b) After how many seconds is the height 48 feet?

415. Find two consecutive integers such that the square of their sum is 25.

416. The product of two consecutive integers is 8 more than twice their sum. Find the integers.

417. Find all possible pairs of consecutive odd integers whose sum is equal to their product decreased by 7.

418. The square of the sum of two consecutive even integers is 52 more than twice their product. Find the integers.

419. The product of two consecutive even integers is 60 more than twice the larger. Find the integers.

420. Find three consecutive odd integers such that four times the sum of all three equals 13 more than the product of the smaller two.

Section 5.7 Solving Quadratic Inequalities

Objective 1 Solve quadratic inequalities and graph their solutions.

Solve the inequality and graph the solutions.

421. $(x-4)(x+1)>0$

422. $(m+1)(m-2)\le 0$

423. $(g+5)(g+4)\ge 0$

424. $(a-6)(a+1)<0$

425. $(r+2)(r+5)<0$

426. $(q-12)(q-1)\ge 0$

427. $(7r+1)(r-1)\le 0$

428. $x(x+6)>0$

429. $2b^2-11b+5>0$

430. $6p^2+5p>6$

431. $r^2-2r-8\le 0$

432. $4k^2-9<0$

433. $2m^2+5m\ge 12$

434. $t^2>81$

435. $x^2\le 36$

436. $k^2-7k+12<0$

437. $2g^2-g>15$

438. $z^2+3z+2\le 0$

439. $r^2+3r-10>0$

440. $3x^2+6x-9\le 0$

CHAPTER 6   RATIONAL EXPRESSIONS

Section 6.1   The Fundamental Property of Rational Expressions

Objective 1   Find the values for which a rational expression is undefined.

Find any values for which the expression is undefined.

1. $\dfrac{11}{2x}$

2. $\dfrac{1}{x-2}$

3. $\dfrac{3x^2}{x+3}$

4. $\dfrac{x-7}{7x^2-49x}$

5. $\dfrac{p+4}{p^2-8p+15}$

6. $\dfrac{9x^2}{x^2+1}$

7. $\dfrac{x}{x^2-36}$

8. $\dfrac{z-4}{z^2+16}$

9. $\dfrac{z}{z^2-4z+4}$

10. $\dfrac{p}{p^2+5p+6}$

11. $\dfrac{x+5}{x^3+3x^2-10x}$

12. $\dfrac{y+6}{y^4-16}$

Objective 2   Find the numerical value of a rational expression.

Find the numerical value of the expression when (a) $x=3$ and (b) $x=-2$.

13. $\dfrac{5x^2-x}{3x}$

14. $\dfrac{-3x+1}{x+1}$

15. $\dfrac{(-5-x)^2}{4x-2}$

16. $\dfrac{2x+8}{x^2-3x+2}$

17. $\dfrac{4x-5}{8+x}$

18. $\dfrac{2x^2}{6x-3}$

19. $\dfrac{2x}{(3x-1)^2}$

20. $\dfrac{x}{x^2-5x+7}$

21. $\dfrac{4x^2}{x^2-x-6}$

22. $\dfrac{x^3}{x^2-1}$

23. $\dfrac{4x-3}{2+x-x^2}$

24. $\dfrac{-4x^2}{2x^2-x+3}$

Note: In the remaining exercises in this chapter, assume that no values of a variable make any denominator equal to zero.

Objective 3   Write rational expressions in lowest terms.

Write the rational expression in lowest terms.

25. $\dfrac{4k}{12k^2}$

26. $\dfrac{6x^2 y^3}{-18x^3 y^4}$

27. $\dfrac{-21ab^4 c^3}{-18ab^2 c^5}$

28. $\dfrac{3p+6}{2p+4}$

29. $\dfrac{z^2 - 4}{z+2}$

30. $\dfrac{b^2 - 6b + 9}{b^2 - 4b + 3}$

31. $\dfrac{4a^2 + 4a + 1}{4a^2 - 1}$

32. $\dfrac{x^2 - x - 6}{x^2 + x - 12}$

33. $\dfrac{y^2 - 1}{1 - y}$

34. $\dfrac{1 - p}{p^2 - 1}$

35. $\dfrac{12k^3 - 12k}{3k^2 + 3k}$

36. $\dfrac{3q^2 - 4qz + z^2}{q^2 - 2qz + z^2}$

37. $\dfrac{6r^2 + rs - 5s^2}{r^2 + 2rs + s^2}$

38. $\dfrac{5x^2 + 3xy - 2y^2}{x^2 + 2xy + y^2}$

39. $\dfrac{vw - 5v + 3w - 15}{vw - 5v - 2w + 10}$

40. $\dfrac{2r^3 + 2rs^2 + 3r^2 s + 3s^3}{4r^3 + 4rs^2 - r^2 s - s^3}$

Objective 4   Recognize equivalent forms of rational expressions.

Write an equivalent form for each of the following rational expressions.

41. $\dfrac{2x+3}{3x-2}$

42. $\dfrac{4a-5}{4-5a}$

43. $-\dfrac{6-p}{d+9}$

44. $-\dfrac{9y+2}{4y-2}$

45. $-\dfrac{6-11y}{6+11z}$

46. $\dfrac{8u-2v}{-(3-2z)}$

47. $\dfrac{5y-3z}{-(3y-5z)}$

48. $\dfrac{-(y-7p)}{4y+8p}$

49. $\dfrac{-(2w-3)}{3-2w}$

Mixed Exercises

Find any values for which the expression is undefined.

50. $\dfrac{-15}{x+5}$

51. $\dfrac{3m+2}{3m^2-m-10}$

52. $\dfrac{y+6}{9y^2+6y+1}$

53. $\dfrac{a-5}{a^2-25}$

54. $\dfrac{2+y}{y^3-36y}$

55. $\dfrac{x}{(x-3)(x^2+4x+4)}$

56. $\dfrac{b-16}{b^2+2b-15}$

57. $\dfrac{y^2+4}{4y^2-25}$

Find the numerical value of each expression when (a) $x=3$ and (b) $x=-2$.

58. $\dfrac{4x^2-x}{2x}$

59. $\dfrac{3x+1}{x+2}$

60. $\dfrac{4x-1}{x^2+2x+5}$

61. $\dfrac{3x+2}{1+x-x^2}$

62. $\dfrac{x+4}{2x-3}$

63. $\dfrac{-3x^2}{x^3-1}$

64. $\dfrac{x^3+8}{x+4}$

65. $-\dfrac{2x}{x^2+1}$

Write the rational expression in lowest terms.

66. $\dfrac{21b^2c^3}{28b^3c^4}$

67. $\dfrac{-66q^2rt^4}{-44qr^3t}$

68. $\dfrac{16r^3-8r^2}{8r^4-4r^3}$

69. $\dfrac{9x^2-9x-108}{2x-8}$

70. $\dfrac{2x^2-3x+2xy-3y}{2x^2-8xy-3x+12y}$

71. $\dfrac{r^2-4r}{4r-r^2}$

72. $\dfrac{2a^2 + ab - b^2}{a^2 + 2ab + b^2}$

73. $\dfrac{z^3 - 27}{z^2 - 9}$

74. $\dfrac{8t^2 + 6t - 9}{16t^2 - 9}$

75. $\dfrac{ab^2 + a - cb^2 - c}{da - dc}$

76. $\dfrac{4x^2 - 28xy + 49y^2}{6x^2 - 17xy - 14y^2}$

77. $\dfrac{m^3 + 1}{m^3 - m^2 + m}$

Writing/Conceptual Exercises

78. Which of the following expressions is equivalent to $\dfrac{3x+6}{x+2}$?

    (a) $2x+4$ (b) 6 (c) 3 (d) $2x+3$

79. Which of the following expressions is equal to -1?

    (a) $\dfrac{x+5}{x-5}$ (b) $\dfrac{x-5}{5-x}$ (c) $\dfrac{x-5}{x+5}$ (d) $\dfrac{x-5}{x-5}$

80. A student was asked in algebra class to find the value of the expression

$$\frac{x^2 - 9}{x - 3}$$

when $x = 3$. She said that the numerator and denominator would both be 0, so the answer is 0. How would you respond?

81. A student who was asked to write the fraction

$$\frac{5x + 10}{10x + 20}$$

in lowest terms, gave 1 as her answer. When her instructor asked her to explain her answer, she said, "That's easy. You just cancel. $5x$ divided by $10x$ is $\dfrac{1}{2}$, and 10 divided by 20 is $\dfrac{1}{2}$, so the answer is $\dfrac{1}{2} + \dfrac{1}{2} = 1$." How should the instructor respond? Show the correct way to write the given expression in lowest terms.

Section 6.2 Multiplication and Division of Rational Expressions

Objective 1 Multiply rational expressions.

Multiply. Write answers in lowest terms.

82. $\dfrac{8m^2n}{3} \cdot \dfrac{21}{mn^2}$

83. $\dfrac{2b+8}{b} \cdot \dfrac{2}{3b+12}$

84. $\dfrac{9-3z}{4} \cdot \dfrac{8}{3z-9}$

85. $\dfrac{x^2+3x+2}{x^2+5x+4} \cdot \dfrac{x^2+10x+24}{x^2+5x+6}$

86. $\dfrac{a^2-a-2}{a^2-4} \cdot \dfrac{a^2-1}{a^2-3a+2}$

87. $\dfrac{6m^2-5m-6}{6m^2+5m-6} \cdot \dfrac{12m^2-17m+6}{12m^2-m-6}$

88. $\dfrac{2x^2-5x-12}{x^2-10x+24} \cdot \dfrac{x^2-9x+18}{4x^2-9}$

89. $\dfrac{y^2-6y+9}{y^2-5y+6} \cdot \dfrac{y^2-4y+4}{y^2-4}$

90. $\dfrac{6x+42}{12} \cdot \dfrac{10}{7x+49}$

91. $\dfrac{2r+2p}{8z} \cdot \dfrac{72z^3}{r^2+rp}$

Objective 2 Divide rational expressions.

Divide. Write answers in lowest terms.

92. $\dfrac{4a^2}{8a} \div \dfrac{16a^4}{3a^3}$

93. $\dfrac{6z^3}{9z} \div \dfrac{z^2}{12z}$

94. $\dfrac{b-3}{16} \div \dfrac{3-b}{8}$

95. $\dfrac{2x+2y}{8z} \div \dfrac{x^2(x+y)^2}{72}$

96. $\dfrac{4m+12}{2m-10} \div \dfrac{m^2-9}{m^2-25}$

97. $\dfrac{4a(a+3)}{2a+1} \div \dfrac{a^2(a+3)}{4a^2-1}$

98. $\dfrac{m^2-n^2}{m^2-1} \div \dfrac{m^2+2mn+n^2}{m^2+m}$

99. $\dfrac{ab-a^2}{a^2-1} \div \dfrac{a-b}{a^2+2a+1}$

100. $\dfrac{2k^2+3k-2}{6k^2-7k+2} \div \dfrac{k^2+k-2}{4k^2-5k+1}$

101. $\dfrac{2z^2-5z-12}{z^2-10z+24} \div \dfrac{4z^2-9}{z^2-9z+18}$

Mixed Exercises

Multiply or divide as indicated. Write answers in lowest terms.

102. $\dfrac{16q^3r}{3} \cdot \dfrac{27}{20qr}$

103. $\dfrac{p^3q^4}{pq^5} \cdot \dfrac{q^2}{p^3}$

104. $\dfrac{26x^3}{27} \div \dfrac{52x}{9}$

105. $\dfrac{p^8}{q^3} \div \dfrac{p^5}{q^2}$

106. $\dfrac{r-s}{14} \cdot \dfrac{10}{s-r}$

107. $\dfrac{6p-27}{7} \div \dfrac{18p-81}{5}$

108. $\dfrac{5r^2+5r}{3r^2-4r} \cdot \dfrac{r^2-r-6}{3r^2+2r-8}$

109. $\dfrac{x^2+4x-21}{x^2-49} \cdot \dfrac{x^2-12x+35}{x^2+6x-27}$

110. $\dfrac{2y^2-y-1}{4y^2-4y-3} \cdot \dfrac{4y^2+12y+9}{2y^2+y-3}$

111. $\dfrac{3k^2+k-2}{2k^2+k-1} \div \dfrac{3k^2+4k-4}{2k^2+7k-4}$

112. $\dfrac{4x+12}{2x-10} \cdot \dfrac{x^2-x-20}{x^2-9}$

113. $\dfrac{m^2-16}{m+3} \cdot \dfrac{m^2-9}{m-4}$

114. $\dfrac{9a-18}{6a+12} \cdot \dfrac{3a+6}{30-15a}$

115. $\dfrac{x^2-4}{x^2-1} \cdot \dfrac{x^2-x}{2x^2+4x}$

116. $\dfrac{y^2+yz-12z^2}{y^2-yz-20z^2} \div \dfrac{y^2-2yz-3z^2}{y^2+yz-30z^2}$

117. $\dfrac{4(b-3)(b+2)}{b^2+3b+2} \div \dfrac{b^2-6b+9}{b^2+4b+4}$

118. $\dfrac{2a^2-3a-5}{3a^2+2a-1} \div \dfrac{2a^2+a-15}{3a^2+5a-2}$

119. $\dfrac{2y^2+21y-11}{8y^2-2} \div \dfrac{y^2+12y+11}{4y^2+18y+8}$

120. $\dfrac{24r^2s^3}{5} \cdot \dfrac{25s^4}{6rs^2}$

121. $\dfrac{3(m-n)}{m} \cdot \dfrac{n}{2(m-n)}$

122. $\dfrac{y^2+7y+10}{3y+6} \cdot \dfrac{4y-4}{y^2+2y-15}$

123. $\dfrac{2(a-1)}{b} \cdot \dfrac{2b}{5(a-1)}$

124. $\dfrac{p+q}{2} \div \dfrac{(p+q)^2}{14}$

125. $\dfrac{5n+25}{10} \div \dfrac{36(n+5)^2}{12}$

126. $\dfrac{6(r+2)}{3(r-1)^2} \div \dfrac{(r+2)^2}{9(r-1)}$

127. $\dfrac{x^2+4x+3}{x+3} \div \dfrac{x^2-4x-5}{x-3}$

128. $\dfrac{z^4-z^2}{z^5-4z^3} \div \dfrac{5z-5}{8z-16}$

129. $\dfrac{2b^2+5b+3}{b^2-b-2} \cdot \dfrac{b^2+9b+20}{2b^2+11b+12}$

130. $\left(\dfrac{12x^2+11x+2}{3x^2+5x+2} \cdot \dfrac{3x^2+2x-1}{12x^2-7x+1}\right) \div \dfrac{4x+1}{3x+2}$

131. $\left(\dfrac{3b^2-2b-5}{b^2-4b+4} \div \dfrac{6b^2-7b-5}{2b^2-3b-2}\right) \div \dfrac{b-1}{b^2-1}$

Section 6.3   The Least Common Denominator

Objective 1   Find the least common denominator for a group of fractions.

Find the LCD for the fractions in the list.

132. $\dfrac{5}{18}, \dfrac{7}{15}$

133. $\dfrac{1}{8}, \dfrac{5}{6}$

134. $\dfrac{7}{12}, \dfrac{11}{40}, \dfrac{3}{20}$

135. $\dfrac{9}{14}, \dfrac{13}{35}, \dfrac{19}{25}$

136. $\dfrac{7}{16y}, \dfrac{9}{10y}$

137. $\dfrac{13}{32b^4}, \dfrac{17}{24b^2}$

138. $\dfrac{5}{21r^3}, \dfrac{8}{3r-15}$

139. $\dfrac{15}{4t-18}, \dfrac{22}{6t-27}$

140. $\dfrac{6}{x-y}, \dfrac{5}{y-x}$

141. $\dfrac{1}{a^2-b^2}, \dfrac{10}{b^2-a^2}$

142. $\dfrac{3}{2a^2-10a}, \dfrac{7}{a^2-6a+5}$

143. $\dfrac{v-1}{3v^4-18v^3}, \dfrac{v+1}{v^2-3v-18}$

Objective 2   Rewrite rational expressions with given denominators.

Write the rational expression on the left with the indicated denominator.

144. $\dfrac{5}{6} = \dfrac{}{42}$

145. $\dfrac{9}{r} = \dfrac{}{4r}$

146. $\dfrac{5m}{8n} = \dfrac{}{32n^4}$

147. $\dfrac{-2k}{9k-18} = \dfrac{}{18k-36}$

148. $\dfrac{-5y}{6y+18} = \dfrac{}{6(y+3)^2}$

149. $\dfrac{2z}{5z-5} = \dfrac{}{5z^2+5z-10}$

150. $\dfrac{7}{r^2+7r} = \dfrac{}{5r(r+7)(r+1)}$

151. $\dfrac{7}{y^2-9} = \dfrac{}{(y+3)^2(y-3)}$

152. $\dfrac{9}{7p-42} = \dfrac{}{21p^3-126p^2}$

153. $\dfrac{t+1}{t^2-3t+9} = \dfrac{}{t^3+27}$

Mixed Exercises

Find the LCD for the list of fractions.

154. $\dfrac{2}{14}, \dfrac{4}{21}, \dfrac{5}{6}$

155. $\dfrac{3}{40}, \dfrac{7}{75}, \dfrac{11}{60}$

156. $\dfrac{1}{3x}, \dfrac{3}{5x}$

157. $\dfrac{5}{9a^2}, \dfrac{1}{6a^3}$

158. $\dfrac{11}{w^3-25w}, \dfrac{14w}{w^2+4w-5}$

159. $\dfrac{5t}{t^2-3t-4}, \dfrac{7t}{t^2-2t-8}$

160. $\dfrac{4z}{2z^2+7z-4}, \dfrac{3}{2z^2-7z+3}$

161. $\dfrac{m+4}{m^3-5m^2}, \dfrac{5-m}{m^2-2m-15}$

162. $\dfrac{8z}{z^4+z^3-6z^2}, \dfrac{3z+1}{z^3-9z}$

163. $\dfrac{p-1}{p^2+3p-18}, \dfrac{p+1}{p^2-6p+9}$

164. $\dfrac{11}{2q^2-q-10}, \dfrac{21}{2q^2-13q+20}$

165. $\dfrac{17r}{9r^2-6r-8}, \dfrac{-13r}{9r^2-9r-4}$

Write the rational expression on the left with the indicated denominator.

166. $\dfrac{3}{8} = \dfrac{}{96}$

167. $\dfrac{15}{8w} = \dfrac{}{72w^2}$

168. $\dfrac{7}{3c-15} = \dfrac{}{18c-90}$

169. $\dfrac{11}{6q-36} = \dfrac{}{42q^2-252q}$

170. $\dfrac{2p}{p-4} - \dfrac{}{p^2-16}$

171. $\dfrac{r+1}{r^2+4r} = \dfrac{}{r^3+r^2-12r}$

172. $\dfrac{12}{k^2+3k} = \dfrac{}{k^3+7k^2+12k}$

173. $\dfrac{3x+2}{x^2-4} = \dfrac{}{2x^3-8x}$

174. $\dfrac{5}{r-7} = \dfrac{}{7-r}$

175. $\dfrac{3p}{2p+7} = \dfrac{}{10p^2+23p-42}$

176. $\dfrac{18}{w^2+9w} = \dfrac{}{w^3+5w^2-36w}$

177. $\dfrac{1}{x+y} = \dfrac{}{x^3y+2x^2y^2+xy^3}$

178. $\dfrac{2k}{k^2+2k-35} = \dfrac{}{(k+7)(k-3)(k-5)}$

179. $\dfrac{36z}{z^2-z-6} = \dfrac{}{(z-3)(z+2)(z+1)}$

180. $\dfrac{4x}{x^2-8x+15} = \dfrac{}{(x-5)(x-3)(x+2)}$

181. $\dfrac{x+y}{x^2+xy+y^2} = \dfrac{}{x^3-y^3}$

Writing/Conceptual Exercises

182. Suppose that you were asked to work the problem

$$\frac{3}{9} + \frac{5}{12},$$

and choose $9 \cdot 12 = 108$ as your common denominator rather than 36. Would you still get the correct answer? Explain.

183. Given the problem

$$\frac{3}{r-s} - \frac{6}{s-r},$$

one student chose $r-s$ as the LCD and the other chose $s-r$. Which one is correct?

184. Suppose that $(3r-5)(2r-1)$ is the least common denominator for two fractions. Is $(5-3r)(1-2r)$ also acceptable as the least common denominator? Explain.

185. If a fraction with denominator $x-5$ is to be rewritten as an equivalent fraction with denominator $x^2 - x - 20$, by what expression must the original fraction be multiplied in both the numerator and denominator?

Section 6.4   Addition and Subtraction of Rational Expressions

Objective 1   Add rational expressions having the same denominator.

Find the sum. Write the answer in the lowest terms.

186. $\dfrac{5x}{x-1} + \dfrac{x}{x-1}$

187. $\dfrac{3}{k-7} + \dfrac{k}{k-7}$

188. $\dfrac{x}{x^2-9} + \dfrac{4x}{x^2-9}$

189. $\dfrac{3t-5}{2t+1} + \dfrac{6t+7}{2t+1}$

190. $\dfrac{b}{b^2-1} + \dfrac{1}{b^2-1}$

191. $\dfrac{2x+1}{x^2+4x+3} + \dfrac{2-x}{x^2+4x+3}$

192. $\dfrac{3}{y^2+y-6} + \dfrac{y}{y^2+y-6}$

193. $\dfrac{4m}{m^2+3m+2} + \dfrac{8}{m^2+3m+2}$

194. $\dfrac{6x}{(x+1)^2} + \dfrac{6}{(x+1)^2}$

195. $\dfrac{y-5}{2y^2-5y-3} + \dfrac{y+6}{2y^2-5y-3}$

196. $\dfrac{2x}{4x^2-9y^2} + \dfrac{3y}{4x^2-9y^2}$

197. $\dfrac{6v}{6v^2+13vw+6w^2} + \dfrac{4w}{6v^2+13vw+6w^2}$

Objective 2   Add rational expressions having different denominations.

Find the sum. Write the answer in lowest terms.

198. $\dfrac{x}{4} + \dfrac{2}{3}$

199. $\dfrac{m}{3p} + \dfrac{2}{7p}$

200. $\dfrac{4}{a^2-1}+\dfrac{6}{a+1}$

201. $\dfrac{-9}{h+1}+\dfrac{h}{1-h^2}$

202. $\dfrac{2y}{y^2+5y+6}+\dfrac{y+1}{y^2+2y-3}$

203. $\dfrac{2m}{m^2-5m+4}+\dfrac{3}{m^2-1}$

204. $\dfrac{m+2}{m}+\dfrac{m+3}{4m}$

205. $\dfrac{6}{x-5}+\dfrac{2}{x+5}$

206. $\dfrac{1}{n^2-4}+\dfrac{3}{2n+4}$

207. $\dfrac{2s+1}{3s^2+10s-8}+\dfrac{3s+5}{2s^2+5s-12}$

208. $\dfrac{4p-1}{2p^2+5p-3}+\dfrac{p+3}{6p^2+p-2}$

209. $\dfrac{z}{z^2-1}+\dfrac{z-1}{z^2+2z+1}$

210. $\dfrac{3}{4x^2-1}+\dfrac{3}{2x^2+5x+2}$

211. $\dfrac{3}{2x^2+9x+9}+\dfrac{-1}{2x^2+7x+6}$

Objective 3   Subtract rational expressions.

Find the difference. Write the answer in lowest terms.

212. $\dfrac{6p}{p-3}-\dfrac{18}{p-3}$

213. $\dfrac{z^2}{z-y}-\dfrac{y^2}{z-y}$

214. $\dfrac{x}{x^2-7x+10}-\dfrac{2}{x^2-7x+10}$

215. $\dfrac{3x}{9x^2-25y^2}-\dfrac{5y}{9x^2-25y^2}$

216. $\dfrac{7}{5}-\dfrac{x}{4}$

217. $\dfrac{4+2k}{5}-\dfrac{2+k}{10}$

218. $\dfrac{5-4s}{8}-\dfrac{2-3s}{6}$

219. $\dfrac{1}{m^2-9}-\dfrac{1}{m+3}$

220. $\dfrac{3}{x-y}-\dfrac{5+y}{y-x}$

221. $\dfrac{-6}{x^2-4}-\dfrac{3}{2x+4}$

222. $\dfrac{6}{2q^2+13q+15}-\dfrac{5}{2q^2+9q-5}$

223. $\dfrac{m}{m^2-1}-\dfrac{1-m}{m^2+2m+1}$

224. $\dfrac{b-1}{b^2+2b-8}-\dfrac{2-3b}{b^2+3b-4}$

225. $\dfrac{1}{n^2-1}-\dfrac{1}{n^2+3n+2}$

Mixed Exercises

Add or subtract as indicated. Write the answer in lowest terms.

226. $\dfrac{x}{x^2+4x+4}+\dfrac{2}{x^2+4x+4}$

227. $\dfrac{5}{m^2-25}-\dfrac{m}{m^2-25}$

228. $\dfrac{2z}{3z+3}+\dfrac{5z}{2z+2}$

229. $\dfrac{a}{a^2-1}+\dfrac{a}{a+1}$

230. $\dfrac{3p+5}{3}-\dfrac{p+2}{6}$

231. $\dfrac{4z}{z^2+3z+2}+\dfrac{2z-1}{z^2+6z+5}$

232. $\dfrac{4}{2-m}-\dfrac{7}{m-2}$

233. $\dfrac{1}{z^2-4}-\dfrac{3}{z^2+5z+6}$

234. $\dfrac{3}{9b^2-16}+\dfrac{2}{3b^2+2b-8}$

235. $\dfrac{4}{2m^2+7m-4}+\dfrac{2}{2m^2-7m+3}$

236. $\dfrac{5r}{r+2s}-\dfrac{3s}{-r-2s}$

237. $\dfrac{5}{2c^2+11c+5}-\dfrac{3}{2c^2-7c-4}$

238. $\dfrac{2p}{p^2-p-12}+\dfrac{p+1}{p^2+p-6}$

239. $\dfrac{z+1}{z-1}-\dfrac{z-1}{z+1}$

240. $\dfrac{4x-1}{2x^2+5x-3}-\dfrac{x+3}{6x^2+x-2}$

241. $\dfrac{6z}{(z-1)^2}-\dfrac{2}{z^2-1}$

242. $\dfrac{5x}{6}-\dfrac{2x}{3}$

243. $\dfrac{3+5t}{8}-\dfrac{3t-2}{12}$

244. $\dfrac{3}{a-2}-\dfrac{6}{2-a}$

245. $\dfrac{4}{x^2-4}+\dfrac{3}{x^2+4x+4}$

246. $\dfrac{2b}{b+6}+\dfrac{3}{2b+12}$

247. $\dfrac{18}{x+2y}-\dfrac{7}{x-2y}$

248. $\dfrac{c+3d}{c^2+2cd+d^2}+\dfrac{c-d}{3d^2+4cd+c^2}$

249. $\dfrac{z+2}{z-2}-\dfrac{z-2}{z+2}$

Writing/Conceptual Exercises

250. What are two possible LCD's that could be used for the sum

$$\frac{8r}{r-5} + \frac{6r}{5+r}?$$

251. If one form of the correct answer to a sum or difference of rational expression is

$$\frac{5y}{y-2},$$

what would be an alternate form of the answer if the denominator is $2-y$?

252. Find the error in the following subtraction.

$$\frac{3}{x-3} - \frac{2}{x-2} = \frac{3(x-2)}{(x-3)(x-2)} - \frac{2(x-3)}{(x-2)(x-3)} \qquad (1)$$

$$= \frac{3x-6-2x-6}{(x-2)(x-3)} \qquad (2)$$

$$= \frac{x-12}{(x-2)(x-3)} \qquad (3)$$

Show how you would correct this error.

253. Which one of the following expressions is *not* equivalent to $\dfrac{3-x}{y-5}$?

(a) $\dfrac{x-3}{y-5}$ (b) $\dfrac{x-3}{5-y}$ (c) $-\dfrac{x-3}{y-5}$ (d) $-\dfrac{3-x}{5-y}$

Section 6.5   Complex Fractions

Objective 1   Simplify a complex fraction by writing it as a division problem (Method 1).

Simplify the complex fraction by writing it as a division problem.

254. $\dfrac{-\dfrac{4}{5}}{\dfrac{7}{10}}$

255. $\dfrac{\dfrac{3}{4}-\dfrac{1}{8}}{\dfrac{1}{2}+\dfrac{5}{8}}$

256. $\dfrac{\dfrac{3a^3b}{c^2}}{\dfrac{9ab^2}{c}}$

257. $\dfrac{\dfrac{r+s}{3}}{\dfrac{r^2-s^2}{6}}$

258. $\dfrac{3+\dfrac{1}{x}}{3-\dfrac{2}{x}}$

259. $\dfrac{5-\dfrac{4}{m}}{\dfrac{2}{m}+1}$

260. $\dfrac{\dfrac{p}{2}-\dfrac{1}{4}}{\dfrac{p}{3}+\dfrac{1}{2}}$

261. $\dfrac{\dfrac{3}{z}+1}{\dfrac{3+z}{2}}$

262. $\dfrac{6+\dfrac{2}{s}}{s+\dfrac{2}{3}}$

263. $\dfrac{\dfrac{3}{p}-p}{\dfrac{3-p^2}{9}}$

264. $\dfrac{\dfrac{r+1}{r}}{\dfrac{r+1}{s}}$

265. $\dfrac{\dfrac{25k^2-m^2}{4k}}{\dfrac{5k+m}{7k}}$

Objective 2   Simplify a complex fraction by multiplying by the least common denominator.

Simplify the complex fraction by multiplying by the LCD.

266. $\dfrac{\dfrac{7}{10}}{\dfrac{11}{15}}$

267. $\dfrac{\dfrac{1}{2}+\dfrac{2}{3}}{\dfrac{3}{5}-\dfrac{9}{10}}$

268. $\dfrac{\dfrac{a}{b}}{\dfrac{a^2}{b}}$

269. $\dfrac{\dfrac{2}{x}}{\dfrac{1-x}{x^2}}$

270. $\dfrac{x-y}{x-\dfrac{y^2}{x}}$

271. $\dfrac{r+\dfrac{1}{r}}{\dfrac{5}{r}+t}$

272. $\dfrac{\dfrac{x-5}{x+5}}{\dfrac{x}{x-5}}$

273. $\dfrac{s+\dfrac{1}{s}}{\dfrac{1}{s}-s}$

274. $\dfrac{\dfrac{15}{10k+10}}{\dfrac{5}{3k+3}}$

275. $\dfrac{\dfrac{1}{h}-3}{\dfrac{1}{2}+h}$

276. $\dfrac{\dfrac{2}{x}+\dfrac{3}{y}}{\dfrac{8}{y}-\dfrac{2}{x}}$

277. $\dfrac{\dfrac{a}{a^2-b^2}+\dfrac{b}{a+b}}{\dfrac{a}{a-b}}$

Mixed Exercises

Simplify the complex fraction. Use either method.

278. $\dfrac{\dfrac{x^2}{y}}{\dfrac{x}{y^2}}$

279. $\dfrac{\dfrac{6}{m}}{\dfrac{1+m}{4m^3}}$

280. $\dfrac{\dfrac{20}{m+1}}{\dfrac{15}{3m+3}}$

281. $\dfrac{\dfrac{a+3}{a}}{\dfrac{1}{a}+\dfrac{1}{5}}$

282. $\dfrac{z + \dfrac{1}{3}}{\dfrac{6z+2}{3z}}$

283. $\dfrac{\dfrac{2}{z-1} + 2}{\dfrac{2}{1-z} - 2}$

284. $\dfrac{\dfrac{1}{y} + y}{\dfrac{y^2 + 1}{4}}$

285. $\dfrac{\dfrac{5}{w}}{\dfrac{1-w}{w^2}}$

286. $\dfrac{\dfrac{1}{v+2} - 3}{\dfrac{1}{v+2} + 3}$

287. $\dfrac{\dfrac{1}{n+5} - 1}{\dfrac{1}{n+5} + 1}$

288. $\dfrac{\dfrac{4}{t+2} + \dfrac{5}{t-1}}{\dfrac{3}{t-1} - \dfrac{1}{t+2}}$

289. $\dfrac{\dfrac{4}{s+3} - \dfrac{2}{s-3}}{\dfrac{5}{s^2 - 9}}$

290. $\dfrac{\dfrac{a+1}{a-1}}{\dfrac{1}{a^2 - 1}}$

291. $\dfrac{\dfrac{1}{z+3} + 2}{\dfrac{1}{z+3} - 2}$

292. $\dfrac{\dfrac{1}{w-2} - \dfrac{1}{w+2}}{\dfrac{1}{w+2} + \dfrac{1}{w^2 - 4}}$

293. $\dfrac{\dfrac{4}{rs} - \dfrac{2}{r^2 s}}{\dfrac{3}{rs^2} - \dfrac{1}{rs}}$

294. $\dfrac{\dfrac{4}{s} - 1}{\dfrac{4}{s} + 3}$

295. $\dfrac{\dfrac{1}{s-3} + 2}{\dfrac{1}{s-3} - 2}$

296. $\dfrac{\dfrac{1}{x-4}}{\dfrac{3}{x^2 - 16}}$

297. $\dfrac{\dfrac{6}{k+1} - \dfrac{5}{k-3}}{\dfrac{3}{k-3} + \dfrac{2}{k+2}}$

298. $\dfrac{\dfrac{5x-20}{x+3}}{\dfrac{x^2-16}{x^2-9}}$

299. $\dfrac{\dfrac{1}{8}+\dfrac{2}{12}}{\dfrac{3}{4}-\dfrac{1}{3}}$

300. $\dfrac{\dfrac{3}{x}+\dfrac{4}{x^2+xy}}{\dfrac{6}{x^2-y^2}}$

301. $\dfrac{\dfrac{8xy^2}{2x+3y}}{\dfrac{y}{4x^2-9y^2}}$

Writing/Conceptual Exercises

302. Which one of the following complex numbers is equivalent to

$$\frac{\dfrac{1}{2}-\dfrac{1}{3}}{\dfrac{1}{4}-\dfrac{1}{5}}?$$

Answer the question mentally, without having to show any work. Explain your reasoning.

(a) $\dfrac{\dfrac{1}{2}+\dfrac{1}{3}}{\dfrac{1}{4}+\dfrac{1}{5}}$

(b) $\dfrac{\dfrac{1}{3}-\dfrac{1}{2}}{\dfrac{1}{4}-\dfrac{1}{5}}$

(c) $\dfrac{\dfrac{1}{3}-\dfrac{1}{2}}{\dfrac{1}{5}-\dfrac{1}{4}}$

(d) $\dfrac{-\dfrac{1}{2}-\dfrac{1}{3}}{\dfrac{1}{4}-\dfrac{1}{5}}$

303. Which one of the choices below is equal to

$$\frac{\dfrac{1}{3}-\dfrac{1}{2}}{\dfrac{1}{4}-\dfrac{1}{5}}?$$

Answer the question mentally, without having to show your work. Explain your reasoning.

(a) $\dfrac{10}{3}$

(b) $-\dfrac{10}{3}$

(c) $\dfrac{3}{10}$

(d) 3

304. For the complex fraction

$$\frac{\dfrac{1}{x^2} - \dfrac{1}{y^2}}{\dfrac{1}{x} - \dfrac{1}{y}},$$

a student multiplied all of the denominators to get $x^3 y^3$ as the common denominator and then multiplied numerator and denominator by $x^3 y^3$. Will this student obtain the correct answer? Explain.

305. If you were a mathematics instructor and found the following work on a student's quiz, would you give the student full credit? Explain.

Problem: Simplify the complex fraction.

$$\frac{10 + \dfrac{4}{z}}{\dfrac{5z + 2}{10}}$$

Solution:

$$\frac{10z \left(10 + \dfrac{4}{z} \right)}{10z \left(\dfrac{5z + 2}{10} \right)} = \frac{100z + 40}{5z^2 + 2z}$$

Section 6.6   Equations Involving Rational Expressions

Objective 1   Distinguish between expressions with rational coefficients and equations with terms that are rational expressions.

Identify as an *expression* or an *equation*. If it is an expression, simplify it. If it is an equation, solve it.

306. $\dfrac{2}{3}x + \dfrac{1}{5}x$

307. $\dfrac{2}{3}x + \dfrac{1}{5}x = 1$

308. $\dfrac{4}{5}y - \dfrac{1}{10}y$

309. $\dfrac{4}{5}y - \dfrac{1}{10}y = 2$

310. $\dfrac{1}{3}z - \dfrac{1}{2}z = 4$

311. $\dfrac{1}{3}z - \dfrac{1}{2}z$

312. $\dfrac{5}{8}r + \dfrac{2}{3}r$

313. $\dfrac{5}{8}r + \dfrac{2}{3}r = \dfrac{7}{8}$

314. $\dfrac{5}{12}s - \dfrac{2}{3}s = \dfrac{1}{4}$

315. $\dfrac{5}{12}s - \dfrac{2}{3}s$

Objective 2   Solve equations with rational expressions.

Solve the equation and check your answers.

316. $\dfrac{26}{z} = 4 + \dfrac{6}{z}$

317. $\dfrac{1}{m} + \dfrac{5}{3m} = \dfrac{2}{3}$

318. $\dfrac{2}{3x} + \dfrac{3}{4x} = \dfrac{17}{24}$

319. $\dfrac{x-2}{5} = \dfrac{x+3}{6}$

320. $\dfrac{3}{n-1} - \dfrac{1}{n} = \dfrac{3}{4}$

321. $\dfrac{4}{x} = \dfrac{x}{3x+16}$

322. $\dfrac{2p+8}{9} = \dfrac{10p+4}{27}$

323. $\dfrac{6+x}{6} - \dfrac{x}{4} = \dfrac{2}{3}$

324. $\dfrac{8}{2m-4} + \dfrac{3}{5m-10} = \dfrac{1}{5}$

325. $\dfrac{2}{z-1} + \dfrac{1}{z+1} - \dfrac{5}{4} = 0$

326. $\dfrac{2}{m-3} + \dfrac{12}{9-m^2} = \dfrac{3}{m+3}$

327. $\dfrac{6}{h^2-4h-5} = \dfrac{5}{h-5} + \dfrac{3}{h+1}$

Objective 3   Solve a formula for a specified variable.

Solve the formula for the specified variable.

328. $N = \dfrac{kF}{d}$ for d

329. $F = \dfrac{k}{d-D}$ for D

330. $S = \dfrac{a}{l-r}$ for r

331. $h = \dfrac{2A}{B+b}$ for b

332. $F = \dfrac{GmM}{d^2}$ for M

333. $\dfrac{1}{R} = \dfrac{1}{R_1} + \dfrac{1}{R_2}$ for R_2

334. $\dfrac{V_1 P_1}{T_1} = \dfrac{V_2 P_2}{T_2}$ for T_2

335. $\dfrac{1}{f} = \dfrac{1}{d_0} + \dfrac{1}{d_1}$ for d_0

336. $F = \dfrac{Gm_1 m_2}{d^2}$ for m_2

337. $S_n = \dfrac{n}{2}(a_1 + a_n)$ for a_n

Mixed Exercises

Solve each equation and check your answers.

338. $\dfrac{9}{y} = 5 - \dfrac{1}{y}$

339. $\dfrac{m-4}{4} = \dfrac{8+m}{16}$

340. $\dfrac{b+7}{8} - \dfrac{b+2}{3} = \dfrac{4}{3}$

341. $\dfrac{z+1}{z-3} - 6 = \dfrac{4}{z-3}$

342. $\dfrac{1}{t+5} - \dfrac{3}{t-5} = \dfrac{10}{25-t^2}$

343. $\dfrac{4}{k} - \dfrac{2}{k+1} = 3$

344. $\dfrac{2}{c^2 - 6c - 16} = \dfrac{7}{c+2} + \dfrac{5}{c-8}$

345. $\dfrac{5a+1}{3a+3} = \dfrac{5a-5}{5a+5} + \dfrac{3a-1}{a+1}$

346. $\dfrac{2}{n^2 + 4n - 5} = \dfrac{5}{n-1} + \dfrac{2}{n+5}$

347. $\dfrac{4}{y+7} - \dfrac{3}{y+6} = \dfrac{1}{y^2 + 13y + 42}$

348. $\dfrac{1}{q^2 + 5q + 6} + \dfrac{1}{q^2 - 2q - 8} = \dfrac{-1}{12q + 24}$

349. $\dfrac{3p}{p^2 + 5p + 6} = \dfrac{5p}{p^2 + 2p - 3} - \dfrac{2}{p^2 + p - 2}$

350. $\dfrac{1}{z^2 + 5z + 6} + \dfrac{1}{12(z+2)} = \dfrac{-1}{z^2 - 2z - 8}$

351. $\dfrac{3x}{x^2 + 5x + 6} - \dfrac{5x}{x^2 + 2x - 3} = \dfrac{-2}{x^2 + x - 2}$

Solve the formula for the specified variable.

352. $I = \dfrac{nE}{R + nr}$ for R

353. $\dfrac{E}{e} = \dfrac{R + r}{r}$ for r

354. $B = \dfrac{3V}{h}$ for V

355. $\dfrac{V_1 P_1}{T_1} = \dfrac{V_2 P_2}{T_2}$ for V_1

356. $F = f\left(\dfrac{v + v_0}{v - v_s}\right)$ for v_s

357. $A = \dfrac{2pf}{b(q + 1)}$ for q

358. $A = \dfrac{1}{2}h(b_1 + b_2)$ for b_1

359. $A = \dfrac{R_1 R_2}{R_1 + R_r}$ for R_1

360. $\dfrac{t - 1}{t + 1} = \dfrac{r}{s}$ for t

361. $C = \dfrac{5}{9}(F - 32)$ for F

362. $A = \dfrac{1}{2}h(B + b)$ for B

363. $\dfrac{1}{a} + \dfrac{1}{b} = \dfrac{1}{c}$ for b

Identify as an *expression* or an *equation*. If it is an expression, simplify it. If it is an equation, solve it.

364. $\dfrac{2}{x + 1} + \dfrac{3}{x - 1}$

365. $\dfrac{8}{x - 1} = \dfrac{6}{x + 1}$

366. $\dfrac{1}{2}x - \dfrac{1}{3}x$

367. $\dfrac{6}{2x} + \dfrac{8}{7x} - \dfrac{5}{4x}$

368. $\dfrac{6}{x - 1} = \dfrac{8}{x - 1} + \dfrac{9}{x + 1}$

369. $\dfrac{x + 7}{4} - \dfrac{x + 6}{5} = \dfrac{x + 5}{6}$

Writing/Conceptual Exercises

370. Which one of the following numbers satisfies

$$\dfrac{z - 5}{3} - \dfrac{z + 1}{6} = 4?$$

(a) 8   (b) 11   (c) -1 (d) 35

371. What values of x cannot possibly be solutions of the equation

$$\frac{x-2}{x+3} = \frac{3}{5x}?$$

(Do not attempt to solve the equation.)

372. What is wrong with the following problem?

"Solve $\frac{3}{x-4} + \frac{5}{x+1}$."

373. If we multiply both sides of the equation

$$\frac{8}{x-3} = \frac{8}{x-3}$$

by $x-3$, we get $8 = 8$. Are all real numbers solutions of this equation? Explain.

374. If you are solving a formula for the letter t, and your steps lead you to the equation

$$tp - 5q = tq + 6p,$$

what would be your next step?

375. If you are solving a formula for the letter z, and your steps lead you to the equation

$$az - bz = cy,$$

what would be your next step?

Section 6.7   Applications of Rational Expressions

Objective 1   Solve problems about numbers using rational expressions.

Solve the problem.

376. One-eighth of a number is five less than one-third of the same number. What is the number?

377. If the same number is added to the numerator and denominator of the fraction 3/5, the value of the resulting fraction is 1/3. Find the number.

378. If the same number is added to the numerator and denominator of the fraction –4/7, the value of the resulting fraction is –9/2. Find the number.

379. If a certain number is added to the numerator and twice that number is subtracted from the denominator of the fraction 2/7, the result is equal to 5. Find the number.

380. In a certain fraction, the numerator is 3 less than the denominator. If 5 is added to both the numerator and the denominator, the resulting fraction is equal to 4/5. Find the original fraction.

381. The sum of a number and its reciprocal is 17/4. Find the number.

382. The difference between a number and its reciprocal is 3/2. Find the number.

383. If three times a number is subtracted from twice its reciprocal, the result is −1.

384. Bjorn and Wei-Li worked as computer analysts. In 1993, Bjorn earned 4/5 as much as Wei-Li. If they earned a total of $58,500, how much did each of them earn?

385. A child takes 5/8 the number of pills that an adult takes for the same illness. Together the child and the adult use 52 pills. Find the number used by the adult.

Objective 2   Solve problems about distance using rational expressions.

Solve the problem.

386. Garrett can row 5 miles per hour in still water. It takes him as long to row 4 miles upstream as 16 miles downstream. How fast is the current?

387. Sonja flew from City A to City B at 200 miles per hour and from City B to city a at 180 miles per hour. The trip at the slower speed took ½ hour longer. Find the distance between the two cities. (Assume there is no wind in either direction.)

388. Anjali traveled to her destination at an average speed of 50 miles per hour. Coming home, her average speed was 30 miles per hour and the trip took 2 hours longer. How far did she travel each way?

389. Jarrod flew his plane 500 miles against the wind in the same time it took him to fly 700 miles with the wind. If the speed of the wind was 40 mils per hour, what was the speed of the plane?

390. A plane travels 380 miles with a 30 mile per hour wind. After flying for the same amount of time on the return trip, the plane is 50 miles short of the terminal. Find the speed of the plane in still air.

391. On Saturday, Ranjan jogged 6 miles. On Monday, jogging at the same speed, it took him 30 minutes longer to cover 10 miles. How fast did Ranjan jog?

392. A plane made the trip from Redding to Los Angeles, a distance of 560 miles, in 1.5 hours less than it took to fly from Los Angeles to Portland, a distance of 1130 miles. Find the rate of the plane. (Assume there is no wind in either direction.)

393. A boat goes 240 miles down river in the same time it can go 160 miles upriver. The speed of the current is 5 miles per hour. Find the speed of the boat in still water.

394. A ship goes 120 miles down river in 2 2/3 hours less than it takes to go the same distance upriver. If the speed of the current is 6 miles per hour, find the speed of the ship.

395. A plane traveling 450 miles per hour can go 1000 miles with the wind in ½ hour less than when traveling against the wind. Find the speed of the wind

Objective 3   Solve problems about work using rational expressions.

Solve the problem.

396. Fran can clean the garage in 3 hours, but it takes Angie 4 hours. How long would it take them to clean the garage if they worked together?

397. Karen can wash the walls in a certain room in 3 hours and Erika can wash these walls in 2 hours. How long would it take them to complete this task if they work together?

398. Evan can install the carpet in a room in 3 hours, but Kendal needs 5 hours. How long will it take them to complete this task if they work together?

399. Mr. Turoso rakes the leaves in 5 hours, but his son Frank needs 9 hours. How long does it take father and son to rake the leaves together?

400. Leone can weed the garden in 1/2 hour, but Grace takes 3/4 hour. How long does it take them to weed the garden if they work together?

401. Kentaro can paint a certain room in 2 1/2 hours, but Kevin needs 3 hours to paint the same room. How long does it take them to paint the room if they work together?

402. Julio can type twice as fast as Cindy. Together, they can type a certain job in 2 hours. How long would it take Julio to type the entire job by himself?

403. Working together, Felice and al can balance the books for a certain company in 3 hours. Working alone, it would take Felice 2/3 as long as Al to balance the books. How long would it take Al to do the job alone?

404. Salvatore and Yolanda can clean a pool together in 4 hours. It takes Yolanda 4/5 as long as Salvatore to do the job alone. How long would it take Yolanda working alone?

405. Marco can seal an asphalt driveway in 5/6 the time it takes Cliff. Working together, it takes them 1 1/2 hours. How long would it have taken Marco working alone?

Objective 4   Solve problems about variation using rational expressions.

Solve the problem.

406. If m varies directly as p, and $m = 20$ when $p = 2$, find m when p is 5.

407. If y varies inversely as x, and $y = 10$ when $x = 3$, find y when x is 12.

408. For a given period of time, the interest earned on an investment varies directly as the interest rate. If the interest is $125 when the rate is 5%, find the interest when the rate is 6 1/2 %.

409. For a specified distance, time varies inversely with speed. If Ramona walks a certain distance on a treadmill in 20 minutes at 4.2 miles per hour, how long will it take her to walk the same distance at 3.5 miles per hour?

410. If the volume is constant, the pressure of a gas in a container varies directly as the temperature, where the temperature is measured in the Kelvin scale, abbreviated as K. If the pressure is 10.00 pounds per square inch at a temperature of 300 K, what is the pressure at a temperature of 325 K?

411. If the temperature is constant, the pressure of a gas in a container varies inversely as the volume of the container. If the pressure is 9 pounds per square foot in a container of 6 cubic feet, what is the pressure in a container of 7.5 cubic feet?

412. The circumference of a circle varies directly as the radius. A circle with a radius of 7 centimeters has a circumference of 43.96 centimeters. Find the circumference if the radius changes to 11 centimeters.

413. The current in a simple electrical circuit varies inversely as the resistance. If the current is 50 amps (an *ampere* is a unit for measuring current) when the resistance is 10 ohms (an *ohm* is a unit for measuring resistance), find the current if the resistance is 5 ohms.

414. For a given height, the area of a triangle varies directly as its base. Find the area of a triangle with a base of 5.2 centimeters, if the area is 9.6 square centimeters when the base is 3.2 centimeters.

415. The force required to compress a spring varies directly as the change in the length of the spring. If a force of 25 pounds is required to compress a spring 10 inches, how much force is required to compress the spring 8 inches?

Mixed Exercises

Solve the problem.

416. In a certain fraction, the denominator is 5 more than the numerator. If 4 is added to both the numerator and the denominator, the value of the resulting fraction is 2/7. Find the original fraction.

417. The Cuyahoga River has a current of 2 miles per hour. Julia can paddle 10 miles downstream in the time it takes her to paddle 2 miles upstream. How fast can Julia paddle?

418. If z varies inversely as t, and $z=12$ when $t=1/7$, find z when $t=2/7$.

419. Dennis can do a job in 4 days. When Dennis and Sue work together, the job takes 2 1/3 days. How long would the job take Sue if she worked along?

420. The area of a circle varies directly as the square of the radius. A circle with a radius of 5 centimeters has an area of 78.5 square centimeters. Find the area if the radius changes to 7 centimeters.

421. Eric can row 6 miles per hour in still water. It takes him as long to row 5 miles upstream as 19 miles downstream. How fast is the current?

422. The distance traveled by Kynan in a car moving at a constant rate varies directly as the time. If Kynan travels 364 miles in 7 hours, how far will he travel in 15 hours?

423. If the reciprocal of a number is subtracted from twice the number, the result is 7/2. Find the number.

424. Machine A can do a certain job in 7 hours, and machine B takes 12 hours. How long will the job take the two machines working together?

425. Lucy drove from Palmyra to Hamilton, a distance of 200 miles, in 1 hour longer than it took her to drive the return trip. If her speed averaged 10 miles per hour faster on the return trip, what was her speed from Palmyra to Hamilton?

426. An apprentice is paid 3/4 the salary of an experienced journeyman. If the total wages paid an apprentice and a journeyman are $56,000, find the amount paid to the journeyman.

427. A cold water faucet can fill a sink in 12 minutes, and a hot water faucet can fill it in 15 minutes. The drain can empty the sink in 25 minutes. If both faucets are on and the drain is open, how long will it take to fill the sink?

428. If a number is added to twice its reciprocal, the result is 9/2. Find the number.

429. If two times a number is added to one-half of its reciprocal, the answer is 13/6. Find the number.

430. Dana drove 24 miles in 6 minutes less than it took Carol to drive the same distance. Find Dan's average speed if his speed is 8 miles per hour faster than Carol's.

431. If n varies inversely as a^2, and $n = 12$ when $a = 1/2$, find n when a is 3.

432. The illumination produced by a light source varies inversely as the square of the distance from the source. If the illumination produced 5 feet from a light source is 60 foot-candles, find the illumination produced 10 feet from the same source.

433. With constant resistance, the power used in a simple electric circuit varies directly as the square of the current. If the power used is 37,500 watts when the current is 50 amps, find the power used if the current is reduced to 40 amps.

434. The resistance of a wire varies directly as the length and inversely as the square of the diameter. If the resistance of a 15 meter length of wire is 10 ohms when the diameter is .03 centimeters, find the resistance of 15 meters of the same type of wire if the diameter is increased to .06 centimeters.

435. A tank can be filled by each of two intake pipes in 15 minutes. A full tank takes 45 minutes to empty. If both intake pipes are on and the drain is open, how long will it take to fill the tank?

436. Latanya averages 10 miles per hour riding her bike to town. If she drives her car at an average speed of 30 miles per our, she gets to town 2 hours faster. How far does she travel to town?

437. Vikram can fly his plane 315 miles against the wind in the same time it takes him to fly 405 miles with the wind. The wind blows at 30 miles per hour. Find the speed of his plane in still air.

438. James and Hazel earned a total of $60,500 last year. If James earned 4/7 as much as Hazel, how much did each of them earn?

439. An inlet pipe can fill a vat of chocolate in 12 hours. An outlet pipe can empty the vat in 15 hours. How long will it take to fill the vat if both pipes are left open?

Writing/Conceptual Exercises

440. If an airplane flies 525 miles per hour in still air and the wind speed is y miles per hour, how fast does the plane fly with the wind? How fast does it fly against the wind?

441. If it takes Jolene 15 hours to paint her apartment, how much of the apartment can she paint in 9 hours?

State whether each of the following situations is an example of direct of inverse variation.

442. The heavier a package you ship, the less the cost per pound.

443. The longer a library book is overdue, the larger the fine you owe.

CHAPTER 7   EQUATIONS OF LINES, INEQUALITIES, AND FUNCTIONS

Section 7.1   Equations of a Line

Objective 1   Write an equation of a line given its slope and y-intercept.

Write an equation in slope-intercept form for the line.

1. $m = 3; b = -2$

2. $m = -1; b = 0$

3. $m = \dfrac{1}{2}; b = -\dfrac{1}{2}$

4. $m = \dfrac{2}{3}; b = 4$

5. $m = 1; b = -1$

6. $m = 0, b = 5$

7. Slope 2; y-intercept $(0, 3)$

8. Slope $\dfrac{3}{4}$; y-intercept $\left(0, -\dfrac{1}{2}\right)$

9. Slope -5; y-intercept $(0, 0)$

10. Slope -1; y-intercept $(0, 1)$

11. Slope 0; y-intercept $(0, -2)$

12. Slope 1; y-intercept $\left(0, \dfrac{4}{3}\right)$

Objective 2   Graph a line given its slope and a point on the line.

Graph the line passing through the given point and having the given slope.

13. $(2, 2); m = \dfrac{1}{3}$

14. $(3, -1); m = 2$

15. $(-3, 3); m = -\dfrac{1}{5}$

16. $(-2, -1); m = -4$

17. $(4, -2); m = -1$

18. $(-3, -2); m = \dfrac{2}{3}$

19. $(2, 4);$ undefined slope

20. $(-2, 1); m = 1$

21. $(1, -3); m = -\dfrac{5}{2}$

22. $(-3, -4); m = 0$

Objective 3   Write an equation of a line given its slope and any point on the line.

Write an equation for the line passing through the given point and having the given slope. Write the equation in the form $Ax + By = C$.

23. $(2, 4)$; $m = 3$

24. $(-2, 1)$; $m = 1$

25. $(-4, -1)$; $m = -\dfrac{1}{3}$

26. $(1, -2)$; $m = -5$

27. $(2, -9)$; $m = -4$

28. $(-4, -1)$; $m = \dfrac{1}{4}$

29. $(-5, 2)$; $m = -\dfrac{3}{4}$

30. $(0, 0)$; $m = 0$

31. $(-3, 2)$; $m = -\dfrac{2}{5}$

32. $(-6, -5)$; undefined slope

33. $(3, -4)$; $m = \dfrac{5}{2}$

34. $(1, -5)$; $m = -1$

35. $\left(\dfrac{3}{2}, -5\right)$; $m = -\dfrac{1}{2}$

36. $(4, 2)$; $m = \dfrac{5}{4}$

Objective 4   Write an equation of a line given two points on the line.

Write an equation for the line passing through the given pair of points. Write the equation in the form
$Ax + By = C$.

37. $(1, 3)$ and $(3, 5)$

38. $(4, -2)$ and $(5, 1)$

39. $(-3, 2)$ and $(2, 0)$

40. $(-3, -5)$ and $(-2, 4)$

41. $(1, -5)$ and $(-2, 10)$

42. $(-3, 1)$ and $(-3, -2)$

43. $(-2, 1)$ and $(3, 11)$

44. $(-2, 3)$ and $(2, 7)$

45. $(3, 6)$ and $(-2, -3)$

46. $(4, 8)$ and $(-6, -3)$

47. $(1, 7)$ and $(-2, -3)$

48. $(2, -5)$ and $(-4, 6)$

49. $\left(\frac{1}{2}, \frac{2}{3}\right)$ and $\left(-\frac{3}{2}, 1\right)$ 50. $\left(-\frac{4}{5}, \frac{1}{8}\right)$ and $\left(-\frac{8}{5}, -\frac{3}{8}\right)$

Objective 5   Find the equation of a line that fits a data set.

The total expenditures (in millions of dollars) for the purchase of memorabilia collectibles is given below. Use the information in the chart to answer questions 51-54.

Year	x	Millions of dollars (y)
1993	0	84
1994	1	101
1995	2	123
1996	3	136
1997	4	160
1998	5	181
1999	6	196

51.   Use the data from 1994 and 1999 to find the slope of the line that approximates this information. Then use the slope to find the equation of the line in slope-intercept form.

52.   To see how well the equation in exercise #51 approximates the ordered pairs (x, y) in the table of data, let $x = 4$ (for 1997) and find y.

53.   Use the data from 1995 and 1998 to find the slope of the line that approximates this information. Then use the slope to find the equation of the line in slope-intercept form.

54.   To see how well the equation in exercise #53 approximates the ordered pairs (x, y) in the table of data, let $x = 3$ (for 1996) and find y.

Mixed Exercises

Write an equation in slope-intercept form for the line.

55. Slope -2; y-intercept $(0, 3)$ 56. $m = -4; b = 5$

57. Slope $\frac{1}{2}$; y-intercept $\left(0, -\frac{2}{3}\right)$ 58. $m = 0; b = -3$

59. Slope 1; y-intercept $(0, 0)$ 60. Slope $\frac{3}{4}$; y-intercept $\left(0, -\frac{4}{3}\right)$

61. Slope $-\dfrac{2}{3}$; y-intercept $(0, 5)$ 62. Slope -3; y-intercept $\left(0, \dfrac{3}{5}\right)$

Graph the line passing through the given point and having the given slope.

63. $(3, 4)$; $m = -2$ 64. $(0, 2)$; $m = \dfrac{3}{4}$

65. $(-2, -3)$; $m = 4$ 66. $(5, 2)$; $m = -\dfrac{1}{3}$

67. $(-2, 4)$; $m = 0$ 68. $(-3, -1)$; undefined slope

Write an equation in the form $Ax + By = C$ for the line.

69. Through $(3, -2)$; $m = \dfrac{1}{2}$ 70. Through $(7, 5)$ and $(6, 3)$

71. Through $(-4, -5)$; $m = 0$ 72. Through $(-1, -4)$ and $(0, 4)$

73. Through $(3, -9)$; $m = -3$ 74. Through $(5, 6)$ and $(5, -6)$

75. Through $(2, -5)$; $m = -\dfrac{1}{2}$ 76. Through $(4, -7)$ and $(-2, -3)$

77. Through $(-12, 20)$; $m = -\dfrac{3}{4}$ 78. Through $(0, 6)$ and $(-18, 0)$

In Exercises 79-83, two views of the same line are shown on a calculator screen. Use the displays at the bottom of the screen to find the equation of the form $y = mx + b$ for the line. Then graph the line on your own calculator to support your answer. Use the standard viewing window.

79.

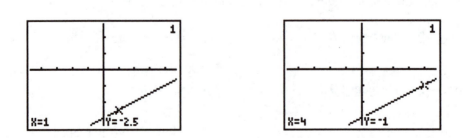

80.

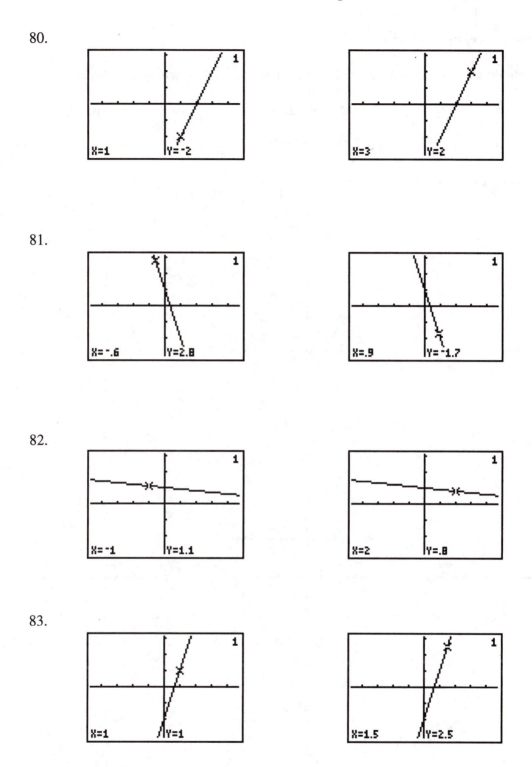

81.

82.

83.

Writing/Conceptual Exercises

Matching the equation with the graph that would *most closely resemble* its graph.

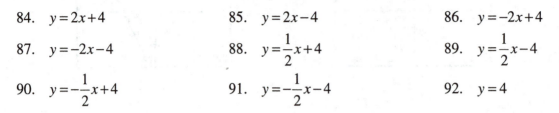

84. $y = 2x + 4$ 85. $y = 2x - 4$ 86. $y = -2x + 4$

87. $y = -2x - 4$ 88. $y = \dfrac{1}{2}x + 4$ 89. $y = \dfrac{1}{2}x - 4$

90. $y = -\dfrac{1}{2}x + 4$ 91. $y = -\dfrac{1}{2}x - 4$ 92. $y = 4$

A.   B.   C.

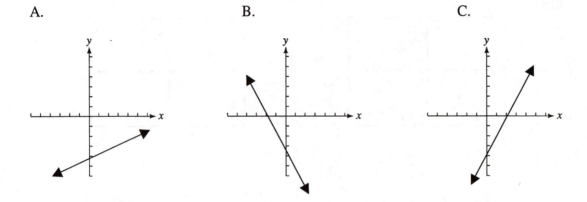

D.   E.   F.

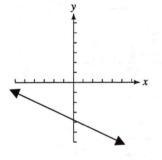
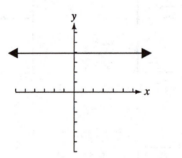
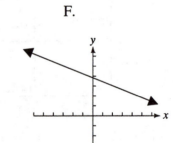

G.   H.   I.

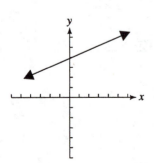
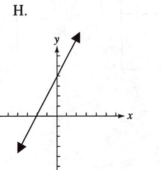
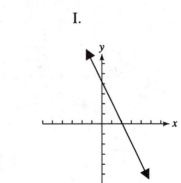

Section 7.2   Graphing Linear Inequalities in Two Variables

Objective 1   Graph $\leq$ or $\geq$ linear inequalities.

Graph the linear inequality.

93. $y \leq x + 4$

94. $y \geq x - 2$

95. $y \geq 3x$

96. $y \leq -\dfrac{2}{5}x + 2$

97. $x + y \geq 2$

98. $x - y \leq -3$

99. $y \geq -1$

100. $2x + 5y \leq -8$

101. $3x - 2y \leq 6$

102. $x + 2 \leq -1$

Objective 2   Graph > or < linear inequalities.

Graph the linear inequality.

103. $y < x - 3$

104. $y > -2x + 4$

105. $x < 2y + 4$

106. $x + 3y < 3$

107. $2x + 5y > -10$

108. $2x - 3y < 6$

109. $5x + 4y > 20$

110. $3x - 2y < 6$

111. $4 - 3y > 2x$

112. $5x - 2y + 10 < 0$

Objective 3   Graph inequalities with a boundary through the origin.

Graph the linear inequality.

113. $y \leq 3x$

114. $y \geq 2x$

115. $x > 4y$

116. $x < 2y$

117. $x < -2y$

118. $x < -4y$

119. $y \le \dfrac{2}{5}x$

120. $y \ge \dfrac{1}{3}x$

121. $3x - 4y \ge 0$

122. $3x - 2y < 0$

Mixed Exercises

Graph the linear inequality.

123. $y \ge -x + 4$

124. $y < 3x - 2$

125. $x \ge -5$

126. $y < 3$

127. $3x + 2y \le -6$

128. $y \ge -\dfrac{1}{2}x$

129. $x < \dfrac{1}{3}y$

130. $-3x + 5y > 15$

131. $x - 4 < 0$

132. $y \le -\dfrac{1}{2}x + 6$

133. $y + 2 \ge 0$

134. $3x - 4y - 12 > 0$

135. $2x \le 3y$

136. $y > \dfrac{2}{3}x - 5$

137. $3x - 2y \le 12$

138. $2x + 5y > 10$

139. $3x - y > 0$

140. $4x + y \ge 2$

In Exercises 141-144, a graphics calculator was used to generate the shaded graphs in choices A, B, C, and D. Match the inequality with the appropriate choice.

A.

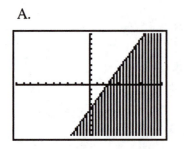

B.

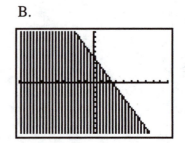

C.

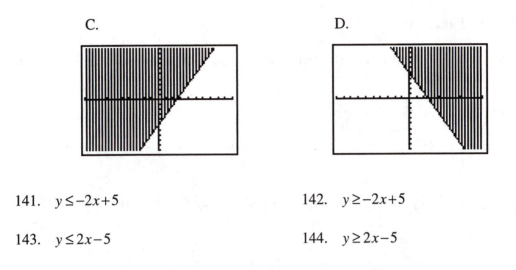

D.

141. $y \leq -2x+5$

142. $y \geq -2x+5$

143. $y \leq 2x-5$

144. $y \geq 2x-5$

In Exercises 145-148, a graphics calculator was used to generate the shaded graphs in choices A, B, C, and D. Match the inequality with the appropriate choice.

A.

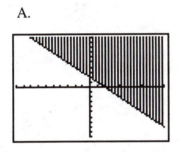

B.

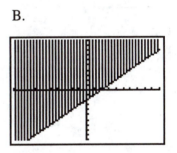

C.

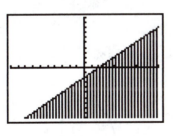

D.

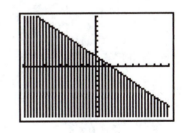

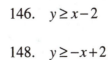

145. $y \leq x-2$

146. $y \geq x-2$

147. $y \leq -x+2$

148. $y \geq -x+2$

Section 7.3   Functions

Objectives 1 and 2   Understand the definition of a relation; understand the definition of a function.

Decide whether the relation is a function.

149. $\{(1, 3), (2, 4), (4, 7), (3, 9)\}$

150. $\{(1, 4), (2, 4), (3, 4), (4, 4)\}$

151. $\left\{\left(3, \dfrac{1}{2}\right), (4, 7), (3, 9), (2, 6)\right\}$

152. $\{(6, 8), (7, 9), (8, 10)\}$

153.

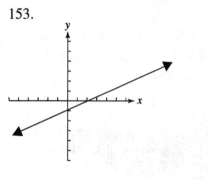

154.

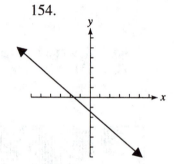

155.

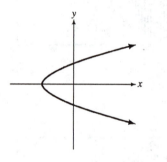

156.

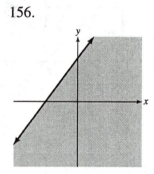

157.

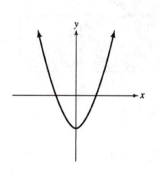

158.

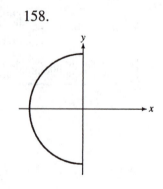

Objective 3   Decide whether an equation defines a function.

Decide whether the equation or inequality defines y as a function of x.

159. $y = 4x - 2$ 160. $y = 8 - 7x$

161. $6x - 7y = 2$ 162. $5x + 11y = 2$

163. $y = x^2 + 7x + 6$ 164. $y = 3 - x^2$

165. $x = y^2 - 2$ 166. $y = -\sqrt[3]{x+1}$

167. $6x - y < 4$ 168. $2x > 3y$

169. $y = \dfrac{1}{x-1}$ 170. $y = \dfrac{-2}{3x-1}$

171. $y = |x-1|$ 172. $y = -|2-x|$

173. $y = -x^3$ 174. $x = |y|$

175. $y > x$ 176. $x = y^4$

177. $x^2 + y^2 = 25$ 178. $4x - 5y = 20$

Objective 4   Find domains and ranges.

Find the domain and the range for the function.

179. $y = 3x - 7$ 180. $y = 9 - x$

181. $y = x^2 + 4$

182. $y = x^2 - 7$

183. $y = 5 - x^2$

184. $y = 2$

185. $y = (x+3)^2$

186. $y = x^3$

187. $y = |x| + 6$

188. $y = |x-3|$

Objective 5   Use $f(x)$ notation.

For the function f, find (a) $f(2)$ and (b) $f(-1)$.

189. $f(x) = 7x - 4$

190. $f(x) = 9x + 11$

191. $f(x) = 6 - 3x$

192. $f(x) = 8 - 5x$

193. $f(x) = x^2 + 4$

194. $f(x) = 3x^2 - 7$

195. $f(x) = -x^2 + 8x - 2$

196. $f(x) = -x - x^2$

197. $f(x) = (x+4)^2$

198. $f(x) = 9x^2 - 6x + 1$

199. $f(x) = -(2x-1)^2$

200. $f(x) = -(x-2)^2$

201. $f(x) = |x-5|$

202. $f(x) = -|3-x|$

For the polynomial function P, find (a) $P(0)$, (b) $P(-2)$, and (c) $P(3)$.

203. $P(x) = 2x^2 + 3x$

204. $P(x) = x^2 - x - 6$

205. $P(x)=x^3$

206. $P(x)=x^3+2x-1$

207. $P(x)=x^4-16$

208. $P(x)=2x^3+x^2-2x+1$

Objective 6   Apply the function concept in a function.

The following table gives the Nielsen television rating information for five random programs during the week of June 14-20, 1999.

Program	Share (x)	Audience in millions (y)
60 Minutes	22.0	9.9
NBA Finals-Game One	21.0	11.4
Friends	18.0	8.8
Frasier	17.0	9.2
Becker	14.0	8.6

http://ytv.yahoo.com/nielsen/

209.   Use the table to write a set or ordered pairs that define a function f.

210.   What is the domain of f?

211.   What is the range of f?

212.   Find $f(21.0)$ and $f(14.0)$.

213.   For what x-value does $f(x)$ equal 9.2?

214.   For what x-value does $f(x)$ equal 8.8?

215.   60 Minutes drew an audience of 11.4. Translate this into actual numbers of viewers.

Mixed Exercises

Decide whether the relation is a function. Give the domain and range.

216. $\{(1, 3), (1, 4), (2, -1), (3, 7)\}$

217. $\{(-1, 2), (0, 5), (1, 8)\}$

218. $\{(2, -2), (3, -3), (4, -4)\}$

219. $\{(6, -3), (4, -2), (2, -1), (0, 0)\}$

220. $\{(0, 4), (3, 2), (0, 0), (3, 5)\}$

221. $\{(1, 1), (1, 2), (1, 7), (1, 0)\}$

Decide whether the equation or inequality defines y as a function of x. If so, give the domain and the range for the function.

222. $y + x = 7$ 223. $x^2 + y^2 = 1$

224. $y \geq 8x - 2$ 225. $y = \dfrac{1}{x}$

226. $y = 3x + 4$ 227. $y = x^2 + 1$

228. $y = |x + 2|$ 229. $y = 3$

For the function f, find (a) $f(-2)$ and (b) $f(4)$.

230. $f(x) = 2x + 5$ 231. $f(x) = 6 - 2x$

232. $f(x) = 3x^2$ 233. $f(x) = x^2 - 2x$

234. $f(x) = \dfrac{4}{x^2 + 1}$ 235. $f(x) = \dfrac{2x + 1}{5}$

In Exercises 236-243, a calculator-generated graph of a relation is shown. Decide whether the relation is a function.

236.

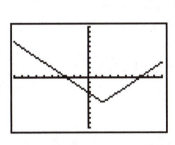

237.

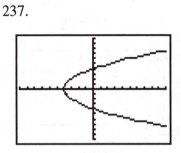

238.

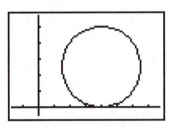

239.

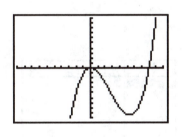

240.

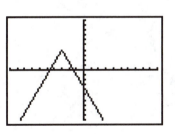

241.

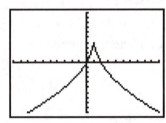

242.

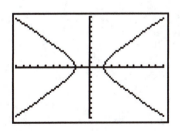

243.

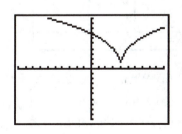

In Exercises 244-248, the table shown was generated by a calculator. The expression y_1 represents $f(x)$. the points represented in the table lie on a straight line.

244.

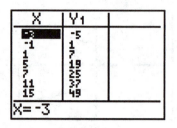

(a) What is $f(1)$?

(b) Find the value of x if $f(x)=7$.

(c) Write the function in the form $y=mx+b$.

245.

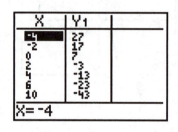

(a) What is $f(2)$?

(b) Find the value of x if $f(x)=-3$.

(c) Write the function in the form $y=mx+b$.

246.

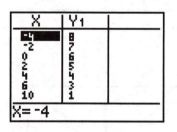

(a) What is $f(6)$?

(b) Find the value of x if $f(x)=6$.

(c) Write the function in the form $y=mx+b$.

247.

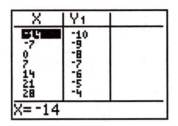

(a) What is $f(7)$?

(b) Find the value of x if $f(x) = -8$.

(c) Write the function in the form $y = mx + b$.

248.

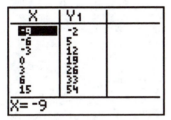

(a) What is $f(6)$?

(b) Find the value of x if $f(x) = 12$.

(c) Write the function in the form $y = mx + b$.

Writing/Conceptual Exercises

249. In your own words, explain the difference between a relation and a function.

250. In your own words, explain the idea of a function.

251. How can you tell from its graph whether a relation is a function?

CHAPTER 8   LINEAR SYSTEMS

Section 8.1   Solving Systems of Linear Equations by Graphing

Objective 1   Decide whether a given ordered pair is a solution of a system.

Decide whether the given ordered pair is a solution of the given system.

1. $(4, 1)$
    $x + 4y = 8$
    $2x - 5y = 1$

2. $(3, -2)$
    $2x + 3y = 1$
    $3x - 2y = 10$

3. $(-3, -2)$
    $5x - 3y = -9$
    $2x + 3y = -12$

4. $(4, -2)$
    $4x + 3y = 10$
    $x - 4y = 12$

5. $(-5, -7)$
    $x - y = 2$
    $4x + y = 27$

6. $(3, -6)$
    $5x + y = 9$
    $2x - 3y = 24$

7. $(-4, -3)$
    $x - y = 1$
    $-2x + 3y = -1$

8. $(1, -2)$
    $-3x + y = -5$
    $4x - 3y = 10$

9. $(-1, -5)$
    $5x - 2y = -5$
    $y = -3x - 8$

10. $(-4, 8)$
    $3x + 2y = 4$
    $y = -2x$

11. $(2, 8)$
    $9x - 2y = 2$
    $8x - y = 8$

12. $(4, -1)$
    $5x + 6y = 14$
    $-4x - 10y = -6$

13. $(-3, -8)$
    $x + 8 = 0$
    $y + 3 = 0$

14. $(1, -2)$
    $3y = x - 7$
    $2y = 3x - 7$

15. $(2, -8)$
    $3x - 2y = 22$
    $-x + y = -10$

Objective 2   Solve linear systems by graphing.

Solve the system by graphing both equations on the same axes.

16. $x + y = 5$
     $3x - y = 3$

17. $2x - y = -6$
     $3x + 2y = -2$

18. $4x - 3y = 6$
     $2x - 3y = 12$

19. $2x + 5y = -5$
    $3x = 5y + 30$

20. $3x - 2y = 8$
    $7x + 2y = 12$

21. $x - y = 0$
    $x + 4 = 2y$

22. $8x + 5y = 40$
    $2x - 3y = -24$

23. $6x - 5y = -5$
    $2x - 5y = 15$

24. $y - 2 = 0$
    $3x - 2y = -16$

25. $2x = y$
    $5x + 3y = 0$

26. $3x - y = -4$
    $2x + y = 1$

27. $2x + 24 = 3y$
    $2y + x = 2$

28. $2x + y = 8$
    $x + 2y = 10$

29. $3x + 4y = 12$
    $y = -x + 3$

30. $2x + y = -6$
    $x + 5 = 0$

Objective 3 Solve special systems by graphing.

Solve the system of equations by graphing both equations on the same axes. If the two equations produce parallel lines, write *no solution*. If the two equations produce the same line, write *infinite number of solutions*.

31. $4x - 3y = 12$
    $6y - 8x = -24$

32. $3x + 3y = 12$
    $x = 4 - y$

33. $3x - 2y = 6$
    $-6x + 4y = 12$

34. $x + 2y = 4$
    $6y = -3x + 18$

35. $4x + y = 8$
    $y = -4x - 8$

36. $5x - 2y = 10$
    $4y - 10x = -20$

37. $3x - y = 2$
    $-6x + 2y = -4$

38. $-4x + y = 6$
    $12x - 3y = 5$

39. $x + y = 3$
    $-x - y = 1$

40. $x - \dfrac{1}{2}y = 5$
    $-2x + y = 1$

41. $2x + y = 3$
    $-4x - 2y = -6$

42. $x + y = 6$
    $2x + 2y = -7$

Objective 4   Identify special systems without graphing.

Without graphing, answer the following questions for each linear system.

(a) Is the system *inconsistent*, are the equations *dependent*, or *neither*?

(b) Is the graph a pair of *intersecting lines*, a pair of *parallel lines*, or *one line*?

(c) Does the system have *one solution, no solution*, or an *infinite number of solutions*?

43. $x + y = 2$
     $x + y = 5$

44. $x - 2y = 5$
     $2x - 4y = 10$

45. $3x + y = 7$
     $6x + 2y = 9$

46. $2x - 3y = 7$
     $4x + 6y = 14$

47. $x + 3y = 2$
     $4x + 12y = 8$

48. $y = 3x - 2$
     $3x - y = 4$

49. $2x - y = 4$
     $x + 3y = 2$

50. $2x - y = 4$
     $2x = y + 3$

51. $3x - 4y = 8$
     $x - 2y = 2$

52. $3x - 9y = 6$
     $y = \dfrac{1}{3}x - \dfrac{2}{3}$

53. $6x + 2y = 12$
     $y = -3x + 4$

54. $4x + 3y = 8$
     $y = -\dfrac{4}{3}x + \dfrac{5}{3}$

Objective 5   Recognize how a graphing calculator is used to solve a linear system.

Use the graphing capabilities of your calculator to find the solutions to the given systems.

55. $2x - y = 0$
    $3x - 2y = -1$

56. $y = -2x$
    $2x - y = 16$

57. $x + y = -2$
    $2x - 3y = 21$

58. $3x - y = 17$
    $3y - x = 4$

59. $2x - 3y = -9$
    $3x + y = 25$

60. $3y = -x$
    $9x + 8y = -76$

61. $\dfrac{1}{3}x - \dfrac{1}{2}y = 2$
    $\dfrac{2}{5}x + \dfrac{1}{2}y = 9$

62. $3x - 2y = 6$
    $\dfrac{3}{2}x - 3 = y$

63. $4x - 2y = 8$
    $y = 2x - 4$

64. $\dfrac{1}{3}x + y = 7$

    $\dfrac{-1}{3}x + 4 = y$

65. $3x = 2y$

    $9x + 4y = 30$

66. $1 + 5x = \dfrac{1}{2}y$

    $1 + \dfrac{1}{4}y = 10x$

Mixed Exercises

Decide whether the given ordered pair is a solution of the given system.

67. $(-1, 2)$

    $3x - y = -5$

    $4x + y = -2$

68. $(-5, 3)$

    $3x + 2y = 9$

    $2x + 5y = 5$

69. $(5, 0)$

    $3x - 2y = 15$

    $-x + 3y = -5$

70. $(0, -4)$

    $2x - y = 4$

    $3x + 2y = 8$

71. $(7, -3)$

    $3x + y = 18$

    $3x + 4y = 9$

72. $\left(\dfrac{3}{2}, -2\right)$

    $2x + 5y = -7$

    $-2x + 3y = -9$

Solve the system of equations by graphing both equations on the same axes. If the system is inconsistent or the equations are dependent, say so.

73. $3x - y = 6$

    $4x + y = 8$

74. $5x + 3y = 30$

    $10x + 6y = 60$

75. $2x + 3y = 6$

    $x - 2y = 10$

76. $7x - 3y = -21$

    $-14x + 6y = -42$

77. $2x + 24 = 3y$

    $2y + x = 2$

78. $3x - 5y = 15$

    $2x - y = -4$

Without graphing, determine whether the system has *one solution, no solution,* or an *infinite number of solutions.*

79. $2x + 5y = -5$

    $-2x = 5y + 30$

80. $3x - 2y = 8$

    $7x + 2y = 12$

81. $x - y = 0$

    $x + 4 = 2y$

82. $8x + 5y = 40$

    $2x - 3y = -24$

83. $6x + 15y = -45$

    $-2x - 5y = 15$

84. $y - 2 = 0$

    $3x - 2y = -16$

85. $2x = y$

$5x + 3y = 0$

86. $4y + 1 = x$

$\dfrac{1}{4}x - y = \dfrac{1}{4}$

87. $y = 3x$

$\dfrac{1}{3}y - \dfrac{2}{3} = x$

88. $x = 5$

$y + 3 = 0$

89. $5x + 2y = 8$

$10x + 8 = -4y$

90. $3x - y = -4$

$2x + y = -1$

91. $y + 2 = 0$

$x - 4 = 0$

92. $4x - 3y = 12$

$6y - 8x = -24$

93. $3x + 3y = 12$

$x = 4 - y$

94. $8x = 6y - 6$

$4x - 3y = 15$

95. $2x + 24 = 3y$

$2y + x = 2$

96. $\dfrac{1}{2}x - \dfrac{1}{4}y = 5$

$-2x + \quad y = 3$

Writing/Conceptual Exercises

97. A student solved a linear system and found the solutions to be $(2, 3)$ and $(5, 5)$. What is wrong with this?

98. Why is the graphing method not the best method for solving a linear system?

99. One student found that a point in the first quadrant had coordinates which were a solution of a linear system. A second student found a similar point in the second quadrant. Both students were correct. What does this tell you about the system?

100. Find a system of equations with the solution $(4, -2)$ and show the graph.

Section 8.2   Solving Systems of Linear Equations by Substitution

Objective 1   Solve linear systems by substitution.

Solve the system by the substitution method. Check your solution.

101. $x + y = 8$

$y = 3x$

102. $3x + 2y = 9$

$y = x + 2$

103. $2x + y = 0$

$y = -x$

104. $x - 2y = 10$

$x = 3 - 5y$

105. $y + 4x = 0$

$x = -11 - 3y$

106. $5x + 2y = 14$

$y = 2x - 11$

107. $x + y = 5$
$5x - 2y = 4$

108. $x - 4y = 12$
$3x - 4y = -4$

109. $-8x + 5y = -13$
$x - y = 5$

110. $3x - 2 = -y$
$y + 2x = -2$

111. $5x + 3y = 14$
$x + y = 4$

112. $3x - 2y = 13$
$x - 5y = 0$

Objective 2   Solve special systems.

Solve the system by the substitution method.

113. $3x - 2y = 6$
$6x - 4y = 7$

114. $x + y = 5$
$-x - y = -5$

115. $4x - 8y = 12$
$x - 2y = 3$

116. $8x - 2y = 4$
$4x - y = 4$

117. $6x + 2y = 12$
$3x + y = 4$

118. $36x + 20y = 12$
$-27x - 15y = -9$

119. $5x + 2y = -8$
$10x + 4y = 6$

120. $48x - 56y = 32$
$21y - 18x = -12$

121. $12x - 18y = 25$
$4x - 6y = 5$

122. $72x - 60y = -12$
$25y - 30x = 5$

123. $x + y = 8$
$\dfrac{x}{2} + \dfrac{y}{2} = 4$

124. $3x - 5y = 12$
$x - \dfrac{5}{3}y = 3$

Objective 3   Solve linear systems with fractions.

In the given systems, begin by clearing fractions and then solve the system by substitution.

125. $\dfrac{x}{4} + \dfrac{y}{4} = 1$
$\dfrac{x}{2} + \dfrac{y}{2} = -1$

126. $\dfrac{7}{3}x + y = 5$
$2x + \dfrac{3}{2}y = 3$

127. $\dfrac{5}{4}x - y = -\dfrac{1}{4}$
$-\dfrac{7}{8}x + \dfrac{5}{8}y = 1$

128. $x + \dfrac{5}{3}y = 11$
$\dfrac{5}{6}x + \dfrac{2}{3}y = \dfrac{29}{6}$

129. $x - \dfrac{7}{5}y = \dfrac{6}{5}$

  $\dfrac{1}{4}x - \dfrac{1}{2}y = \dfrac{1}{6}$

130. $\dfrac{3}{14}x - \dfrac{1}{7}y = 1$

  $-\dfrac{1}{2}x = \dfrac{1}{4}y$

131. $\dfrac{x}{2} - \dfrac{2y}{3} = \dfrac{1}{3}$

  $x = \dfrac{4y}{3} + \dfrac{2}{3}$

132. $\dfrac{1}{20}x + \dfrac{1}{15}y = -\dfrac{1}{6}$

  $\dfrac{1}{6}x + 1 = 0$

133. $\dfrac{3}{4}x + \dfrac{5}{8}y = -9$

  $\dfrac{5}{6}x - \dfrac{3}{7}y = -10$

134. $\dfrac{5x}{4} + \dfrac{2y}{3} = -4$

  $\dfrac{2x}{3} - \dfrac{3y}{2} = 9$

Mixed Exercises

Solve the system by the substitution method.

135. $3x + 4y = -10$
  $2x + 3y = -6$

136. $3x - 2y = 5$
  $x - 5 = 0$

137. $5x + 2y = 16$
  $y + 2 = 0$

138. $6x + 5y = 8$
  $x + 7 = 0$

139. $3x - 6y = 1$
  $x = 2y + 3$

140. $6x + 8y = 10$
  $4y = 5 - 3x$

141. $4x + 3y = 10$
  $3x + 5y = 2$

142. $2x - 5y = 4$
  $3x - 4y = -1$

143. $y = 3x + 1$
  $2x - y = -2$

144. $18x - 14y = 0$
  $7y - 9x = 0$

145. $15x - 10y = 5$
  $-12x + 8y = -4$

146. $10 + 8x = 6y$
  $20 + 12y = 16x$

147. $13x - 39y = 0$
  $x = 3y$

148. $9x - 15y = 21$
  $-12x + 20y = 28$

149. $3x - 7y = 12$
  $-9x + 21y = 0$

150. $6x = 2y - 8$
$5y = 20 + 15x$

151. $3x - 8y = 9$
$x = \dfrac{8}{3}y + 3$

152. $2x + y = -8$
$x + \dfrac{1}{2}y = -4$

153. $\dfrac{x}{2} - \dfrac{y}{3} = \dfrac{1}{2}$
$\dfrac{x}{4} - \dfrac{y}{12} = \dfrac{3}{4}$

154. $\dfrac{x}{2} + \dfrac{y}{3} = 1$
$\dfrac{x}{4} - \dfrac{y}{6} = \dfrac{7}{2}$

155. $\dfrac{9}{2}x - \dfrac{3}{4}y = 3$
$-\dfrac{3}{4}x + \dfrac{1}{8}y = -\dfrac{1}{2}$

156. $\dfrac{8}{3}x - y = \dfrac{7}{5}$
$\dfrac{1}{4}x + \dfrac{3}{8}y = \dfrac{9}{40}$

157. $\dfrac{1}{3}x - \dfrac{1}{4}y = 5$
$\dfrac{1}{2}x + \dfrac{1}{3}y = -1$

158. $\dfrac{1}{2}x - \dfrac{3}{4}y = -7$
$-\dfrac{1}{8}x + \dfrac{5}{2}y = 11$

Solve each equation in the system for y. Then use the intersection feature on your graphics calculator to find the coordinates of the point of intersection.

159. $3x - y = -2$
$2x + y = 7$

160. $9x + 2y = -10$
$-7x + 3y = 26$

161. $3x + 6y = 23$
$-2x - 12y = -18$

162. $-9x + 14y = 93$
$21x + 16y = 75$

Writing/Conceptual Exercises

In Exercises 163 through 166, which variable would be easier to solve for in the first step of the substitution method? In which equation would you solve for it? Why?

163. $3x + y = 8$
$4x - 7y = 9$

164. $7x - y = 10$
$2x + 3y = 4$

165. $5x - 9y = 14$
$x - 5y = 7$

166. $9x + 7y = 6$
$2x + 4y = -10$

Section 8.3   Solving Systems of Linear Equations by Elimination

Objective 1   Solve linear systems by elimination.

Solve the system by the addition method. Check your answers.

167. $x + y = 4$
      $x - y = 2$

168. $3x - y = 7$
      $2x + y = 3$

169. $x - 8y = -8$
      $-x + y = -6$

170. $2x - y = 4$
      $3x + y = 11$

171. $4x + 2y = 8$
      $6x - 2y = -13$

172. $3x - 4y = 8$
      $x + 4y = -2$

173. $8x + 2y = 6$
      $3x - 2y = -6$

174. $-2x + 5y = 3$
      $2x + y = -9$

175. $x - 3y = -1$
      $-x + 4y = 3$

176. $3x - 5y = -16$
      $x + 5y = 8$

177. $5x + 8y = 2$
      $3x - 8y = 14$

178. $3x + 2y = -1$
      $-3x + y = 31$

Objective 2   Multiply when using the elimination method.

Solve the system by the addition method. Check your answers.

179. $6x + 7y = 4$
      $2x - 3y = 12$

180. $8x + 6y = 22$
      $4x - y = -1$

181. $4x - 3y = 8$
      $3x - 2y = 6$

182. $3x + 5y = -19$
      $2x - y = -4$

183. $x - 4y = 12$
      $x + 6y = -8$

184. $6x + y = -2$
      $3x - 4y = 35$

185. $2x + y = 3$
      $-3x + y = -22$

186. $4x - 5y = -17$
      $3x + 2y = -7$

187. $4x - 9y = -5$
      $3x + 2y = 5$

188. $3x + y = 1$
      $2x + 3y = -11$

189. $2x + 2y = 4$
      $3x - 5y = -2$

190. $3x - 5y = 0$
      $2x + 3y = 19$

Objective 3   Use an alternative method to find the second value in a solution.

Solve the system by using the addition method twice. Check your answer.

191. $2x - y = 6$
 $x + 2y = 8$

192. $3x + 2y = 6$
 $9x + 4y = 24$

193. $4x - 5y = -22$
 $-16x + 9y = 110$

194. $6x - 5y = -22$
 $9x + 2y = 5$

195. $x + 5y = 8$
 $x - 3y = -1$

196. $-4y = 2x - 1$
 $y + 2x = 7$

197. $2x + 3y = 5$
 $3x - 2y = -4$

198. $3x + 5y = 0$
 $5x - 4y = 1$

199. $5x + 2y = -1$
 $2x - 5y = 1$

200. $2x - 6y = 3$
 $-5x + 2y = 0$

201. $4x - 7y = -8$
 $3x + 2y = 23$

202. $3x + 2y = -1$
 $5x + 4y = -1$

Objective 4   Use the elimination method to solve special systems.

Solve the system by the addition method.

203. $3x - 2y = 6$
 $6x - 4y = 7$

204. $x + y = 5$
 $-x - y = -5$

205. $4x - 8y = 12$
 $x - 2y = 3$

206. $8x - 2y = 4$
 $4x - y = 4$

207. $6x + 2y = 12$
 $3x + y = -6$

208. $24x - 15y = 30$
 $-16x + 10y = -20$

209. $5x + 2y = -8$
 $10x + 4y = 6$

210. $48x - 56y = 32$
 $21y - 18x = -12$

211. $12x - 18y = 25$
 $4x - 6y = 5$

212. $72x - 60y = -12$
 $25y - 30x = 5$

213. $18x - 14y = 0$
 $7y - 9x = 0$

214. $15x - 10y = 5$
 $-12x + 8y = 4$

Mixed Exercises

Solve the system by the addition method. Check your answers.

215. $x + y = 9$
 $x - y = 3$

216. $2x - y = -13$
 $3x + y = 8$

217. $-8x + 3y = -1$
 $2x - 7y = -31$

218. $15x + 7y = -45$
     $-3x + 5y = 9$

219. $13x - 39y = 0$
     $x = 3y$

220. $9x - 15y = 21$
     $-12x + 20y = 28$

221. $14x - 5y = 2$
     $-11x + 8y = -26$

222. $4x = 3 - 9y$
     $7y + 9 + 5x = 0$

223. $6x - 5y = -22$
     $9x + 2y = 5$

224. $5x = 5 - 4y$
     $5y = 7 - 6x$

225. $6x = 2y - 8$
     $5y = 20 + 15x$

226. $5x + 7y = -1$
     $-5x + 13y = 16$

227. $5x - 2y = 8$
     $x + 2y = 16$

228. $x - y = 5$
     $x + y = 7$

229. $-3x + 5y = -2$
     $3x + 2y = 16$

230. $5x - 6y = 9$
     $3x + 6y = 7$

231. $3x + y = 2$
     $6x - y = -5$

232. $15x - 3y = 8$
     $21x + 3y = 10$

233. $-2x + 9y = 5$
     $2x - 3y = -2$

234. $6x - 4y = 4$
     $12x + 8y = 12$

235. $3x - 4y = 12$
     $-6x + 8y = -3$

236. $5x + 4y = -7$
     $-3x + 5y = 19$

237. $3x - 4y = 8$
     $4x + 5y = -10$

238. $3x + 2y = 10$
     $4x - 3y = -15$

239. $-2x - 4y = -12$
     $y = 3x - 4$

240. $5x = 15 - 10y$
     $3x = 9 + 2y$

241. $5x - 7y = 17$
     $3x - 12 = 6y$

242. $2x + 12 = -3y$
     $7y - 30 = 5x$

243. $3x - 4 = -3y - 1$
     $2x + y = -1$

244. $x - 2y = 6$
     $-\dfrac{1}{2}x + y = -3$

Writing/Conceptual Exercises

245. Only one of the following systems does not require that we multiply one or both equations by a constant in order to solve the system by the addition method. Which one is it?

   (a) $4x + y = 7$
           $2x - 3y = 6$

   (b) $-5x + 2y = 8$
             $4x - 2y = 1$

(c) $3x + 2y = 5$
       $2x - 3y = 9$

(d) $x + y = 5$
       $2x - 2y = 10$

246. For the system

$$4x + 10y = 3$$
$$5x - 25y = 18,$$

if we were to multiply the first (top) equation by -5, by what number would we have to multiply the second (bottom) equation in order to:

(a) eliminate the x terms when solving by the addition method?

(b) eliminate the y terms when solving by the addition method?

247. Without actually solving the system, explain why

$$x - y = 12$$
$$x - y = 10$$

can have no solution.

248. On an algebra quiz, Mischa and Shala solved the following system by the addition method:

$$3x + 2y = 10$$
$$x - y = -5.$$

After multiplying the second equation by 2 and adding the result to the first equation, both of them obtained the equation $5x = 0$. Mischa wrote down that the system has no solution, while Shala wrote on her paper that the system has an infinite number of solutions. Who is correct? Explain.

Section 8.4   Applications of Linear Systems

Objective 1   Solve problems about unknown numbers.

Use a system of equations to solve the problem.

249. The sum of two numbers is 74. Their difference is 16. Find the numbers.

250. Find two numbers whose sum is -103 and whose difference is 53.

251. The difference between two numbers is 9. If the larger is one more than three times the smaller, find the numbers.

252. The sum of two numbers is 12. If three times the larger is added to twice the smaller, the result is 33. Find the numbers.

253. The difference between two numbers is 11. If two times the smaller is added to one-half the larger, the result is 33. Find the numbers.

254. Two neighboring towns, Rockland and Brookdale, have a combined population of 8085. There are 1243 more people living in Rockland than in Brookdale. Find the population in each town.

255. There are a total of 53 students in the two second grade classes at Hastings School. If Ms. Ferrari has 3 more students in her class than Mrs. Baylies, find the number of students in each class.

256. A rope 58 centimeters long is cut into two pieces with one piece three times as long as the other. Find the length of each piece.

257. The perimeter of a rectangular room is 40 feet. The length is 4 feet less than three times the width. Find the dimensions of the room.

258. The perimeter of a triangular pennant is 116 centimeters. If two sides are of equal length, and the third side is 20 centimeters longer than each of the equal sides, what are the lengths of the three sides?

Objective 2   Solve problems about quantities and their costs.

Use a system of equations to solve the problem.

259. Admission prices at a football game were $8 for adults and $4 for children. The total receipts for the game were $52,000. Tickets were sold to 8000 people. How many adults and how many children attended the game?

260. The receipts from a concert were $2450. The price for a regular ticket was $7 and the student tickets were half the regular price. If 400 tickets were sold, how many of each type were there?

261. The cashier at Rex's Drug Store has some $10 bills and some $20 bills. The total value of the money is $1290. If there is a total of 77 bills, how many of each type are there?

262. A postal clerk has 1250 stamps in his drawer that are worth a total of $287.50. If there are only 25¢ and 15¢ stamps, how many of each kind are there?

263. There were 411 tickets sold for a soccer game, some for students and some for non-students. Student tickets cost $1.25 and non-student tickets cost $3.50 each. The total receipts were $763.50. How many of each type were sold?

264. The cashier at Taco Bell has some $5 bills and some $20 bills. The total value of the money is $1350. If the number of twenties is equal to twice the number of fives, how many of each type are there?

265. The total receipts for a basketball game were $2432.50. There were 423 tickets sold, some for children and some for adults. If the adult tickets cost $7.50 and the children's tickets cost $5, how many of each type were there?

266. Twice as many general admission tickets to a basketball game were sold as reserved seat tickets. General admission tickets cost $10 and reserved seat tickets cost $15. If the total value of both kinds of tickets was $26,250, how many tickets of each kind were sold?

267. Salina has $10,000 to invest, part at 5% and part at 8%. She wants the income from simple interest on the two investments to total $710 annually. How much should she invest at each rate?

268. Ms. Oberle has $12,000 to invest at 7% and 9%. She wants the income from simple interest on the two investments to total $1000 yearly. How much should she invest at each rate?

Objective 3   Solve problems about mixtures.

Use a system of equations to solve the problem.

269. A merchant wishes to mix coffee worth $6 a pound with coffee worth $3 a pound to get 45 pounds of a mixture worth $4 a pound. How many pounds of the $6 and the $3 coffee will be needed?

270. A grocer wishes to blend candy selling for $1.60 a pound with candy selling for $2.50 a pound to get a mixture that will be sold for $1.90 a pound. How many pounds of the $1.60 and the $2.50 candy should be used to get 30 pounds of the mixture?

271. How many barrels of olives worth $40 a barrel must be mixed with olives worth $65 a barrel to get 50 barrels worth $53 a barrel?

272. How many pounds of $4.20 a pound candy should be mixed with candy worth $3.50 a pound to get 14 pounds of candy worth $3.90 a pound?

273. Roberta Candies sells caramels that cost $3.65 per pound mixed with creams that cost $3.25 per pound. How much of each kind of candy is in a pound of the mixture if it costs $3.49?

274. How many liters of a 35% solution should be mixed with a 55% solution to get 40 liters of a 43% solution?

275. A 90% antifreeze solution is to be mixed with a 75% solution to make 30 liters of an 80% solution. How many liters of the 90% and 75% solutions should be used?

276. A pharmacist needs 15 liters of 20% alcohol solution. He has only 15% alcohol solution and 30% alcohol solution on hand to make the mixture. How many liters of each solution should he combine to make the mixture?

277. A 10% solution of antifreeze is mixed with a 50% solution of antifreeze to get 100 liters of a 22% solution. How many liters of each solution are used?

278. A 25% alcohol solution is mixed with a 40% alcohol solution to get 60 liters of a 30% alcohol solution. How many liters of each solution are used?

Objective 4   Solve problems about rate or speed.

Use a system of equations to solve the problem.

279. Two cars start from positions 300 miles apart and race toward each other. They meet after 3 hours. Find the average speed of each car if one car travels 30 miles per hour faster than the other.

280. Two trains start from positions 500 miles apart and travel toward each other. They meet after 2 ½ hours. Find the average speed of each train if one train travels 20 miles per hour faster than the other.

281. At the beginning of a fundraising walk, Barry and Dave are 30 miles apart. If they leave at the same time and walk in the same direction, Barry would overtake Dave in 15 hours. If they walked toward each other, they would meet in 3 hours. What are their speeds?

282. Two cars leave from the same place and travel in the same direction. One car travels 1 3/8 times as fast as the other. After 4 hours they are 60 miles apart. Find the speed of each car.

283. Gareth left Somerset traveling to Akron 210 miles away at the same time as Gila left Akron traveling to Somerset. They met after 2 hours. If Gila was traveling 1.1 times as fast as Gareth, what were their speeds?

284. Mr. Burton left Louisville at noon on the same day that Mr. Quigley left Louisville at 1 P.M. Both were traveling in the same direction. At 5 P.M., Mr. Quigley was 62 miles behind Mr. Burton. If Mr. Burton was traveling 2 miles per hour faster than Mr. Quigley, what were their speeds?

285. It takes a boat 1 hour to go 16 miles downstream and 8 hours to return. Find the speed of the current and the speed of the boat in still water.

286. A plane can travel 300 miles per hour with the wind and 230 miles per hour against the wind. Find the speed of the wind and the speed of the plane in still air.

287. It takes a kayak 1½ hours to go 24 miles downstream and 4 hours to return. Find the speed of the current and the speed of the kayak in still water.

288. Two planes left Atlanta traveling in opposite directions. Plane A left 15 minutes before plane B. After plane B had been flying for 1 hour, the planes were 860 miles apart. What were the speeds of the two planes if plane A was flying 40 miles per hour faster than plane B?

Mixed Exercises

Use a system of equations to solve the problem.

289. The perimeter of a rectangle is 68 centimeters. The length is 7 centimeters more than twice the width. Find the dimensions of the rectangle.

290. Stefan plans to buy 10 ties with exactly $132. If some ties cost $12, and the others cost $15, how many of each price should he buy?

291. Rosita has 14 bills in her wallet worth $95 altogether. If the wallet contains only $5 and $10 bills, how many bills of each denomination does she have?

292. Two airplanes start out together and travel in opposite directions. At the end of 4 hours, they are 4240 kilometers apart. If one plane travels 60 kilometers per hour faster than the other, what are their speeds?

293. A 30% acid solution is to be mixed with a 50% acid solution to get 100 milliliters of a 35% acid solution. How many milliliters of 30% solution are needed?

294. Emil invested $50,000 in two accounts last year. One account earned 6% and the other earned 7% annual interest. If the total income last year from these accounts was $3350, how much was invested in each account?

295. It takes Carla's boat ½ hour to go 8 miles downstream and 1 hour to make the return trip upstream. Find the speed of the current and the speed of Carla's boat in still water.

296. A merchant wishes to make 150 pounds of coffee blend that can be sold for $4 per pound. The blend will be a mixture of coffee worth $6 per pound and coffee worth $3 per pound. How many pounds of each kind of coffee should be used in the mixture?

297. Andy leaves Wilmont, driving by car to Southdale, which is 168 kilometers away. At the same time, Ralph, riding his bicycle, leaves Southdale cycling toward Wilmont. Andy is traveling 40 kilometers per hour faster than Ralph. They pass each other 1½ hours later. What are their speeds?

298. How many liters of water should be added to a 25% antifreeze solution to get 30 liters of a 20% solution?

299. The sum of two numbers is –17. If one number is one less than three times the other, find the numbers.

300. The sum of two numbers is –3. If one number is tripled, the result is 23 more than the other number. Find the original numbers.

301. Two angles are complementary. The larger angle measures 15° more than twice the smaller. Find the measures of the angles.

302. Two angles are supplementary. The difference between their measures is 56°. Find the measures of the angles.

303. A bank teller has 83 bills of $1 and $5 denominations. How many of each type does he have if the total value of the money is $215?

304. The box office is selling reserved seat tickets and general admission tickets for $10 and $5 respectively. The total value of the tickets sold is $630. If there were 103 tickets sold, how many of each type are there?

305. Mr. Rivera has twice as much money invested at 14% as he has at 11%. If his yearly income from the investment is $780, how much does he have invested at each rate?

306. Mr. Salazar has half as much money invested at 9% as he has at 12%. If his yearly income from the two investments is $1650, how much does he have invested at each rate?

307. Two cars leave the same town at the same time traveling in opposite directions. After 5 hours they are 475 miles apart. Find the average speed of each car if the average speed of one is 5 miles per hour greater than the average speed of the other.

308. John hiked up to Lookout Point from the ranger station in 2 hours. The trip down along the same route took him 1 1/2 hours. Find the distance from the ranger station to Lookout Point if his average speed going was 5 miles per hour less than twice his average speed returning.

309. A freight train leaves a station traveling west. One hour later a passenger train leaves the same station traveling in the same direction. The passenger train overtakes the freight train in three hours. Find the average speed of the passenger train if the average speed of the passenger train is 15 miles per hour more than the average speed of the freight train.

310. Jorge has a coin collection containing pennies, nickels, and dimes. He has 432 coins altogether. The total value of the coins is $32.28. If he has the same number of pennies as nickels, how many coins of each denomination does he have?

311. Traveling with the current a boat can travel the same distance in 2 hours as it can traveling against the current in 5 hours. If the average speed of the boat in still water is 4 miles per hour more than the speed of the current, find the average speed of the boat in still water.

312. Shamim and Rohit live 460 miles apart. They start driving toward each other at the same time and meet in 4 hours. Rohit's average speed is 5 miles per hour more than Shamim's average speed. Find Shamim's average speed.

Section 8.5   Solving Systems of Linear Inequalities

Objective 1 and 2   Solve systems of linear inequalities by graphing; Recognize how a graphing calculator is used to solve a system of linear inequalities.

Graph the solutions of each system of linear inequalities. Then use the graphing capability of your calculator to verify your answers.

313. $4x + 5y \le 20$
     $y \le x + 3$

314. $3x + 5y \ge 15$
     $y \ge x - 2$

315. $x + y \le 3$
     $5x - y \ge 5$

316. $x + y \ge 3$
     $x - 2y \le 4$

317. $7x + 3y \ge 21$
     $x - y \le 6$

318. $2x + 5y \le 10$
     $x - y \ge 4$

319. $3x - y \le 6$
     $3y - 6 \le 2x$

320. $x + 2y \ge -4$
     $5x \le 10 - 2y$

321. $3x - y > 3$
     $4x + 3y < 12$

322. $6x - y > 6$
     $2x + 5y < 10$

323. $3x - 4y < 12$
     $y > -4$

324. $4x - 3y > 12$
     $x < 4$

325. $4x - 3y < 9$
     $y > 2$

326. $5x - 2y \le 10$
     $y \le -2$

327. $2x - y \ge 4$
     $5y + 15 \ge -3x$

328. $4x - y \le 4$
     $7y + 14 \ge -2x$

329. $x - y \le 2$
     $y \le 2$

330. $3x - 2y < 8$
     $x < 4$

331. $x - 4y < 6$

    $x > 2y$

332. $x < 2y + 3$

    $0 < x + y$

333. $2x + 3y \leq 6$

    $y \leq x$

    $x \leq 2$

334. $4x - y \leq 8$

    $x + 2y \leq 4$

    $x \geq -2$

335. $x \geq 2$

    $x \leq 6$

    $y \geq 1$

    $y \leq 5$

336. $x - 2y \geq 6$

    $x + y \leq 6$

    $x \geq 0$

    $y \geq 0$

Writing/Conceptual Exercises

337. Explain the difference between infinitely many solutions to a system of linear inequalities and any ordered pair being a solution.

338. What determines whether to draw a solid boundary line or a dashed boundary line in solving a system of linear inequalities?

339. Can a system of linear inequalities have solutions in all quadrants? If so, graph the solution of such a system.

340. Can a system of linear inequalities have solutions in only one quadrant? If so, graph the solution of such a system.

CHAPTER 9 ROOTS AND RADICALS

Section 9.1 Evaluating Roots

Objective 1 Find square roots.

Find all square roots of the number.

1. 81

2. 169

3. 256

4. 400

5. 324

6. 576

7. 361

8. 1225

9. $\dfrac{49}{64}$

10. $\dfrac{121}{144}$

11. $\dfrac{169}{625}$

12. $\dfrac{196}{289}$

Find the square root.

13. $\sqrt{100}$

14. $-\sqrt{144}$

15. $-\sqrt{1}$

16. $\sqrt{36}$

17. $\sqrt{225}$

18. $-\sqrt{900}$

19. $\sqrt{10,000}$

20. $-\sqrt{529}$

21. $\sqrt{\dfrac{64}{225}}$

22. $-\sqrt{\dfrac{169}{441}}$

23. $\sqrt{\dfrac{256}{361}}$

24. $-\sqrt{\dfrac{2500}{6400}}$

Objective 2 Decide whether a given root is rational, irrational, or not a real number.

Tell whether the square root is *rational, irrational,* or *not a real number.*

25. $\sqrt{144}$

26. $\sqrt{3}$

27. $-\sqrt{16}$

28. $\sqrt{-25}$

29. $\sqrt{-95}$

30. $-\sqrt{43}$

31. $\sqrt{1600}$

32. $\sqrt{200}$

33. $\sqrt{\dfrac{4}{49}}$

34. $-\sqrt{\dfrac{625}{484}}$

35. $\sqrt{2.5}$

36. $\sqrt{-.9}$

Objective 3   Find decimal approximations for irrational square roots.

Use a calculator to find a decimal approximation for the root. Round answers to the nearest thousandth.

37. $\sqrt{45}$ 38. $\sqrt{31}$ 39. $-\sqrt{15}$ 40. $-\sqrt{90}$

41. $-\sqrt{150}$ 42. $\sqrt{300}$ 43. $-\sqrt{131}$ 44. $\sqrt{245}$

45. $\sqrt{832}$ 46. $-\sqrt{1100}$ 47. $\sqrt{12,500}$ 48. $-\sqrt{26,358}$

Objective 4   Use the Pythagorean formula.

Find the length of the unknown side of the right triangle with sides a, b, and c, where c is the hypotenuse. If necessary, round your answer to the nearest thousandth.

49. $a = 8,\ b = 15$ 50. $a = 28,\ b = 21$

51. $c = 35,\ a = 25$ 52. $c = 25,\ b = 24$

53. $c = 17,\ a = 16$ 54. $a = 3,\ b = 8$

Use the Pythagorean formula to solve the problem. If necessary, round your answer to the nearest thousandth.

55. The hypotenuse of a right triangle measure 10 centimeters and one leg measures 4 centimeters. How long is the other leg?

56. Two sides of a rectangle measure 7 centimeters and 11 centimeters. How long are the diagonals of the rectangle?

57. A diagonal of a rectangle measures 29 inches. The length of the rectangle is 21 inches. Find the width of the rectangle.

58. A diagonal of a rectangle measures 2.5 meters. The width of the rectangle is 1.5 meters. Find the length of the rectangle.

59. A ladder 25 feet long leans against a wall. The foot of the ladder is 7 feet from the base of the wall. How high on the wall does the top of the ladder rest?

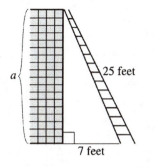

60. Laura is flying a kite on 60 feet of string. How high is the kite above her hand (vertically) if the horizontal distance between Laura and the kite is 36 feet?

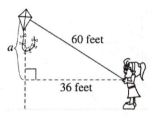

61. Kevin started to drive due south at the same time Lydia started to drive due west. Lydia drove 21 miles in the same time that Kevin drove 28 miles. How far apart were they at that time?

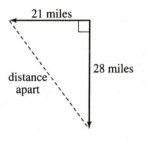

62. A cable from the top of a pole 15 feet tall is pulled taut and attached to the ground 8 feet from the base of the pole. How long is the cable?

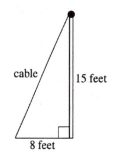

63. The foot of a loading ramp 9 feet long is placed 7 feet from the base of a platform. The top of the ramp rests on the platform. How high is the platform?

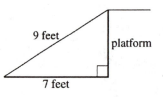

64. A plane flies due east for 35 miles and then due south until it is 37 miles from its starting point. How far south did the plane fly?

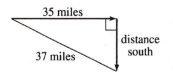

Objective 5   Use the distance formula.

Find the distance between each pair of points.

65. $(6,6),(3,8)$

66. $(6,2),(9,1)$

67. $(3,4),(0,-2)$

68. $(5,-6),(-5,6)$

69. $(-5,-1),(4,-7)$

70. $(0,0),(4,-3)$

71. $(-7,-6),(-7,6)$

72. $(4,0),(0,4)$

73. $(2,-2),(-6,1)$

74. $(5,-8),(-2,1)$

75. $(3,-2),(2,3)$

76. $(7,-11),(9,2)$

Objective 6   Find the roots.

Find each root that is a real number.

77. $\sqrt[3]{27}$

78. $\sqrt[3]{-125}$

79. $-\sqrt[5]{1}$

80. $\sqrt[3]{-1000}$

81. $\sqrt[3]{-216}$

82. $-\sqrt[3]{-8}$

83. $-\sqrt[5]{243}$

84. $-\sqrt[3]{-64}$

85. $\sqrt[4]{-81}$

86. $\sqrt[3]{-343}$

87. $-\sqrt[4]{625}$

88. $-\sqrt[5]{-32}$

89. $-\sqrt[6]{-1}$

90. $\sqrt[5]{-100,000}$

91. $-\sqrt[4]{-1}$

92. $\sqrt[3]{512}$

Mixed Exercises

Find each root that is a real number.

93. $\sqrt{196}$

94. $\sqrt[7]{-1}$

95. $\sqrt{-144}$

96. $\sqrt[4]{256}$

97. $-\sqrt[6]{64}$

98. $-\sqrt[4]{-81}$

99. $-\sqrt{\dfrac{225}{289}}$

100. $-\sqrt[3]{-\dfrac{27}{125}}$

Write *rational, irrational,* or *not a real number* to describe the number. If the number is rational, give its exact value. If the number is irrational, give a decimal approximation to the nearest thousandth. Use a calculator as necessary.

101. $\sqrt{225}$

102. $\sqrt{7}$

103. $-\sqrt{49}$

104. $\sqrt{75}$

105. $-\sqrt{\dfrac{81}{169}}$

106. $\sqrt{\dfrac{361}{441}}$

107. $\sqrt{8.5}$

108. $-\sqrt{.09}$

Find the length of the unknown side of the right triangle with sides a, b, and c, where c is the hypotenuse. If necessary, round your answer to the nearest thousandth.

109. $a = 10$, $b = 24$ 110. $b = 5$, $c = 9$

111. $a = 1$, $c = 12$ 112. $c = 41$, $a = 40$

113. $a = 5$, $b = 8$ 114. $b = 8$, $c = 11$

115. $a = 9$, $c = 17$ 116. $b = 6$, $c = 11$

Writing/Conceptual Exercises

Answer the question.

117. How many fourth roots does 0 have?

118. How many real number fourth roots does any positive number have?

119. How many real number fourth roots does any negative number have?

120. Which of the following numbers have rational square roots?

     (a) 5   (b) 49   (c) -9 (d) 121

121. Which of the following numbers have irrational square roots?

     (a) $\dfrac{25}{36}$ (b) $\dfrac{1}{2}$ (c) 15   (d) $-\dfrac{4}{9}$

What must be true about b for the statement to be true?

122. $\sqrt[3]{b}$ is a negative number.   123. $-\sqrt[3]{b}$ is a negative number.

124. $\sqrt[3]{-b}$ is a positive number.   125. $-\sqrt[3]{b}$ is a positive number.

126. $\sqrt[3]{b}$ is zero.

Section 9.2   Multiplication and Division of Radicals

Objective 1   Multiply radicals.

Use the product rule for radicals to find the product.

127. $\sqrt{5} \cdot \sqrt{3}$ 128. $\sqrt{12} \cdot \sqrt{3}$

129. $\sqrt{7} \cdot \sqrt{7}$ 130. $\sqrt{2} \cdot \sqrt{32}$

131. $\sqrt{6} \cdot \sqrt{7}$ 132. $\sqrt{6} \cdot \sqrt{24}$

133. $\sqrt{23} \cdot \sqrt{23}$ 134. $\sqrt{5} \cdot \sqrt{20}$

135. $\sqrt{13} \cdot \sqrt{11}$ 136. $\sqrt{6} \cdot \sqrt{r}, r > 0$

137. $\sqrt{2x} \cdot \sqrt{7}, x > 0$ 138. $\sqrt{18a} \cdot \sqrt{2b}, a > 0, b > 0$

Objective 2   Simplify radicals using the product rule.

Simplify the radical. Assume that all variables represent nonnegative real numbers.

139. $\sqrt{24}$ 140. $\sqrt{72}$ 141. $\sqrt{500}$ 142. $-\sqrt{150}$

143. $-\sqrt{56}$ 144. $\sqrt{288}$ 145. $-2\sqrt{54}$ 146. $3\sqrt{1000}$

147. $\sqrt{y^3}$ 148. $\sqrt{p^2 q^4}$ 149. $\sqrt{80a^5}$ 150. $\sqrt{125r^7 s^6}$

Find the product and simplify.

151. $\sqrt{5} \cdot \sqrt{15}$ 152. $\sqrt{3} \cdot \sqrt{20}$ 153. $\sqrt{6} \cdot \sqrt{12}$

154. $\sqrt{30} \cdot \sqrt{10}$ 155. $\sqrt{11} \cdot \sqrt{22}$ 156. $\sqrt{18} \cdot \sqrt{24}$

Objective 3   Simplify radical quotients using the quotient rule.

Use the quotient rule and product rule, as necessary, to simplify the expression.

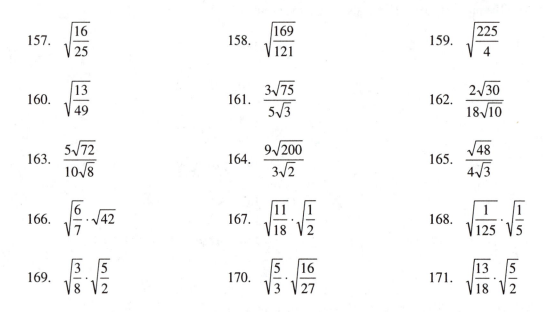

157. $\sqrt{\dfrac{16}{25}}$

158. $\sqrt{\dfrac{169}{121}}$

159. $\sqrt{\dfrac{225}{4}}$

160. $\sqrt{\dfrac{13}{49}}$

161. $\dfrac{3\sqrt{75}}{5\sqrt{3}}$

162. $\dfrac{2\sqrt{30}}{18\sqrt{10}}$

163. $\dfrac{5\sqrt{72}}{10\sqrt{8}}$

164. $\dfrac{9\sqrt{200}}{3\sqrt{2}}$

165. $\dfrac{\sqrt{48}}{4\sqrt{3}}$

166. $\sqrt{\dfrac{6}{7}} \cdot \sqrt{42}$

167. $\sqrt{\dfrac{11}{18}} \cdot \sqrt{\dfrac{1}{2}}$

168. $\sqrt{\dfrac{1}{125}} \cdot \sqrt{\dfrac{1}{5}}$

169. $\sqrt{\dfrac{3}{8}} \cdot \sqrt{\dfrac{5}{2}}$

170. $\sqrt{\dfrac{5}{3}} \cdot \sqrt{\dfrac{16}{27}}$

171. $\sqrt{\dfrac{13}{18}} \cdot \sqrt{\dfrac{5}{2}}$

Objective 4   Use the product and quotient rules to simplify higher roots.

Simplify the expression.

172. $\sqrt[3]{24}$

173. $\sqrt[3]{81}$

174. $\sqrt[3]{128}$

175. $\sqrt[4]{243}$

176. $\sqrt[5]{64}$

177. $\sqrt[4]{320}$

178. $-\sqrt[3]{\dfrac{64}{27}}$

179. $-\sqrt[3]{-\dfrac{1}{216}}$

180. $\sqrt[3]{\dfrac{8}{125}}$

181. $\sqrt[3]{\dfrac{1728}{1000}}$

182. $-\sqrt[4]{\dfrac{16}{81}}$

183. $\sqrt[4]{\dfrac{625}{256}}$

184. $\sqrt[3]{3} \cdot \sqrt[3]{9}$

185. $\sqrt[5]{25} \cdot \sqrt[5]{125}$

186. $\sqrt[4]{3} \cdot \sqrt[4]{27}$

Mixed Exercises

Find the product and simplify if possible.

187. $\sqrt{11} \cdot \sqrt{11}$

188. $\sqrt{13} \cdot \sqrt{15}$

189. $\sqrt{3} \cdot \sqrt{75}$

190. $\sqrt{10} \cdot \sqrt{50}$

191. $\sqrt{13} \cdot \sqrt{26}$

192. $\sqrt{79} \cdot \sqrt{79}$

Simplify the expression. Assume that all variables represent positive numbers.

193. $\sqrt{99}$

194. $\sqrt{245}$

195. $-\sqrt{175}$

196. $\sqrt{45} \cdot \sqrt{15}$

197. $\sqrt{28} \cdot \sqrt{63}$

198. $\sqrt{9x} \cdot \sqrt{3x}$

199. $-\sqrt{\dfrac{144}{225}}$

200. $\dfrac{28\sqrt{33}}{7\sqrt{11}}$

201. $\sqrt[5]{-\dfrac{243}{32}}$

202. $\sqrt{\dfrac{50a}{a^3}}$

203. $\sqrt{\dfrac{5}{3}} \cdot \sqrt{\dfrac{27}{125}}$

204. $\sqrt{\dfrac{40}{63}} \cdot \sqrt{\dfrac{7}{10}}$

205. $\sqrt{75a^3b}$

206. $\sqrt[3]{40}$

207. $\sqrt[5]{-64}$

208. $\sqrt[3]{2} \cdot \sqrt[3]{4}$

209. $\sqrt{98x^6 y^7}$

210. $\sqrt[4]{243}$

Writing/Conceptual Exercises

Decide whether the statement is *true* or *false*.

211. $\sqrt{4 \cdot 25} = \sqrt{4} \cdot \sqrt{25}$

212. $\sqrt{4 + 25} = \sqrt{4} + \sqrt{25}$

213. $\sqrt{(-10)^2} = -10$

214. $\sqrt[3]{(-10)^3} = -10$

215. Which of the following radicals are simplified according to the guidelines given in the textbook?

    (a) $\sqrt{54}$ (b) $\sqrt{19}$ (c) $\sqrt{21}$ (d) $\sqrt{72}$

216. Which of the following radicals are simplified according to the guidelines given in the textbook?

(a) $\sqrt[3]{9}$ (b) $\sqrt[3]{16}$ (c) $\sqrt[3]{125}$ (d) $\sqrt[3]{20}$

Section 9.3   Addition and Subtraction of Radicals

Objective 1   Add and subtract radicals.

Add or subtract, as indicated.

217. $-4\sqrt{5} + 9\sqrt{5}$

218. $7\sqrt{11} - 5\sqrt{11}$

219. $\sqrt[3]{10} + \sqrt[3]{10}$

220. $5\sqrt{7} - 4\sqrt{7}$

221. $\sqrt{5} + \sqrt{3}$

222. $5\sqrt{3} - 2\sqrt{3} + 7\sqrt{3}$

223. $2\sqrt{6} - 4\sqrt{6} + \sqrt{6}$

224. $5\sqrt{5} + 3\sqrt{5} - 4\sqrt{5}$

225. $9\sqrt[4]{12} + 6\sqrt[4]{12} - 2\sqrt[4]{12}$

226. $6\sqrt[5]{4} - 2\sqrt[5]{4} + \sqrt[5]{4}$

Objective 2   Simplify radical sums and differences.

Simplify and add or subtract terms wherever possible. Assume that all variables represent nonnegative real numbers.

227. $\sqrt{8} + 3\sqrt{2}$

228. $2\sqrt{32} + \sqrt{2}$

229. $3\sqrt{200} - 6\sqrt{18}$

230. $3\sqrt{11} - 5\sqrt{44}$

231. $5\sqrt{24} - 2\sqrt{54} + 3\sqrt{20}$

232. $-5\sqrt{7} - 2\sqrt{28} + 4\sqrt{28}$

233. $\dfrac{1}{5}\sqrt{50} - \dfrac{3}{2}\sqrt{12}$

234. $\dfrac{\sqrt{8}}{2} - \dfrac{5}{4}\sqrt{32}$

235. $\dfrac{5}{6}\sqrt{72} - \dfrac{3}{4}\sqrt{48}$

236. $\dfrac{3}{8}\sqrt{128} - \dfrac{3}{2}\sqrt{8}$

237. $6\sqrt{8z} - 3\sqrt{18z}$

238. $3\sqrt{96w} - 2\sqrt{6w}$

239. $3\sqrt{20x} - \sqrt{500x}$

240. $\sqrt[4]{162} - \sqrt[4]{32}$

241. $2\sqrt[4]{243} - 3\sqrt[4]{48}$

242. $4\sqrt[3]{r^5} + 3\sqrt[3]{27r^5}$

Objective 3   Simplify radical sums involving multiplication.

Perform the indicated operations. Assume that all variables represent nonnegative real numbers.

243. $\sqrt{5} \cdot \sqrt{3} + 3\sqrt{15}$

244. $4\sqrt{7} \cdot \sqrt{6} - 2\sqrt{42}$

245. $7\sqrt{14} - 2\sqrt{7} \cdot \sqrt{2}$

246. $8\sqrt{35} + 4\sqrt{5} \cdot \sqrt{7}$

247. $3\sqrt{x} \cdot \sqrt{4} + \sqrt{4x}$

248. $3\sqrt{2y} \cdot \sqrt{18} - 5\sqrt{6y}$

249. $10x\sqrt{5} - 3\sqrt{15x} \cdot \sqrt{3x}$

250. $11\sqrt{5w} \cdot \sqrt{30w} - 8w\sqrt{24}$

251. $\sqrt{6k^2} \cdot \sqrt{3k} + k\sqrt{2k}$

252. $3\sqrt{7y} \cdot \sqrt{35} - 2\sqrt{80y}$

Mixed Exercises

Simplify and perform the indicated operations. Assume that all variables represent nonnegative real numbers.

253. $3\sqrt{19} - 2\sqrt{19}$

254. $3\sqrt[3]{11} + 5\sqrt[3]{11} - 7\sqrt[3]{11}$

255. $6\sqrt{3} + 2\sqrt{12} - 3\sqrt{27}$

256. $\sqrt{80} + 2\sqrt{45}$

257. $4\sqrt{63m} - 3\sqrt{7m}$

258. $3\sqrt[3]{16} - 4\sqrt[3]{54}$

259. $\sqrt{7} \cdot \sqrt{11} - 3\sqrt{7m}$

260. $\sqrt{5r^2} \cdot \sqrt{40} - \sqrt{8r^2}$

261. $\dfrac{3}{4}\sqrt{48} - \dfrac{1}{2}\sqrt{12}$

262. $\dfrac{5}{8}\sqrt{32} + \dfrac{3}{10}\sqrt{50}$

263. $6\sqrt{18} + 2\sqrt{48} - 6\sqrt{28}$

264. $3\sqrt[3]{128} + 2\sqrt[3]{250}$

265. $3\sqrt{13} - \sqrt{13} + 5\sqrt{52}$

266. $2\sqrt{18} - 5\sqrt{32} + 7\sqrt{162}$

267. $6\sqrt[3]{135} + 3\sqrt[3]{40}$

268. $-3\sqrt[4]{243} - 2\sqrt[4]{48}$

269. $3\sqrt{6} - 8\sqrt{6} - 5\sqrt{24}$

270. $\sqrt{98} - 2\sqrt{8} + \sqrt{32}$

271. $\sqrt[3]{625} + \sqrt[3]{135} - \sqrt[3]{40}$

272. $-2\sqrt[3]{81} + \sqrt[3]{24}$

273. $3\sqrt{8z} + 3\sqrt{2z} + \sqrt{32z}$

274. $\sqrt[3]{81} + \sqrt[3]{24} + \sqrt[3]{192}$

275. $\sqrt{48y} + \sqrt{12y} + \sqrt{27y}$

276. $2\sqrt[3]{16r} + \sqrt[3]{54} - \sqrt[3]{16r}$

277. $3\sqrt{48} + 5\sqrt{27}$

278. $\sqrt{100x} - \sqrt{9x} + \sqrt{25x}$

Writing/Conceptual Exercises

279. Write an equation showing how the distributive property is used to justify the statement

$$4\sqrt{7} + 8\sqrt{7} = 12\sqrt{7}.$$

280. Write an equation showing how the distributive property is used to justify the statement

$$11\sqrt[3]{3} - 4\sqrt[3]{3} = 7\sqrt[3]{3}.$$

Despite the fact that $\sqrt{25}$ and $\sqrt[3]{8}$ are radicals that have different root indexes, they can be added to obtain a single term:

$$\sqrt{25} + \sqrt[3]{8} = 5 + 2 = 7.$$

281. Make up a similar sum of radicals that leads to an answer of 12.

282. Make up a similar difference of radicals that leads to an answer of 8.

Section 9.4   Rationalizing the Denominator

Objective 1   Rationalize denominators with square roots.

Rationalize the denominator.

283. $\dfrac{2}{\sqrt{5}}$

284. $\dfrac{-4}{\sqrt{13}}$

285. $\dfrac{-2}{\sqrt{6}}$

286. $\dfrac{5}{\sqrt{15}}$

287. $\dfrac{2}{\sqrt{2}}$

288. $\dfrac{25}{\sqrt{10}}$

289. $\dfrac{-22}{\sqrt{11}}$

290. $\dfrac{-8}{\sqrt{12}}$

291. $\dfrac{\sqrt{3}}{\sqrt{7}}$

292. $\dfrac{\sqrt{5}}{\sqrt{8}}$

293. $\dfrac{\sqrt{6}}{\sqrt{12}}$

294. $\dfrac{4}{\sqrt{28}}$

295. $\dfrac{\sqrt{3}}{\sqrt{75}}$

296. $\dfrac{\sqrt{5}}{\sqrt{125}}$

297. $\dfrac{25\sqrt{5}}{\sqrt{250}}$

298. $\dfrac{6\sqrt{3}}{\sqrt{96}}$

Objective 2   Write radicals in simplified form.

Perform the indicated operations and write all answers in simplest form. Rationalize all denominators. Assume that all variables represent positive real numbers.

299. $\sqrt{\dfrac{1}{5}}$

300. $\sqrt{\dfrac{6}{7}}$

301. $\sqrt{\dfrac{3}{8}}$

302. $\sqrt{\dfrac{32}{27}}$

303. $\sqrt{\dfrac{63}{50}}$

304. $\sqrt{\dfrac{3}{2}} \cdot \sqrt{\dfrac{5}{6}}$

305. $\sqrt{\dfrac{8}{21}} \cdot \sqrt{7}$

306. $\sqrt{6} \cdot \sqrt{\dfrac{11}{30}}$

307. $\sqrt{\dfrac{1}{5}} \cdot \sqrt{\dfrac{6}{35}}$

308. $\dfrac{\sqrt{6y}}{\sqrt{x}}$

309. $\sqrt{\dfrac{7r}{s}}$

310. $\sqrt{\dfrac{3a^2b}{5}}$

311. $\dfrac{\sqrt{k^2m}}{\sqrt{k^5}}$

312. $\sqrt{\dfrac{df^3}{d^4}}$

313. $\sqrt{\dfrac{20a^3b}{6a}}$

Objective 3   Rationalize denominators with cube roots.

Rationalize the denominator. Assume that all variables in the denominator represent nonzero real numbers.

314. $\dfrac{\sqrt[3]{6}}{\sqrt[3]{9}}$

315. $\dfrac{\sqrt[3]{9}}{\sqrt[3]{4}}$

316. $\sqrt[3]{\dfrac{1}{16}}$

317. $\sqrt[3]{\dfrac{5}{2}}$

318. $\sqrt[3]{\dfrac{7}{5}}$

319. $\sqrt[3]{\dfrac{5}{3}}$

320. $\sqrt[3]{\dfrac{8}{25}}$

321. $\dfrac{6}{\sqrt[3]{36}}$

322. $\sqrt[3]{\dfrac{1}{24}}$

323. $\sqrt[3]{\dfrac{3}{80}}$

324. $\sqrt[3]{\dfrac{1}{162}}$

325. $\sqrt[3]{\dfrac{4}{5r^2}}$

326. $\sqrt[3]{\dfrac{2s}{9r}}$

327. $\sqrt[3]{\dfrac{v}{36w^2}}$

328. $\sqrt[3]{\dfrac{8t}{25u}}$

Mixed Exercises

Simplify the expression. Assume that all variables represent positive real numbers.

329. $\dfrac{63}{\sqrt{7}}$

330. $\dfrac{-24}{\sqrt{6}}$

331. $\dfrac{2\sqrt{3}}{\sqrt{2}}$

332. $\sqrt{\dfrac{5}{3}}$

333. $\sqrt{\dfrac{11}{20}}$

334. $\dfrac{6\sqrt{7}}{\sqrt{14}}$

335. $\sqrt[3]{\dfrac{2}{7}}$

336. $\sqrt{\dfrac{2}{5}} \cdot \sqrt{\dfrac{15}{22}}$

337. $\sqrt[3]{\dfrac{5}{6x}}$

338. $\sqrt{\dfrac{27}{98}} \cdot \sqrt{\dfrac{1}{3}}$

339. $\sqrt{\dfrac{72qt^5}{3t^6}}$

340. $\sqrt[3]{\dfrac{7x^2}{81y^2}}$

341. $\dfrac{-14}{\sqrt{27}}$

342. $\sqrt[3]{\dfrac{r}{98}}$

343. $\sqrt{\dfrac{27}{48}}$

344. $\sqrt[3]{\dfrac{t^{15}}{x}}$

345. $\sqrt[3]{\dfrac{7}{36}}$

346. $\dfrac{3\sqrt{2y}}{\sqrt{t^3}}$

347. $\sqrt{\dfrac{162x^4}{t^5}}$

348. $\dfrac{7}{\sqrt{75}}$

349. $\sqrt{\dfrac{5}{8x^2}}$

350. $\sqrt{\dfrac{19}{32}}$

351. $\sqrt[3]{\dfrac{c^3}{d^2}}$

352. $\dfrac{3\sqrt{2}}{\sqrt{t}}$

Writing/Conceptual Exercises

353. Which one of the following would be an appropriate choice for multiplying the numerator and denominator of $\dfrac{\sqrt[3]{5}}{\sqrt[3]{7}}$ in order to rationalize the denominator?

   (a) $\sqrt[3]{5}$ (b) $\sqrt[3]{7}$ (c) $\sqrt[3]{49}$ (d) $\sqrt[3]{2}$

354. Which one of the following would be an appropriate choice for multiplying the numerator and denominator of $\dfrac{\sqrt[3]{4y}}{\sqrt[3]{10z}}$ in order to rationalize the denominator?

   (a) $\sqrt[3]{10z}$ (b) $\sqrt[3]{100z}$ (c) $\sqrt[3]{10z^2}$ (d) $\sqrt[3]{100z^2}$

355. Which one of the following would be an appropriate choice for multiplying the numerator and denominator of $\dfrac{\sqrt[3]{5s}}{\sqrt[3]{9r^2}}$ in order to rationalize the denominator?

   (a) $\sqrt[3]{3r}$ (b) $\sqrt[3]{9r^2}$ (c) $\sqrt[3]{3r^2}$ (d) $\sqrt[3]{25s^2}$

356. What would be your first step in simplifying the radical $\sqrt[3]{\dfrac{7}{11}}$? What property would you be using in this step?

Section 9.5   Simplifying Radical Expressions

Objective 1   Simplify products of radical expressions.

Simplify the expression.

357. $\sqrt{7}\left(\sqrt{5}-\sqrt{3}\right)$

358. $\sqrt{10}\left(\sqrt{5}+\sqrt{6}\right)$

359. $\sqrt{5}\left(\sqrt{15}+2\sqrt{3}\right)$

360. $\sqrt{7}\left(3\sqrt{11}-2\sqrt{7}\right)$

361. $\left(2\sqrt{5}+\sqrt{3}\right)\left(\sqrt{2}+\sqrt{7}\right)$

362. $\left(\sqrt{5}-\sqrt{11}\right)\left(\sqrt{3}+\sqrt{2}\right)$

363. $\left(\sqrt{6}-\sqrt{5}\right)\left(2\sqrt{6}+\sqrt{10}\right)$

364. $\left(4\sqrt{2}-3\sqrt{3}\right)\left(\sqrt{2}+4\sqrt{3}\right)$

365. $\left(\sqrt{5}+\sqrt{10}\right)\left(\sqrt{5}-\sqrt{10}\right)$

366. $\left(2\sqrt{3}-\sqrt{7}\right)\left(2\sqrt{3}+\sqrt{7}\right)$

367. $\left(2\sqrt{3}-\sqrt{5}\right)^2$

368. $\left(\sqrt{10}-\sqrt{11}\right)^2$

369. $\left(3+4\sqrt{5}\right)^2$

370. $\left(3\sqrt{2}+2\sqrt{3}\right)^2$

Objective 2   Simplify quotients of radical expressions.

Rationalize the denominator.

371. $\dfrac{2}{3-\sqrt{2}}$

372. $\dfrac{5}{2+\sqrt{5}}$

373. $\dfrac{5}{4-\sqrt{3}}$

374. $\dfrac{4+\sqrt{3}}{\sqrt{3}}$

375. $\dfrac{\sqrt{5}}{\sqrt{3}-1}$

376. $\dfrac{2}{\sqrt{5}+\sqrt{3}}$

377. $\dfrac{9}{\sqrt{5}+\sqrt{7}}$

378. $\dfrac{6}{\sqrt{3}+\sqrt{10}}$

379. $\dfrac{\sqrt{7}+\sqrt{2}}{\sqrt{3}-\sqrt{2}}$

380. $\dfrac{\sqrt{6}+1}{\sqrt{2}-4}$

381. $\dfrac{\sqrt{2}+\sqrt{3}}{\sqrt{7}-\sqrt{2}}$

382. $\dfrac{\sqrt{10}+3}{\sqrt{10}-3}$

Objective 3   Write radical expressions with quotients in lowest terms.

Write the quotient in lowest terms.

383. $\dfrac{8-2\sqrt{3}}{2}$

384. $\dfrac{6+2\sqrt{7}}{12}$

385. $\dfrac{5\sqrt{3}+15}{5}$

386. $\dfrac{4\sqrt{2}+8}{14}$

387. $\dfrac{9+6\sqrt{2}}{12}$

388. $\dfrac{4\sqrt{7}-6}{10}$

389. $\dfrac{4+\sqrt{8}}{2}$

390. $\dfrac{3+\sqrt{27}}{6}$

391. $\dfrac{8\sqrt{6}-12}{16}$

392. $\dfrac{14+7\sqrt{3}}{28}$

393. $\dfrac{135\sqrt{3}+45}{15}$

394. $\dfrac{72\sqrt{5}-16\sqrt{2}}{24}$

Mixed Exercises

Simplify the expression.

395. $\sqrt{6}\left(\sqrt{5}-\sqrt{11}\right)$

396. $\left(4-\sqrt{6}\right)\left(2\sqrt{5}+1\right)$

397. $\left(\sqrt{11}+\sqrt{2}\right)^2$

398. $\left(7+\sqrt{13}\right)\left(7-\sqrt{13}\right)$

399. $\dfrac{1}{4-\sqrt{11}}$

400. $\dfrac{\sqrt{5}+2}{3-\sqrt{2}}$

401. $\dfrac{\sqrt{5}-1}{\sqrt{7}+1}$

402. $\dfrac{-5\sqrt{5}}{3-\sqrt{15}}$

403. $\dfrac{\sqrt{7}-\sqrt{3}}{\sqrt{2}}$

404. $\dfrac{\sqrt{7}+\sqrt{3}}{\sqrt{7}-\sqrt{3}}$

405. $\left(5\sqrt{15}+2\sqrt{5}\right)\left(3\sqrt{15}-\sqrt{5}\right)$

406. $\left(2\sqrt{5}+5\sqrt{2}\right)^2$

407. $\left(\sqrt{2}-\sqrt{12}\right)^2$

408. $\left(\sqrt{10}+\sqrt{3}\right)\left(\sqrt{6}-\sqrt{11}\right)$

409. $\left(5+\sqrt{2}\right)\left(5-\sqrt{2}\right)$

410. $\dfrac{26}{\sqrt{11}+\sqrt{2}}$

Write the quotient in lowest terms.

411. $\dfrac{4-2\sqrt{7}}{8}$

412. $\dfrac{20-10\sqrt{10}}{4}$

413. $\dfrac{6\sqrt{10}-9}{3}$

414. $\dfrac{5\sqrt{11}-15}{10}$

415. $\dfrac{65\sqrt{2} - 52}{13}$

416. $\dfrac{19\sqrt{6} - 133\sqrt{3}}{95}$

417. $\dfrac{7 - \sqrt{98}}{14}$

418. $\dfrac{2 - \sqrt{12}}{6}$

419. $\dfrac{16 - 12\sqrt{72}}{24}$

420. $\dfrac{9 + 3\sqrt{60}}{12}$

Section 9.6   Solving Equations with Radicals

Objective 1   Solve equations with radicals.

Solve the equation.

421. $\sqrt{w} = 9$

422. $\sqrt{q} - 1 = 0$

423. $8 - \sqrt{y} = 6$

424. $\sqrt{b + 16} = 6$

425. $\sqrt{2q + 6} = 4$

426. $\sqrt{2z - 5} = 0$

427. $\sqrt{2 + 3k} = 2\sqrt{k}$

428. $\sqrt{6r + 15} = 3\sqrt{r}$

429. $\sqrt{3x + 3} = 3\sqrt{x}$

430. $\sqrt{t + 6} - 5 = 0$

431. $\sqrt{5t + 6} = \sqrt{6t - 1}$

432. $\sqrt{4x + 18} = \sqrt{3x + 15}$

Objective 2   Identify equations with no solutions.

Solve the equation if it has a solution.

433. $\sqrt{y} = -10$

434. $8 - \sqrt{p} = 0$

435. $\sqrt{z} + 7 = 0$

436. $0 = 5 - \sqrt{r}$

437. $\sqrt{y + 3} = -1$

438. $\sqrt{k - 5} + 7 = 0$

439. $\sqrt{x + 2} + 6 = 0$

440. $\sqrt{2m + 13} = 3\sqrt{m + 3}$

441. $\sqrt{4-3n} = 2\sqrt{-n}$

442. $\sqrt{3p-2} = \sqrt{4p+6}$

443. $r = \sqrt{r^2 - 6r + 12}$

444. $s = \sqrt{s^2 + 4s + 4}$

445. $b = \sqrt{b^2 - 3b + 15}$

446. $\sqrt{d^2 + 4d + 12} + d = 0$

Objective 3   Solve equations that require squaring a binomial.

Find all solutions for the equation.

447. $\sqrt{5y+1} = y + 1$

448. $\sqrt{m+5} = m + 3$

449. $\sqrt{x-5} + 5 = x$

450. $\sqrt{x+13} - 1 = x$

451. $t + 3 = \sqrt{t^2 + 2t + 5}$

452. $q - 1 = \sqrt{q^2 + 8q + 11}$

453. $5\sqrt{p} - 4 = p + 2$

454. $3\sqrt{a} - 1 = a + 1$

455. $\sqrt{x} - 2 = x - 8$

456. $\sqrt{y+21} = y + 1$

457. $\sqrt{b-2} = b - 4$

458. $3\sqrt{p+1} = p + 1$

459. $2\sqrt{c+2} = c + 3$

460. $6\sqrt{t} - 5 = t + 3$

461. $3\sqrt{x+13} = x + 9$

462. $\sqrt{2x} - x + 4 = 0$

463. $\sqrt{t+2} - \sqrt{t-3} = 1$

464. $\sqrt{a-1} + 1 = \sqrt{a+4}$

465. $\sqrt{3k+7} = 2 - \sqrt{k+1}$

466. $\sqrt{x+21} = 3 + \sqrt{x}$

467. $\sqrt{3x+3} = \sqrt{2x+3} + 1$

468. $\sqrt{2-r} + \sqrt{r+11} = 5$

Mixed Exercises

Find all solutions for the equation.

469. $\sqrt{z-8} = 5$

470. $\sqrt{2q+3} = 4$

471. $\sqrt{3r+7} = 0$

472. $\sqrt{6p-7} + 2 = 0$

473. $3\sqrt{r} = \sqrt{8r+16}$

474. $\sqrt{5y-5} = \sqrt{4y+1}$

475. $k = \sqrt{k^2 - 2k + 10}$

476. $\sqrt{2x+1} = x - 7$

477. $\sqrt{3n+4} = 2\sqrt{n}$

478. $\sqrt{m} + 6 = 5$

479. $\sqrt{2x+2} = \sqrt{3x-5}$

480. $3t = \sqrt{9t^2 - 6t + 12}$

481. $\sqrt{2x+2} = \sqrt{3x-5}$

482. $\sqrt{3m+3} = \sqrt{5m-1}$

483. $p = \sqrt{p^2 - 3p - 12}$

484. $\sqrt{3m+3} = \sqrt{5m-1}$

485. $\sqrt{3x-4} = x - 10$

486. $2r = \sqrt{4r^2 + 5r - 30}$

487. $\sqrt{4x+13} = 2x - 1$

488. $\sqrt{6x+7} - 1 = x + 1$

489. $\sqrt{7y-5} - 3 = 0$

490. $\sqrt{x-3} = x - 3$

491. $\sqrt{7y+15} = 2y + 3$

492. $\sqrt{a-4} + \sqrt{a+4} = 4$

493. $\sqrt{3x-1} + 5 = 0$

494. $\sqrt{5c+6} = \sqrt{c+3} + 3$

495. $\sqrt{3q-8} = q - 2$

496. $\sqrt{k+10} = 2 - \sqrt{2k+19}$

497. $\sqrt{25-8x} = x - 2$

498. $\sqrt{7r+8} - \sqrt{r+1} = 5$

Writing/Conceptual Exercises

499. How can you tell that the equation $\sqrt{2x+1} = -9$ has no real number solution without performing any algebraic steps?

500. Explain why the equation $x^2 = 81$ has two real number solutions, while the equation $\sqrt{x} = 9$ has only one real number solution.

501.   A student is told that he *must* check his solutions to the equation

$$\sqrt{r-2} = r-4.$$

The student said that he doesn't see why this is necessary because he is sure he didn't make any mistakes in solving the equation. How would you respond?

502. Explain why the equation

$$z = \sqrt{z^2 + 3z + 4}$$

cannot have a negative solution.

Section 9.7   Fractional Exponents

Objective 1   Define and use $a^{1/n}$.

Evaluate the expression.

503. $16^{1/2}$	504. $25^{1/2}$	505. $169^{1/2}$	506. $225^{1/2}$
507. $8^{1/3}$	508. $27^{1/3}$	598. $64^{1/3}$	510. $125^{1/3}$
511. $216^{1/3}$	512. $512^{1/3}$	513. $729^{1/3}$	514. $1000^{1/3}$
515. $16^{1/4}$	516. $81^{1/4}$	517. $625^{1/4}$	518. $1296^{1/4}$
519. $32^{1/5}$	520 $243^{1/5}$	521. $64^{1/6}$	522. $128^{1/7}$

Objective 2   Define and use $a^{m/n}$.

Evaluate the expression.

523. $16^{3/2}$	524. $25^{3/2}$	525. $36^{3/2}$	526. $64^{3/2}$
527. $8^{2/3}$	528. $8^{4/3}$	529. $27^{2/3}$	530. $27^{4/3}$
531. $64^{2/3}$	532. $125^{2/3}$	533. $216^{2/3}$	534. $512^{2/3}$
535. $8^{-2/3}$	536. $25^{-3/2}$	537. $36^{-3/2}$	538. $64^{-3/2}$
539. $16^{-3/4}$	540. $27^{-2/3}$	541. $27^{-4/3}$	542. $1000^{-2/3}$

Objective 3   Use rules for exponents with fractional exponents.

Simplify the expression. Write the answer in exponential form with only positive exponents. Assume that all variables represent positive numbers.

543. $5^{1/2} \cdot 5^{3/2}$

544. $6^{2/3} \cdot 6^{4/3}$

545. $7^{5/8} \cdot 7^{3/8}$

546. $4^{8/7} \cdot 4^{6/7}$

547. $8^{-5/8} \cdot 8^{-3/8}$

548. $9^{-1/2} \cdot 9^{-3/2}$

549. $\dfrac{25^{3/4}}{25^{1/4}}$

550. $\dfrac{8^{-1/2}}{8^{1/2}}$

551. $\dfrac{6^{4/5}}{6^{-3/5}}$

552. $\left(r^4 s^{2/3}\right)^{3/2}$

553. $\left(9^{2/3}\right)^3$

554. $\left(4^{3/5}\right)^{10}$

555. $\dfrac{12^{-5/8}}{12^{1/2}}$

556. $\left(\dfrac{4}{9}\right)^{3/2}$

557. $\left(\dfrac{64}{27}\right)^{-2/3}$

558. $\left(\dfrac{a^{2/3}}{b^{1/4}}\right)^{12}$

559. $\left(\dfrac{r^{1/2}}{s^{-1/2}}\right)^{2/3}$

560. $\dfrac{y^{1/2} \cdot y^{-3/2}}{y^2}$

561. $\dfrac{r^{-3/4} \cdot r^{-1/4}}{r}$

562. $\dfrac{z^{-2/3} \cdot z^{7/3}}{z^{2/3}}$

563. $\left(\dfrac{c^6}{x^3}\right)^{2/3}$

564. $\dfrac{8^{3/5} \cdot 8^{-8/5}}{8^{-2}}$

565. $\left(\dfrac{x^{1/4}}{x^{-3/4}}\right)^2$

566. $\left(a^{-1}\right)^{1/2}\left(a^{-3}\right)^{-1/2}$

Objective 4   Use fractional exponents to simplify radicals.

Simplify the radical by first writing it in exponential form. Give the answer as an integer or a radical in simplest form. Assume that all variables represent nonnegative numbers.

567. $\sqrt[4]{7^2}$

568. $\sqrt[6]{3^3}$

569. $\sqrt[4]{4^2}$

570. $\sqrt[6]{9^3}$

571. $\sqrt[9]{125^3}$

572. $\sqrt[8]{81^2}$

573. $\sqrt[6]{p^2}$

574. $\sqrt[6]{z^3}$

575. $\sqrt[6]{r^2}$

576. $\sqrt[8]{a^2}$

577. $\sqrt[8]{16^2}$

578. $\sqrt[4]{100^2}$

Mixed Exercises

Evaluate the expression.

579. $625^{1/4}$

580. $16^{-3/2}$

581. $\sqrt[8]{81^2}$

582. $\left(3^{2/3}\right)^6$

583. $\left(\dfrac{8}{125}\right)^{4/3}$

584. $27^{4/3}$

585. $1024^{1/10}$

586. $64^{-3/2}$

587. $\left(\dfrac{1}{8}\right)^{1/3}$

588. $32^{3/5}$

589. $\sqrt[9]{27^6}$

590. $36^{3/2}$

591. $100{,}000^{1/5}$

592. $\sqrt[9]{64^6}$

593. $343^{1/3}$

594. $100^{-3/2}$

595. $9^{5/7} \cdot 9^{9/7}$

596. $36^{-3/2}$

Simplify the expression. Write the answer in exponential form with only positive exponents. Assume that all variables represent positive numbers.

597. $\dfrac{y^{-2}}{y^{-1/2}y^{3/2}}$

598. $\dfrac{x^{2/3} \cdot x^{-7/3}}{x^{4/3}}$

599. $\left(\dfrac{x^{3/4}}{y^{2/3}}\right)^{24}$

600. $\dfrac{12^{-1/4}}{12^{5/8}}$

601. $\left(r^{2/3}s^{3/4}\right)^{1/2}$

602. $\left(\dfrac{a^{1/3}}{b^{1/2}}\right)^{1/4}$

Writing/Conceptual Exercises

Decide which one of the four choices is *not* equal to the given expression.

603. $121^{1/2}$ (a) 121^5 (b) $\sqrt{121}$ (c) -11 (d) 11

604. $32^{2/5}$ (a) 4 (b) 2^2 (c) $\sqrt[5]{1024}$ (d) $\dfrac{1}{4}$

605. $1000^{-1/3}$ (a) -10 (b) $\dfrac{1}{10}$ (c) $\dfrac{1}{\sqrt[3]{1000}}$ (d) $.1$

606. $-8^{4/3}$ (a) -16 (b) $-\sqrt[3]{8^4}$ (c) $\dfrac{1}{16}$ (d) -2^4

CHAPTER 10 QUADRATIC EQUATIONS

Section 10.1 Solving Quadratic Equations by the Square Root Property

Objective 1 Solve equations of the form x^2 = a number.

Solve the equation by using the square root property. Express all radicals in simplest form. Give only real number solutions.

1. $x^2 = 100$

2. $y^2 = 225$

3. $r^2 = 400$

4. $z^2 = 12.25$

5. $a^2 = 15$

6. $k^2 = 63$

7. $q^2 = 32$

8. $t^2 = 28$

9. $w^2 = 72$

10. $c^2 + 81 = 0$

11. $d^2 - 150 = 0$

12. $3t^2 + 9 = 27$

Objective 2 Solve equations of the form $(ax+b)^2$ = a number.

Solve the equation by using the square root property. Express all radicals in simplest form.

13. $(x+3)^2 = 4$

14. $(x-5)^2 = 49$

15. $(y+2)^2 = 36$

16. $(y-7)^2 = 25$

17. $(x-5)^2 = 0$

18. $(y+3)^2 = 0$

19. $(m+5)^2 = 1$

20. $(n+4)^2 = 8$

21. $(p-3)^2 = 50$

22. $(q-7)^2 = 45$

23. $(r+5)^2 = 15$

24. $(3s-2)^2 = 27$

Mixed Exercises

Solve the equation by using the square root property. Express all radicals in simplest form. Give only real number solutions.

25. $m^2 = 64$

26. $(x-2)^2 = 9$

27. $p^2 = -121$

28. $(y-5)^2 = 25$

29. $s^2 = 54$

30. $(q-3)^2 - 5 = 0$

31. $(a+12)^2 = -4$

32. $(q+7)^2 = 0$

33. $(5k+6)^2 = 81$

34. $(6p+8)^2 = 56$

35. $(10m-5)^2 + 1 = 0$

36. $s^2 - 80 = 0$

37. $m^2 = 3.24$

38. $t^2 - 30.25 = 0$

39. $(3f+7)^2 = 27$

40. $p^2 = \dfrac{50}{169}$

41. $\left(\dfrac{1}{2}z+2\right)^2 = 49$

42. $\left(\dfrac{1}{3}r-4\right)^2 = 20$

43. $x^2 = \dfrac{4}{25}$

44. $z^2 = \dfrac{121}{64}$

45. $h^2 = \dfrac{40}{289}$

46. $y^2 = -\dfrac{49}{81}$

47. $m^2 = 1.96$

48. $t^2 - 3.24 = 0$

49. $(5m+4)^2 = 75$

50. $(3p-7)^2 - 18 = 0$

51. $(2p+1)^2 - 1 = 0$

52. $(7p-2)^2 = -9$

53. $\left(\dfrac{1}{4}x+2\right)^2 = 16$

54. $\left(\dfrac{1}{5}x-4\right)^2 = 20$

Writing/Conceptual Exercises

Decide whether the statement is *true* or *false*. If it is false, tell why.

55. If $k = 0$, then $x^2 = k$ has exactly one real solution.

56. If k is a positive integer, then $x^2 = k$ must have one rational solution and one irrational solution.

57.   If k is a prime number, $x^2 = k$ has two rational solutions.

58.   If k is a negative integer, $x^2 = k$ has two irrational solutions.

59.   The equation $-x^2 = -9$ has no real solutions.

60.   Which one of these equations has two real number solutions?

(a) $(x+3)^2 = 0$ (b) $(x-5)^2 = 16$

(c) $(x+6)^2 = -1$ (d) $x^2 + 4 = 0$

61.   Which one of these equations has exactly one real number solutions?

(a) $(2x-5)^2 = 20$ (b) $(3x+1)^2 = 0$

(c) $(4x-2)^2 = -5$ (d) $(x-5)^2 = 25$

62.   Which one of these equations has no real number solution?

(a) $(x+5^2)-4 = 0$ (b) $(x-3)^2 + 5 = 5$

(c) $(x-2)^2 + 2 = 0$ (d) $(x+4)^2 - 1 = 8$

Section 10.2   Solving Quadratic Equations by Completing the Square

Objective 1   Solve quadratic equations by completing the square when the coefficient of the squared term is 1.

Solve the equation by completing the square.

63. $x^2 + 8x = 9$ 64. $y^2 - 12y = -27$

65. $p^2 + 10p = 0$ 66. $k^2 + 2k = 8$

67. $m^2 - 4m = -3$ 68. $r^2 + 12r = -24$

69. $r^2 - 8r = -15$ 70. $t^2 - 2t = 5$

71. $x^2 + 4x - 8 = 0$ 72. $w^2 - 3w + 1 = 0$

73. $b^2 - 5b - 5 = 0$ 74. $y^2 + 6y + 8 = 0$

Objective 2   Solve quadratic equations by completing the square when the coefficient of the squared term is not 1.

Solve the equation by completing the square. Give only real number solutions.

75. $2x^2 + 4x = 0$

76. $3m^2 + 12m = 63$

77. $2x^2 - 6x - 27 = 0$

78. $9r^2 = 6r + 5$

79. $3z^2 - 6z - 2 = 0$

80. $2p^2 - 4p - 5 = 0$

81. $6a^2 - 8a = -3$

82. $3b^2 - 2b = 1$

83. $2x^2 - 5x - 3 = 0$

84. $2n^2 + 7n + 5 = 0$

85. $-r^2 + 2r = -1$

86. $-y^2 + 4y + 21 = 0$

Objective 3   Simplify an equation before solving.

Simplify the equation and then solve by completing the square. Give only real number solutions.

87. $4p^2 = 4p + 7$

88. $4x - 3 = x^2 + 2x$

89. $4y^2 - 2y = 2y - 7$

90. $6y^2 + 3y = 4y^2 - y + 3$

91. $(a + 1)(a + 3) = 2$

92. $(b - 1)(b + 7) = 9$

93. $(c + 3)(c + 7) = 5$

94. $(s - 3)(s + 1) = 1$

95. $(p + 5)(p - 3) = 9$

96. $(j + 3)(j + 2) = 3$

Mixed Exercises

Solve the equation by completing the square. Give only real number solutions.

97. $q^2 - 6q = 0$

98. $h^2 - 14h + 24 = 0$

99. $2m^2 + 4m = -7$

100. $4y^2 + 4y - 3 = 0$

101. $k^2 + 5k - 3 = 0$

102. $-x^2 + 4 = 2x$

103. $2z^2 = 8z + 5 - 4z^2$

104. $4p - 3 = p^2 + 2p$

105. $-x^2 + 6x = 4$

106. $m^2 - 4m + 8 = 6m$

107. $y^2 - 8y + 16 = 0$

108. $d^2 + 20d - 21 = 0$

109. $z^2 + 3z + \dfrac{5}{4} = 0$

110. $c^2 - c - \dfrac{3}{2} = 0$

111. $2r^2 - 4r - 7 = 0$

112. $3t^2 + t - 4 = 0$

113. $4p^2 - 6p + 11 = 0$

114. $8q^2 + 4q = 3$

115. $(x - 4)(x + 2) = -11$

116. $\left(a + \dfrac{1}{2}\right)\left(a + \dfrac{1}{2}\right) = \dfrac{9}{4}$

117. $(t - 5)(t + 2) = -11$

118. $(z + 1)(z + 2) = 3$

119. $4t^2 - 12t + 9 = 0$

120. $(y - 5)(y + 3) = 20$

Writing/Conceptual Exercises

121. What is the first step that you would perform in order to solve the equation

$$3x^2 + 6x = 8$$

by completing the square?

122. Why is it not possible to solve the equation

$$x^3 + 2x = 4$$

by completing the square?

123. Suppose that you solve a quadratic equation by completing the square and obtain the solutions -4 and 5. Could this equation have been solved by factoring? If so, write a quadratic equation with these solutions.

124. Suppose that you solve a quadratic equation by factoring and obtain the solutions $3 + \sqrt{5}$ and $3 - \sqrt{5}$. Could this equation have been solved by factoring?

Section 10.3   Solving Quadratic Equations by the Quadratic Formula

Objective 1   Identify the values of a, b, and c in quadratic equation.

Write the equation in the standard form $ax^2 + bx + c = 0$ if it is not in this form already. Then identify the values of a, b, and c. Do not actually solve the equation.

125. $3x^2 + 2x + 1 = 0$

126. $4x^2 - 3x + 2 = 0$

127. $x^2 + 5x = 0$

128. $5x^2 - 3 = 0$

129. $7q^2 + 3q - 2 = 0$

130. $3d^2 = d - 1$

131. $m^2 = 2m + 7$

132. $3n^2 - 4 = 5n$

133. $7p^2 = 5$

134. $2p = -4p^2 + 3$

135. $x(x+1) = 5$

136. $(z+1)(z+4) = -3$

137. $(z-2)(z+2) = 4z$

138. $3 - [(r+2)(r-3)] = 2r$

Objective 2   Use the quadratic formula to solve quadratic equations.

Use the quadratic formula to solve the equation. Give only real number solutions.

139. $x^2 + 3x + 1 = 0$

140. $x^2 + 7x - 30 = 0$

141. $3m^2 + 5m - 1 = 0$

142. $2k^2 - 2k - 3 = 0$

143. $n^2 + n - 5 = 0$

144. $3x^2 + 3x = 0$

145. $y^2 = -15$

146. $p^2 - 7 = 0$

147. $x^2 - 2x + 1 = -5$

148. $2x^2 = 3x - 2$

149. $(2f - 1)^2 = 3$

150. $(2x - 3)(2x + 1) = 0$

151. $g^2 = 8g + 2$

152. $5k^2 + k + 2 = 0$

153. $(2a - 3)^2 = 2$

154. $(x - 6)(3x - 2) = 0$

155. $3y(y - 2) = 4y + 3$

156. $5p(p + 3) = 3p^2 - 7$

Objective 3 Solve quadratic equations with only one solution.

Use the quadratic formula to solve the equation.

157. $x^2 + 6x + 9 = 0$

158. $y^2 - 12y + 36 = 0$

159. $16a^2 + 8a + 1 = 0$

160. $m^2 + 25 = 10m$

161. $c^2 + 64 = 16c$

162. $4q^2 - 12q + 9 = 0$

163. $n^2 + 49 = 14n$

164. $d^2 = 20d - 100$

165. $25r^2 = 20r - 4$

166. $100p^2 + 60p + 9 = 0$

167. $64s^2 - 80s + 25 = 0$

168. $144t^2 + 168t + 49 = 0$

Objective 4 Solve quadratic equations with fractions.

Use the quadratic formula to solve the equation. Give only real number solutions.

169. $\dfrac{1}{2}p^2 + p = \dfrac{5}{2}$

170. $\dfrac{1}{6}y^2 + \dfrac{1}{2}y - \dfrac{1}{3} = 0$

171. $\dfrac{1}{4}q^2 - \dfrac{1}{3}q - \dfrac{1}{8} = 0$

172. $-\dfrac{1}{4}x^2 + 1 = \dfrac{1}{2}x$

173. $\dfrac{1}{4}t^2 - \dfrac{1}{5}t + \dfrac{1}{20} = 0$

174. $b^2 + \dfrac{4}{3}b - \dfrac{2}{3} = 0$

175. $r^2 + \dfrac{2}{3}r - \dfrac{2}{3} = 0$

176. $\dfrac{1}{3}x^2 - \dfrac{1}{2}x = 1$

177. $\dfrac{1}{2}y^2 + \dfrac{1}{3}y = \dfrac{3}{2}$

178. $k^2 + \dfrac{8}{3}k - \dfrac{1}{3} = 0$

179. $\dfrac{1}{3}x^2 = \dfrac{7}{15}x + \dfrac{2}{5}$

180. $\dfrac{1}{4}y^2 + \dfrac{1}{3}y = \dfrac{7}{12}$

Objective 5   Solve an applied problem that leads to a quadratic equation.

Solve the problem. Round answers to the nearest tenth if necessary.

181. A certain projectile is located at a distance of

$$d = 3t^2 - 6t + 1$$

feet from its starting point after t seconds. How many seconds will it take the projectile to travel 10 feet?

182. The time t in seconds for a car to skid 48 feet is given (approximately) by

$$48 = 64t - 16t^2.$$

Solve this equation for t. Are both answers reasonable?

183. The amount A that P dollars invested at a rate of interest r will amount to in two year is

$$A = P(1 + r)^2.$$

At what interest rate will $100 grow to $121 in two years?

184. Use the formula for the distance traveled by a falling object,

$$d = 16t^2,$$

where d is the distance the object falls in t seconds, to solve the following problem.

One expert at marksmanship can hold a silver dollar above his head, drop it, draw his gun, and shoot the coin as it passes waist level. If the coin falls about 4 feet, estimate the time that elapses between the dropping of the coin and the shot.

185. If a ball is thrown into the air from ground level with an initial velocity of 32 feet per second, its height s (in feet) after t seconds is given by the formula

$$s = -16t^2 + 32t.$$

After how many seconds will the ball return to the ground?

186. If an object is thrown upward from ground level with an initial velocity of 80 feet per second, its height s (in feet) after t seconds is given by the formula

$$s = -16t^2 + 80t.$$

After how many seconds will the object reach a height of 100 feet?

187. If an object is thrown upward from a height of 32 feet with an initial velocity of 48 feet per second, then its height h (in feet) after t seconds is given by the formula

$$h = -16t^2 + 48t + 32.$$

At what times will it be 50 feet above the ground?

188. A rule for estimating the number of board feet of lumber that can be cut from a log depends on the diameter of the log. To find the diameter d needed to get x board feet of lumber, use the formula

$$\left(\frac{d-4}{4}\right)^2 = x.$$

Find the diameter needed to get 20 board feet of lumber.

189. The commodities market is very unstable; money can be made or lost quickly on investments in soybeans, wheat, pork bellies, and so on. Suppose that an investor kept track of his total profit, P (in thousands of dollars), at time t, measured in months, after he began investing, and found that his profit was given by the formula

$$P = 4t^2 - 24t + 32.$$

Find the times at which he broke even on his investment.

190. It takes Mario 2 hours longer to mow the lawn than Ethan. Working together, they can mow the lawn in 5 hours. How long does it take each of them working alone?

Mixed Exercises

Write the equation in the standard form $ax^2 + bx + c = 0$ if it is not in this form already. Then identify the values of a, b, and c. Do not actually solve the equation.

191. $4r^2 + 5r - 6 = 0$

192. $z^2 - 12z = 0$

193. $4q^2 = 4$

194. $(x+2)(x-3) = 0$

195. $2p(3p-4)=5$ 196. $(2r+1)(5r-3)=4r$

Use the quadratic formula to solve the equation. Give only real number solutions.

197. $x^2+5x+5=0$ 198. $5x^2+5x-2=0$

199. $y^2=13-12y$ 200. $2z^2+12z+5=0$

201. $4k^2+2k-1=0$ 202. $x^2=18$

203. $z^2-18z+81=0$ 204. $9r^2+48r+64=0$

205. $\dfrac{1}{3}x^2+\dfrac{4}{3}x=-1$ 206. $\dfrac{1}{16}z^2+\dfrac{3}{8}z=-\dfrac{1}{2}$

207. $-9r^2=-2r+3$ 208. $5z^2=2z+3$

209. $(x-2)(2x+3)=4$ 210. $49g^2-126g+81=0$

211. $.4x^2+2x=.3$ 212. $.09y^2+.24y+.16=0$

213. $\dfrac{1}{20}z^2=\dfrac{1}{5}z+\dfrac{1}{8}-\dfrac{1}{10}z$ 214. $\dfrac{1}{2}x^2+\dfrac{1}{4}x-\dfrac{15}{4}=0$

Solve the problem. Round answers to the nearest tenth if necessary.

215. If a ball is thrown in to the air from ground level with an initial velocity of 9.8 meters per second, its height h (in meters) after t seconds is given by the formula

$$h=-4.9t^2+9.8t.$$

After how many seconds will the ball return to the ground?

216. If an object is thrown upward from a height of 20 feet with an initial velocity of 64 feet per second, then its height s (in feet) after t seconds is given by the formula

$$s=-16t^2+64t+20.$$

At what times will it be 68 feet above the ground?

217. Penny and Richard have found that the profit P (in dollars) from their bagel shop is given by the formula

$$P = -10x^2 + 100x + 300,$$

where x is the number of units of bagels sold daily. How many units should be sold for a profit of $510?

218. Ranjana rowed her boat across Lake Bend and back in a total of 3 hours. If her rate returning was 2 miles per hour slower than her rate going, and if the distance each way was 7 miles, find her rate going.

Writing/Conceptual Exercises

219. If we apply the quadratic formula and find that the value of $b^2 - 4ac$ is positive, what can we conclude about the solutions?

220. A student reasoned as follows: "An applied problem can only have one answer. Therefore, whenever an applied problem leads to a quadratic equation with two solutions, I will have to discard one of the solutions." How would you respond to this student?

221. Suppose that a problem asks you to find the height of a triangle, and that the problem leads to a quadratic equation. If h represents the height of the triangle, which of the following solutions to the equation cannot possibly be an answer to the problem? Explain.

   (a) $h = 3 + \sqrt{10}$ (b) $h = 8\dfrac{3}{4}$ (c) $h = 6.5$ (d) $h = 3 - \sqrt{10}$

222. A student solved an equation by the quadratic formula and ended up with

$$x = \frac{3+5}{2}.$$

Is this an acceptable final answer? If not, show how to complete the work?

Section 10.4   Complex Numbers

Objective 1   Write complex numbers as multiples of i.

Write the number as a multiple of i.

223. $\sqrt{-25}$ 224. $\sqrt{-81}$ 225. $\sqrt{-100}$

226. $\sqrt{-169}$

227. $\sqrt{-7}$

228. $\sqrt{-11}$

229. $\sqrt{-48}$

230. $\sqrt{-72}$

231. $\sqrt{-288}$

232. $\sqrt{-1000}$

233. $\sqrt{-75}$

234. $\sqrt{-128}$

Objective 2   Add and subtract complex numbers.

Add or subtract, as indicated.

235. $(2+7i)+(4+9i)$

236. $(1-2i)+(-3+11i)$

237. $(6-i)+(-9-4i)$

238. $(-11-5i)+(-7+12i)$

239. $(-8+13i)+2$

240. $(5-8i)-(1+4i)$

241. $(3+7i)-(-11-2i)$

242. $(-8-i)-(3-5i)$

243. $(7-4i)-(-1+2i)$

244. $(-3+i)-9$

Objective 3   Multiply complex numbers.

Find the product.

245. $5(2-7i)$

246. $2i(5-i)$

247. $(2+7i)(1-3i)$

248. $(5-3i)(2+i)$

249. $(2+7i)(2-7i)$

250. $(1-9i)(1+9i)$

251. $(8+i)(7-i)$

252. $(5+8i)(4-2i)$

253. $(10+5i)(10-5i)$

254. $(-4+6i)(3+2i)$

Objective 4   Write complex number quotients in standard form.

Write the quotient in standard form.

255. $\dfrac{2}{1-i}$

256. $\dfrac{3}{2+i}$

257. $\dfrac{4-i}{1+i}$

258. $\dfrac{7-2i}{3+i}$

259. $\dfrac{2+3i}{2+5i}$

260. $\dfrac{3-4i}{2+3i}$

261. $\dfrac{2+7i}{i}$

262. $\dfrac{3-2i}{i}$

263. $\dfrac{2-i}{4i}$

264. $\dfrac{3+5i}{-2i}$

265. $\dfrac{i}{4+5i}$

266. $\dfrac{-i}{-1+i}$

Objective 5   Solve quadratic equations with complex number solutions.

Solve the quadratic equation for complex solutions by using the square root property. Write solutions in standard form.

267. $(t+4)^2 = -1$

268. $(r-2)^2 = -25$

269. $(2p-1)^2 = -16$

270. $(3z-2)^2 = -72$

271. $(5s+6)^2 = -50$

272. $(2x-5)^2 = -8$

Solve the quadratic equation for complex solutions by the quadratic formula. Write solutions in standard form.

273. $k^2 - 2k + 2 = 0$

274. $y^2 - 10y + 41 = 0$

275. $q^2 + 2q + 3 = 0$

276. $3x^2 - 2x + 4 = 0$

277. $2s^2 = -3s - 2$

278. $3x = 2x^2 + 5$

Mixed Exercises

Write the number as a multiple of i.

279. $\sqrt{-36}$

280. $\sqrt{-400}$

281. $\sqrt{-625}$

282. $\sqrt{-8}$

283. $-\sqrt{-50}$

284. $-\sqrt{-12}$

Perform the indicated operation.

285. $(2-6i)+(7+9i)$

286. $(18+27i)+(-9-6i)$

287. $(4+5i)-(-8-2i)$

288. $(36+i)+(1-36i)$

289. $(26-3i)-(-4+27i)$

290. $16-(2-2i)$

291. $-4(-3+i)$

292. $-3i(2+5i)$

293. $(4-2i)(5-i)$

294. $(9-4i)(3+7i)$

295. $(8-3i)(8+3i)$

296. $(12-3i)(12+3i)$

297. $\dfrac{6+2i}{i}$

298. $\dfrac{5-4i}{-i}$

299. $\dfrac{14+5i}{4-i}$

300. $\dfrac{2i}{1+i}$

301. $\dfrac{-4+33i}{2+3i}$

302. $\dfrac{13-13i}{3-2i}$

Find the complex solutions of the quadratic equation. Write solutions in standard form.

303. $(y+1)^2 = -25$

304. $(z+4)^2 = -1$

305. $(3k-4^2)= -49$

306. $(2k-3)^2 = -45$

307. $(7t+1)^2 = -27$

308. $z^2 - 4z + 13 = 0$

309. $p^2 - 6p + 34 = 0$

310. $r^2 + 4r + 9 = 0$

311. $4r^2 - r + 1 = 0$

312. $5t^2 - 5t = -2$

Writing/Conceptual Exercises

Decide whether the statement is *true* or *false*. If it is false, tell why.

313. $-i^2 = 1$

314. $\dfrac{1}{i} = i$

315. The conjugate of $3 - 4i$ is $-3 + ri.$

316. The conjugate of i is $-i.$

317. The conjugate of a binomial is another name for the opposite of the binomial.

318. To find the quotient

$$\frac{2-5i}{2+5i},$$

multiply the numerator and denominator by 2 +5i.

319. Every complex number is an imaginary number.

320. Some imaginary numbers, such as $i\sqrt{3},$ are irrational numbers.

Section 10.5   Graphing Quadratic Equations in Two Variables

Objectives 1 and 2   Graph quadratic equations of the form $y = ax^2 + bx + c$ $(a \neq 0)$; use the vertex formula and then graph a parabola.

Sketch the graph of the equation and give the coordinates of the vertex.

321. $y = -x^2$

322. $y = x^2 + 2$

323. $y = x^2 - 3$

324. $y = -2x^2 + 1$

325. $y = x^2 - 6x + 11$

326. $y = -x^2 + 6x - 13$

327. $y = x^2 + 2x - 2$

328. $y = x^2 + x - 2$

329. $y = (x - 2)^2$

330. $y = (x + 4)^2$

Objective 3 Use a graph to determine the number of real solutions of quadratic equation.

Decide from the graph how many real number solutions the corresponding equation has. Find any real solutions from the graph.

331.

332.

333.

334.

335.

336.

337.

338.

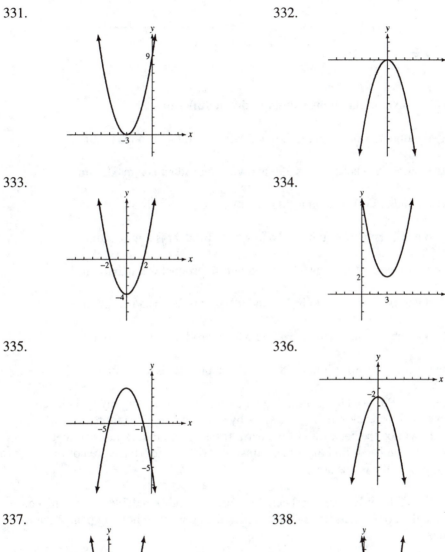

339.   340.

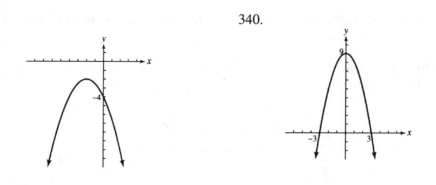

Objective 4   Solve applications using quadratic functions.

341. Find two numbers whose sum is 126 and whose product is a maximum.

342. Find two numbers whose sum is 390 and whose product is a maximum.

343. Find two numbers whose sum is 2000 and whose product is a maximum.

344. Find two numbers whose sum is 50 and whose product is a maximum.

345. Find two numbers whose sum is 816 and whose product is a maximum.

346. Find two numbers whose sum is 478 and whose product is a maximum.

347. Find two numbers whose sum is 24 and whose product is a maximum.

348. Find two numbers whose sum is 78 and whose product is a maximum.

349. The Lambeck Poultry farm is building a holding pen for its egg-laying chickens. The fence will be enclosed on three sides by chicken wire. The fourth side of the enclosure will be the back wall of the barn. If Farmer Lambeck has 816 feet of chicken wire on hand, find the dimensions that will maximize the area of the holding pen. What is this area?

350. In Exercise #349, if Farmer Lambeck purchases an additional 784 feet of chicken wire, find the dimensions that will maximize the area of the holding pen. What is this area?

Mixed Exercises

Sketch the graph of the equation and give the coordinates of the vertex.

351. $y = \dfrac{1}{2}x^2$ 352. $y = -x^2 - 1$

353. $y = 2x^2 - 4$ 354. $y = -x^2 + 6x - 9$

355. $y = -x^2 + 4x - 1$ 356. $y = x^2 + 8x + 14$

357. $y = 2x^2 + 4x$ 358. $y = -x^2 - 3x + 1$

359. $y = -(x+1)^2$ 360. $y = 9 - x^2$

Decide from the graph how many real number solutions the corresponding equation has. Find any real solutions from the graph.

361.

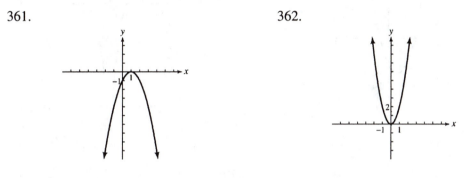

362.

363.

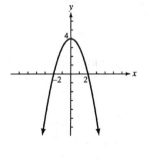

364.

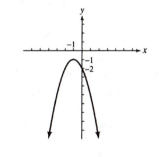

365.

366.

367.

368.

369.

370.

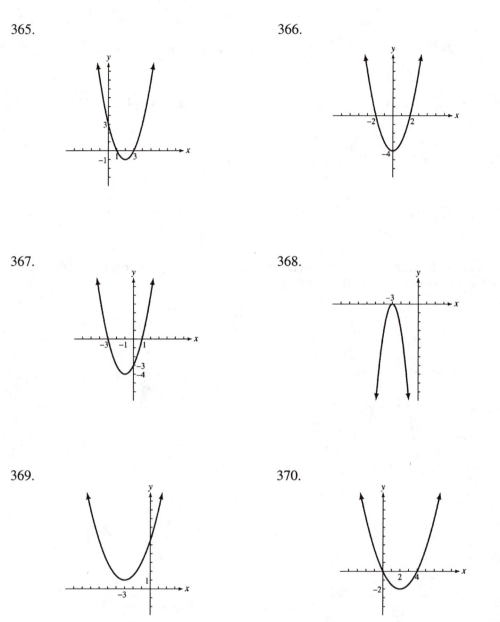

Solve the given quadratic equation by letting the left side be y_1 and using the root-finding capabilities of your graphics calculator.

371. $x^2 - 4x + 1 = 0$

372. $x^2 + 3x - 1 = 0$

373. $2x^2 + 5x - 4 = 0$

374. $3x^2 - 7x - 2 = 0$

Appendix A Review of Decimals and Percents

Objective 1 Add and subtract decimals.

Add or subtract, as indicated.

1. 23.5
    + 42.01

2. 406.13
    + 2.406

3. 593.125
    + 47

4. 231.06
    − .125

5. 456.127
    − 32.125

6. 298.407
    − 231.03

7. 462.95
    28.01
    + 32.0004

8. 351.8
    − 2.704

9. 120.593
    + 117.53

10. $10.78 + .85 + .625$

11. $.222 + .43 + 349.094$

12. $231.7 − 5.039$

13. $2.747 + 6.92 + .556$

14. $50.375 − 15.091 + 1.03$

15. $397.12 + 42.985$

16. $7.003 − 5.8$

17. $275.45 − 24.005$

18. $73.82 + 1852.883 + 1.004$

19. $375.125 − 281.27$

20. $45.829 + 634.51 − 83.4$

Objective 2 Multiply and divide decimals.

Multiply or divide, as indicated.

22. 48.96×9.72

22. 42.1×3.9

23. 8.17×3.6

24. 3.751×8.6

25. 27.8×2.4

26. $3.57 \times .034$

27. $9.4 \times .03$

28. $.007 \times 2.1$

29. $.67 \times .02$

30. $11.9 \times .07$

31. $32.8 \div .08$

32. $46.23 \div 2.3$

33. $.034 \div .08$

34. $5.8962 \div .3$

35. $1.168 \div .002$

36. $51.8373 \div 7.89$

37. $.59 \div 2.5$

38. $67.2 \div .128$

39. $1041.96 \div 45.7$ 40. $2300.25 \div 92.01$

Objective 3   Convert percents to decimals and decimals to percents.

Convert the following percents to decimals.

41. 83%	42. 45%	43. 10%	44. 28%
45. 500%	46. 300%	47. 147%	48. 2%
49. 5%	50. 8%	51. .3%	52. 5.9%
53. .25%	54. 3.7%	55. .58%	56. 32%
57. 63%	58. 721%	59. 99%	60. 100%

Convert the following decimals to percents.

61. .25	62. .3	63. .45	64. .1
65. .78	66. .004	67. .135	68. .983
69. .46	70. .59	71. .473	72. .833
73. .0015	74. .0027	75. 1.05	76. 2.03
77. 4.12	78. 8.5	79. .403	80. .059

Objective 4   Find percentages by multiplication.

Answer each of the following. Round your answers to the nearest hundredth, when necessary.

81. What is 20% of 50?	82. Find 36% of 25.
83. What is 40% of 80?	84. Find 5% of 30.
85. What is 6% of 78?	86. What is 53% of 40?
87. Find 200% of 75.	88. What is 135% of 520?
89. What is .3% of 280?	90. Find 72.75% of 3012.

91. What percent of 85 is 45?

92. 800 is what percent of 1700?

93. What percent of 50 is 250?

94. 65 is what percent of 75?

95. What percent of 8 is .5?

96. 280 is what percent of 1350?

97. What percent of 90 is 28?

98. 3 is what percent of 148?

99. What percent of 420 is 680?

100. 925 is what percent of 3242?

101. A retailer has $27,000 invested in her business. She finds that she earns 13% per year on this investment. How much money is she earning per year?

102. Glenn Reed recently bought a house for $153,000. He expects to earn 9% per year on the purchase price. How many dollars per year will he earn?

103. Barker's Bank pays 9.1% interest per year. What is the annual interest on an account of $4500?

104. Smith's Warehouse is having a sale this week of 20% off any purchase. Find the discount on a purchase of $250.

105. On a recent trip, a traveler figured that the trip totaled 1200 miles with 45% of the trip along the coast. How many miles of the trip were along the coast?

106. If the original price of a shirt is $23, find the price of the shirt if it is reduced by 15%.

107. A family of five with a monthly income of $2700 spends 88% of its income and saves the rest. Find the *annual* savings for this family.

108. If the original price of a book is $16.80, find the sale price after a reduction of 35%.

109. How much annual interest is earned if $28,000 is invested at a rate of 8.7% per year?

110. A furniture store has a price increase on all of its merchandise. The price of a desk originally selling for $340 is marked up by 15%. Find the new price of the desk.

Appendix B   Sets

Objective 1   List the elements of a set.

List the elements of each set.

1. The set of all natural numbers between $\frac{1}{2}$ and $\frac{16}{3}$

2. The set of all integers between 9 and 17, inclusive

3. The set of letters of the alphabet between M and T

4. The set of letters of the alphabet between Q and Y

5. The set of even positive integers

6. The set of all multiples of 4

7. The set of all integers that are neither positive nor negative

8. The set of birds with five legs

9. The set of days of the week

10.   The set of all negative multiples of 3

Objective 2   Learn the vocabulary and symbols used to discuss sets.

Let $R = \{5, 8\}$, $S = \{5, 6, 7, 8\}$, and $T = \{6, 8, 10, 12, 14\}$. Write *true* or *false* for each statement.

11. $5 \in S$

12. $8 \in T$

13. $8 \in R$

14. $5 \notin T$

15. $\{5\} \in R$

16. $\{10\} \in T$

17. $\{8, 5\} = R$

18. $\{8, 6, 5, 7\} = S$

19. $\{0\} = \emptyset$

20. $\{\ \} = \emptyset$

Objective 3   Decide whether a given set is finite or infinite.

List the elements of each set, if possible. Decide whether each set is finite or infinite.

21. The set of all natural numbers

22. The set of all integers between –3 and 3

23. The set of all positive multiples of 10

24. The set of all natural numbers greater than or equal to 6

25. The set of all natural numbers less than 4

26. The set of all integers between $-\dfrac{5}{3}$ and $\dfrac{13}{4}$

27. The set of all natural numbers less than 1

28. The set of all rational numbers between 0 and 1

29. The set of all irrational numbers less than 5

Objective 4   Decide whether a given set is a subset of another set.

Let $U = \{a, b, c, d, e, f, g, h\}$
   $A = \{a, c, d, e\}$
   $B = \{e, f, g\}$
   $C = \{d\}$
   $D = \{e, f, g, h\}$.

Tell whether each of the following is *true* or *false*.

31. $A \subseteq U$ 32. $B \subseteq A$ 33. $C \subseteq A$ 34. $B \subseteq D$

35. $A \subseteq A$ 36. $D \subseteq A$ 37. $\emptyset \subseteq C$ 38. $\{e, h\} \subseteq B$

39. $\{d\} \subseteq C$ 40. $\{d, e, f\} \subseteq B$

Objective 5 and 6   Find the complement of a set; find the union and intersection of two sets.

Let $U = \{1, 2, 3, 4, 5, 6, 7, 8, 9, 10\}$
   $A = \{1, 2, 5, 7, 8\}$
   $B = \{2, 5, 8, 10\}$
   $C = \{5, 9, 10\}$
   $D = \{1, 3, 5, 7, 9\}$.

List elements in the following sets.

41. A' 42. B' 43. C' 44. D'

45. $A \cap B$ 46. $B \cup C$ 47. $A \cup B$ 48. $B \cap C$

49. $A \cup C'$ 50. $B' \cap C$ 51. $A' \cap B'$ 52. $B' \cup C'$

53. $A \cup A'$ 54. $B' \cap A'$ 55. $A \cup \emptyset$ 56. $B' \cap \emptyset$

57. $B \cap D'$ 58. $C' \cup D'$ 59. $D \cap \emptyset$ 60. $D' \cup \emptyset$

ANSWERS
TO
ADDITIONAL EXERCISES

CHAPTER 1   THE REAL NUMBER SYSTEM

Section 1.1

Objective 1

1. Prime   2. Composite   3. Neither   4. Composite   5. Prime

6. Composite   7. Composite   8. Prime   9. Prime   10. Composite

11. Composite 12. Prime   13. $2 \cdot 2 \cdot 7$ 14. $2 \cdot 2 \cdot 2 \cdot 2 \cdot 3$ 15. $2 \cdot 2 \cdot 2 \cdot 2 \cdot 2 \cdot 2$

16. $2 \cdot 3 \cdot 13$ 17. $2 \cdot 2 \cdot 3 \cdot 7$ 18. $2 \cdot 2 \cdot 2 \cdot 13$ 19. $2 \cdot 2 \cdot 3 \cdot 3 \cdot 5$ 20. $2 \cdot 2 \cdot 3 \cdot 13$

21. $2 \cdot 2 \cdot 7 \cdot 7$ 22. $2 \cdot 2 \cdot 3 \cdot 19$ 23. 37 24. $2 \cdot 3 \cdot 7 \cdot 13$

Objective 2

25. $\dfrac{5}{6}$ 26. $\dfrac{3}{5}$ 27. $\dfrac{5}{12}$ 28. $\dfrac{14}{5}$ 29. $\dfrac{8}{25}$ 30. $\dfrac{3}{2}$

31. $\dfrac{1}{12}$ 32. $\dfrac{5}{6}$ 33. $\dfrac{4}{9}$ 34. $\dfrac{2}{7}$ 35. $\dfrac{4}{9}$ 36. $\dfrac{15}{2}$

Objective 3

37. 1 38. $\dfrac{2}{7}$ 39. $\dfrac{3}{2}$ 40. 15 41. 3 42. $\dfrac{8}{15}$

43. $\dfrac{2}{5}$ 44. $\dfrac{8}{7}$ 45. $\dfrac{22}{3}$ or $7\dfrac{1}{3}$ 46. $\dfrac{95}{4}$ or $23\dfrac{3}{4}$

47. $\dfrac{424}{15}$ or $28\dfrac{4}{15}$ 48. $\dfrac{161}{8}$ or $20\dfrac{1}{8}$ 49. 2 50. $\dfrac{4}{3}$

51. $\dfrac{5}{7}$ 52. 1 53. $\dfrac{2}{13}$ 54. 3 55. $\dfrac{1}{88}$ 56. $\dfrac{16}{7}$

57. $\dfrac{12}{5}$ 58. $\dfrac{13}{24}$ 59. $\dfrac{11}{4}$ or $2\dfrac{3}{4}$ 60. 4

Objective 4

61. $\dfrac{2}{5}$ 62. $\dfrac{5}{4}$ 63. 1   64. $\dfrac{25}{36}$ 65. $\dfrac{9}{10}$ 66. $\dfrac{188}{225}$

67. $\dfrac{19}{8}$ 68. $\dfrac{61}{48}$ 69. $\dfrac{41}{36}$ 70. $\dfrac{17}{2}$ *or* $8\dfrac{1}{2}$

71. $\dfrac{121}{10}$ *or* $12\dfrac{1}{10}$ 72. $\dfrac{79}{8}$ *or* $9\dfrac{7}{8}$ 73. $\dfrac{2}{3}$ 74. $\dfrac{3}{11}$

75. $\dfrac{1}{2}$ 76. $\dfrac{13}{45}$ 77. $\dfrac{5}{24}$ 78. $\dfrac{1}{6}$ 79. $\dfrac{1}{12}$ 80. $\dfrac{11}{36}$

81. $\dfrac{103}{144}$ 82. $\dfrac{67}{9}$ *or* $7\dfrac{4}{9}$ 83. $\dfrac{17}{2}$ *or* $8\dfrac{1}{2}$ 84. $\dfrac{31}{24}$ *or* $1\dfrac{7}{24}$

Objective 5

85. $2\dfrac{7}{8}$ ft   86. $7\dfrac{17}{24}$ yd   87. $3\dfrac{1}{4}$ bushels   88. 38 face cords

89. $18\dfrac{2}{3}$ lb   90. 8 sofas   91. $\dfrac{1}{16}$ inch   92. 12 cakes

93. $2\dfrac{5}{8}$ yd   94. $8\dfrac{1}{4}$ hr

Objective 6

95. $\dfrac{43}{125}$ 96. $\dfrac{219}{250}$ 97. $\dfrac{587}{1000}$

98. (a) \$156.6 million more   (b) \$2.5137 billion

(c) \$21.6 million more

Mixed Exercises

99. $3 \cdot 3 \cdot 7$ 100. $2 \cdot 2 \cdot 2 \cdot 2 \cdot 7$ 101. 41   102. $2 \cdot 3 \cdot 5 \cdot 37$

103. $2 \cdot 2 \cdot 2 \cdot 2 \cdot 3 \cdot 3 \cdot 5$ 104. $2 \cdot 5 \cdot 7 \cdot 13$ 105. $\dfrac{13}{15}$ 106. $\dfrac{1}{12}$

107. $\dfrac{12}{5}$ 108. $\dfrac{3}{7}$ 109. $\dfrac{37}{26}$ 110. $\dfrac{2}{3}$ 111. $\dfrac{3}{4}$ 112. $\dfrac{31}{90}$

113. $\dfrac{367}{60}$ or $6\dfrac{7}{60}$ 114. $\dfrac{97}{24}$ or $4\dfrac{1}{24}$ 115. $\dfrac{55}{4}$ or $13\dfrac{3}{4}$

116. 2 117. $\dfrac{2}{3}$ 118. $\dfrac{99}{20}$ or $4\dfrac{19}{20}$ 119. $\dfrac{35}{6}$ or $5\dfrac{5}{6}$

120. $\dfrac{59}{20}$ or $2\dfrac{19}{20}$ 121. $\dfrac{11}{36}$ 122. $\dfrac{2}{3}$ 123. $8\dfrac{7}{12}$ yd

124. $2\dfrac{11}{12}$ bushels

Writing/Conceptual Exercises

125. 6

126. He added the numerators and added the denominators. The two given fractions have different denominators, so we need to get a common denominator in order to add them. The LCD is 30.

$$\frac{3}{5}+\frac{5}{6}=\frac{3 \cdot 6}{5 \cdot 6}+\frac{5 \cdot 5}{6 \cdot 5}$$
$$=\frac{18}{30}+\frac{25}{30}$$
$$=\frac{43}{30} \ or \ 1\frac{13}{30}$$

127. You don't need a common denominator to multiply fractions, only to add or subtract them.

128. (b)

Section 1.2

Objective 1

129. 81   130. 10,000   131. 512   132. 729   133. $\frac{121}{144}$

134. $\frac{1}{256}$ 135. $\frac{27}{343}$ 136. $\frac{625}{1296}$ 137. .49   138. .000008

139. .3136   140. 6.859   141. 1296   142. 324   143. 100,000

144. 1331

Objective 2

145. 26   146. 25   147. 13   148. 20   149. 40

150. 2   151. $\frac{19}{12}$ 152. $\frac{1}{16}$ 153. 2.44   154. 5

Objective 3

155. 66   156. 21   157. 65   158. 40   159. 24

160. 48   161. 192   162. 0   163. 450   164. 297

165. $\frac{59}{13}$ 166. $\frac{87}{4}$ 167. 3   168. 6

Objective 4

169. False   170. True   171. False   172. True   173. False

174. True   175. True   176. True   177. False   178. True

179. True   180. False

Objective 5

181. $10 = 13 - 3$ 182. $14 < 20$ 183. $12 \neq 7$ 184. $22 > 19$

185. $20-8>10$ 186. $19<32$ 187. $6\geq0$ 188. $8+18>15$

189. $3>10\div5$ 190. $17<3\cdot10$

Objective 6

191. $7>5$ 192. $11\leq12$ 193. $\dfrac{1}{100}>0$ 194. $\dfrac{1}{3}<\dfrac{1}{2}$ 195. $\dfrac{3}{4}>\dfrac{2}{3}$

196. $\dfrac{2}{5}\geq\dfrac{2}{7}$ 197. $.47<.48$ 198. $.922\geq.921$ 199. $1\geq0$ 200. $.01<.1$

Objective 7

201. \$140 million   202. \$126 million   203. Indians and Mariners

204. \$335 million

Mixed Exercises

205. $\dfrac{243}{1024}$ 206. 91.125   207. 67   208. 65   209. $\dfrac{1}{4}$

210. 270   211. 512   212. 2   213. $\dfrac{16}{25}$ 214. $\dfrac{16}{81}$

215. 16   216. 14   217. 6   218. 18   219. 1

220. $\dfrac{1}{20}$ 221. False   222. True   223. False   224. True

225. False   226. False   227. $8=13-5$ 228. $14<30$ 229. $10\neq11$

230. $30-6>5$ 231. $99\leq101$ 232. $.231>.213$ 233. $\dfrac{2}{3}\geq\dfrac{7}{12}$ 234. $\dfrac{98}{99}<\dfrac{99}{100}$

Writing/Conceptual Exercises

235. False; to evaluate $4+6\cdot2$, multiply first, then add, obtaining $4+12=16$. To evaluate $(4+6)\cdot2$, add first, then multiply, obtaining $10\cdot2=20$.

236. False; the expression 4^6 means that there are 6 factors of 4, not 7. (It does not mean to multiply 4 by 6 more factors of 4.)

$$4^6 = 4 \cdot 4 \cdot 4 \cdot 4 \cdot 4 \cdot 4$$

237. True   238. 3

239. Answers will vary. One example is

$$3 \cdot 6 > 12 - 2.$$

240. Answers will vary. One example is

$$\frac{50}{5} \leq 12 - 4.$$

This statement is false because it simplifies to the false statement $10 \leq 8$. We can change the inequality symbol to obtain either of the following true statements:

$$\frac{50}{5} \geq 12 - 4$$

$$\frac{50}{5} > 12 - 4.$$

Section 1.3

Objective 1

241. 12   242. 80   243. 128   244. 256   245. 7

246. 8   247. $\dfrac{16}{5}$ 248. $\dfrac{17}{2}$ 249. $\dfrac{1}{2}$ 250. $\dfrac{31}{20}$

251. $\dfrac{26}{27}$ 252. $\dfrac{21}{17}$

Objective 2

253. $x + 8$ 254. $12x$ 255. $13 - x$ 256. $10(x+4)$ 257. $2x - 15$

258. $2x - 7$ 259. $10x + 11$ *or* $11 + 10x$ 260. $8x - 7$ 261. $\dfrac{x+4}{2x}$

262. $\dfrac{2}{3}x - \dfrac{1}{1}x$

Objective 3

263. Yes   264. Yes   265. Yes   266. No   267. Yes   268. No

269. Yes   270. No   271. Yes   272. No   273. No   274. Yes

Objective 4

275. $x+4=10;\ 6$ 276. $x-3=5;\ 8$ 277. $3x+5=23;\ 6$ 278. $5+x=11;\ 6$

279. $5x=40;\ 8$ 280. $3x=2+2x;\ 2$ 281. $\dfrac{10}{x}=3+x;\ 2$ 282. $\dfrac{20}{x}=5;\ 4$

283. $\dfrac{x}{6}=0;\ 0$ 284. $\dfrac{x}{10}=1;\ 10$

Objective 5

285. Expression   286. Equation   287. Equation   288. Expression

289. Expression   290. Equation   291. Expression   292. Equation

293. Equation   294. Expression

Mixed Exercises

295. 9   296. 132   297. 1   298. 3   299. $\dfrac{59}{41}$

300. 20   301. $20-x$ 302. $15x$ 303. $\dfrac{x}{100}$ 304. $12(x+3)$

305. $\dfrac{3x}{(x-2)}$ 306. $6+12x \ or \ 12x+6$ 307. Yes   308. Yes

309. No   310. Yes   311. No   312. Yes   313. $6x=30;\ 5$

314. $\dfrac{63}{x} = 7$; 9   315. $4(x+5)=32$; 3   316. $30-2x=4x$; 5

317. Equation   318. Expression   319. Expression   320. Equation

Writing/Conceptual Exercises

321. In the expression $4x^3$, only x is cubed, not 4. In other words, there are three factors of x but only one factor of 4. On the other hand, in $4x \cdot 4x \cdot 4x = (4x)^3$, there are three factors of $4x$, or, equivalently, three factors of 4 and three factors of x.

$$4x \cdot 4x \cdot 4x = 64x^3 \ne 4x^3$$

322. In the expression $3y^2$, only y is to be squared, not 3y. If we were to multiply 5 by 3 and then square the result, we would be evaluating the expression $(3y)^2$.

323. Answers will vary. Two such pairs are $m=8$, $n=0$ and $m=4$, $n=1$.

324. No, in this case "and" is part of a phrase that signifies division. Although "and" often signifies addition, it does not in this case.

Section 1.4

Objective 1

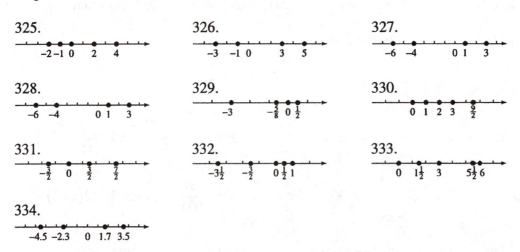

325.

326.

327.

328.

329.

330.

331.

332.

333.

334.

Objective 2

335. Natural numbers, whole numbers, integers, rational numbers, real numbers

336. Integers, rational numbers, real numbers   337. Rational numbers, real numbers

338. Whole numbers, integers, rational numbers, real numbers

339. Rational numbers, real numbers   340. Rational numbers, real numbers

341. Rational numbers, real numbers   342. Irrational numbers, real numbers

343. Rational numbers, real numbers   344. Irrational numbers, real numbers

Objective 3

345. −15   346. −10   347. −.820   348. −2.01   349. $-\dfrac{4}{5}$

350. $\dfrac{5}{12}$ 351. True   352. True   353. False   354. False

Objective 4

355. −25   356. 15   357. −6.5   358. $-\dfrac{5}{8}$ 359. $\dfrac{9}{11}$

360. $2\dfrac{5}{8}$ 361. 0   362. 9.99   363. −7   364. 8

365. 123   366. 0   367. −15   368. −20   369. 7

370. −1   371. 7.52   372. −.7   373. $\dfrac{1}{6}$ 374. $1\dfrac{1}{4}$

375. $-2\dfrac{5}{8}$ 376. $|5|$ 377. $-|2|$ 378. $-|10|$ 379. $-|-8|$

380. $|18-15|$

Objective 5

381. Allen   382. Monroe   383. 1753   384. 635

Mixed Exercises

385. Integers, rational numbers, real numbers

386. Natural numbers, whole numbers, integers, rational numbers, real numbers

387. Rational numbers, real numbers   388. Rational numbers, real numbers

389. Integers, rational numbers, real numbers   390. Irrational numbers, real numbers

391. True	392. True	393. False	394. False	395. True
396. True	397. $-1\frac{3}{5}$	398. $2\frac{1}{3}$	399. -14	400. $-\frac{10}{17}$

401. -45	402. -11	403. -37	404. 4	405. 3
406. -8	407. -4	408. -2	409. -11	410. -1.2

411. $\left|-6\right|$ 412. $\left|-22\right|$ 413. $-\left|-9\right|$ 414. $\left|12-5\right|$

Writing/Conceptual Exercises

415. True   416. False   417. True   418. False

For Exercises 419-422, answers will vary. Three numbers that satisfy each condition are given here.

419. $-10, -5, -2$ 420. $-5, 0, \frac{2}{3}$ 421. $-10, -\frac{5}{2}, 0$ 422. $\sqrt{2}, \pi, \sqrt{7}$

Section 1.5

Objective 1

423. 11	424. 16	425. -7	426. -14	427. -18	428. -12
429. -10	430. -30	431. -47	432. -35	433. -87	434. -100

Objective 2

435. 3	436. -1	437. 0	438. $-\frac{2}{5}$	439. $\frac{1}{6}$	440. $-\frac{4}{35}$

441. $2\frac{3}{8}$ 442. 6.6   443. -4.15 *or* -4.150 444. -17 445. 21

446. -29

Objective 3

447. -10 448. -7 449. -12 450. -14 451. 1   452. 4

453. 4   454. -6.2 455. 7.3   456. $\frac{3}{5}$ 457. $-\frac{1}{30}$ 458. $5\frac{7}{8}$

459. 3   460. -6 461. -3 462. -1 463. -4 464. 14

465. $\frac{4}{5}$ 466. $-\frac{23}{18}$ 467. -18 468. -14

Objective 4

469. 0   470. -18 471. -36 472. -6 473. 3   474. -13.8

475. $-\frac{17}{24}$ 476. $-\frac{19}{18}$ 477. $\frac{4}{5}$ 478. $\frac{3}{8}$

Objective 5

479. $-8+15$; 7   480. $-15+20$; 5   481. $-3+11$; 8

482. $-8+3+5$; 0   483. $[-2+(-3)]+10$; 5   484. $(-2+6)+8$; 12

485. $15+(-20+2)$; -3 486. $-8+(-9)+(-11)$; -28

487. $[-5+(-8)]+12$; -1 488. $[20+(-8)]+(-10)$; 2

489. $-9-4$; -13 490. $10-(-5)$; 15   491. $-6-(-8)$; 2

492. $[-9-(-3)]-4$; -10 493. $-4-8$; -12 494. $-2-9$; -11

495. $-4-(-2-6)$; 4   496. $-3-(-4-1)$; 2   497. $[2+(-3)]-(-6)$; 5

498. $[-4+(-2)]-(-10)$; 4   499. 12 yd gained   500. $244

501. 4566 ft   502. $23   503. 30° F   504. 260 ft   505. 5 yd lost

506. $26   507. $21   508. −102.1° C   509. 3175 ft   510. -$63.26

511. 41° F   512. 1604 m   513. −$516.34   514. $1155   515. 135° F

516. $4517   517. 73 points

518. −300   519. +150   520. −0.15   521. +0.30

Mixed Exercises

522. −5   523. −7   524. −13   525. −25   526. −6   527. 1

528. 3   529. −2   530. 1.6   531. $-\dfrac{2}{15}$ 532. $\dfrac{17}{20}$ 533. −1

534. −2   535. 14   536. 0   537. −6   538. −24   539. 9

540. −30   541. $[5+(-8)]+6$; 3   542. $[4+(-13)]+(-8)$; −17

543. $(-3)+(-9)+(-12)$; −24   544. $[-2+(-8)]+20$; 10

545. 197° F   546. $69.81   547. −10   548. 4   549. 36   550. 0

551. 12.3   552. −15.1   553. 13   554. 7   555. $-\dfrac{2}{3}$ 556. $\dfrac{17}{16}$

557. −7   558. 0   559. 5   560. −14.3   561. 1   562. 7

563. −11   564. −3   565. $6-(-10)$; 16   566. $(-5+12)-6$; 1

567. $[10-(-6)]-7$; 9   568. $[-7-(-5)]-(-2)$; 0   569. 10.931   570. −$25.03

Writing/Conceptual Exercises

571. (a)   572. (c)   573. (d)   574. (a)   575. Negative

576. Positive   577. Positive   578. Negative   579. Positive   580. Positive

Section 1.6

Objective 1

581. –12   582. –120   583. –144   584. –143   585. –640   586. –45

587. $-\dfrac{2}{15}$ 588. $-\dfrac{7}{12}$ 589. $-\dfrac{3}{13}$ 590. -7.5 591. –22   592. –13.12

Objective 2

593. 15   594. 72   595. 169   596. 1000   597. 625   598. 30

599. $\dfrac{4}{5}$ 600. $\dfrac{1}{14}$ 601. $\dfrac{1}{6}$ 602. 2.73   603. 1.02   604. 3.33

Objective 3

605. −24, −12, −8, −6, −4, −3, −2, −1, 1, 2, 3, 4, 6, 8, 12, 24

606. −12, −6, −4, −3, −2, −1, 1, 2, 3, 4, 6, 12

607. −40, −20, −10, −8, −5, −4 −2, −1, 1, 2, 4, 5, 8, 10, 20, 40

608. −13, −1, 1, 13   609. −7, −1, 1, 7

610. −10, −5, −2, −1, 1, 2, 5, 10

611. −36, −18, −12, −9, −6, −4, −3, −2, −1, 1, 2, 3, 4, 6, 9, 12, 18, 36

612. −39, −13, −3, −1, 1, 3, 13, 39   613. −23, −1, 1, 23

614. −50, −25, −10, −5, −2, −1, 1, 2, 5, 10, 25, 50

615. −43, −1, 1, 43

616. −30, −15, −10, −6, −5, −3, −2, −1, 1, 2, 3, 5, 6, 10, 15, 30

Objective 4

617. 5   618. –2   619. –2   620. –9   621. 20   622. 0   623. –3

624. Undefined 625. $\dfrac{9}{10}$ 626. $-\dfrac{4}{3}$ 627. $-\dfrac{1}{2}$ 628. 2

Objective 5

629. –20 630. –5 631. –64 632. 1 633. 4 634. -18

635. 76 636. 108 637. 0 638. –10 639. –38 640. 62

641. –10 642. –7 643. –9 644. 8 645. Undefined 646. 0

647. $-\dfrac{5}{6}$ 648. $\dfrac{3}{4}$ 649. $\dfrac{9}{7}$ 650. $-\dfrac{2}{3}$ 651. 4.1 652. –2.5

Objective 6

653. –7 654. 6 655. 36 656. 5 657. 39 658. 0

659. –3 660. 17 661. 77 662. 37 663. –1 664. $-\dfrac{11}{3}$

Objective 7

665. $4(-3)+5;\ -7$ 666. $(-8)(3)+4;\ -28$ 667. $5-11(-4);\ 49$

668. $-8-(-7)(4);\ 20$ 669. $2[14+(-6)]+(-2);\ 14$

670. $100-\dfrac{3}{10}[50-(-10)];\ 82$ 671. $.40[6+(-2)];\ 1.6$

672. $.85[32-(-8)];\ 34$ 673. $\dfrac{2}{3}[16+(-10)]-(-24);\ 28$

674. $-7+\dfrac{7}{8}(-2-6);\ -14$ 675. $\dfrac{-72}{-12};\ 6$ 676. $\dfrac{90}{35+(-5)};\ 3$

677. $-11+\dfrac{49}{-7};\ -18$ 678. $8-\dfrac{-21}{3};\ 15$ 679. $-12-\dfrac{-18}{-9};\ -14$

680. $\dfrac{30(-8)}{5-(-10)};\ -16$ 681. $\dfrac{-8(6)}{-3+15};\ -4$ 682. $\dfrac{11+(-7)}{-8-(-6)};\ -2$

683. $\dfrac{30-(-6)}{-3(4)}$; -3 684. $\dfrac{-16}{4}+\dfrac{25}{-5}$; -9

Objective 8

685. $9x=-45$; -5 686. $\dfrac{x}{(-2)}=-3$; 6   687. $x+(-4)=1$; 5

688. $x-(-10)=12$; 2   689. $\left(\dfrac{2}{3}\right)x=-4$; -6 690. $-1x=4$; -4

691. $\dfrac{x}{(-3)}=1$; -3 692. $x-8=-5$; 3   693. $-5x=10$; -2

694. $7x=x$; 0

Mixed Exercises

695. -72 696. 180   697. 77   698. -49 699. -8 700. 18

701. 35   702. 13   703. 4   704. 26   705. -14 706. 50

707. 27   708. 0   709. $-15, -5, -3, -1, 1, 3, 5, 15$

710. $-35, -7, -5, -1, 1, 5, 7, 35$

711. $-18, -9, -6, -3, -2, -1, 1, 2, 3, 6, 9, 18$

712. $-79, -1, 1, 79$ 713. $-6, -3, -2, -1, 1, 2, 3, 6$

714. $-17, -1, 1, 17$ 715. $-25, -5, -1, 1, 5, 25$

716. $-49, -7, -1, 1, 7, 49$ 717. $.30(18-12)$; 1.8

718. $-15-(-7)(6)$; 27   719. $-9+2(-8+6)$; -13 720. $(-6)(30)-8(-10)$; -100

721. 1   722. -16 723. 0   724. Undefined   725. -1

726. $\dfrac{6}{5}$ 727. -3 728. 14   729. -4 730. 3

731. Undefined

732. 16

733. $\dfrac{100}{-16+(-9)}$; -4

734. $\dfrac{-24(5)}{12-(-3)}$; -8

735. $\dfrac{-10+(-20)}{3-(-3)}$; -5

736. $\dfrac{-30}{-5}+\dfrac{100}{-25}$; 2

737. $\dfrac{90}{35+(-5)}$; 3

738. $\dfrac{-3-5}{-5+3}$; 4

739. $\dfrac{1}{2}x=-3$; -6

740. $\dfrac{x}{-2}=3$; -6

741. $\dfrac{15}{x}=-3$; -5

742. $\dfrac{x+(-3)}{2}=-4$; -5

Writing/Conceptual Exercises

743. False   744. False   745. True   746. False   747. False   748. True

749. (c)   750. (d)   751. Positive   752. Negative   753. Negative

754. Positive   755. Positive   756. Negative

Section 1.7

Objective 1

757. $y+9=9+y$

758. $5(3)=3(5)$

759. $-3\left(\dfrac{1}{4}\right)=\dfrac{1}{4}(-3)$

760. $3m=m\cdot3$

761. $-7(p+9)=(p+9)(-7)$

762. $10\left(\dfrac{1}{4}\cdot\dfrac{1}{3}\right)=\left(\dfrac{1}{4}\cdot\dfrac{1}{3}\right)(10)$

763. $(pq)(-5)=(-5)(pq)$

764. $-3+18=18+(-3)$

765. $3+[11+(-9)]=[11+(-9)]+3$

766. $-3(4+z)=(4+z)(-3)$

Objective 2

767. $m(np)=(mn)(p)$

768. $(3\cdot5)(-6)=3[5(-6)]$

769. $[-3+(-5)]+y=-3+(-5+y)$

770. $(2m)(-3)=(2)[m(-3)]$

771. $3(ab)=(3a)b$

772. $[2\cdot(-3)]\cdot 5=(2)\cdot[(-3)\cdot 5]$

773. $(-12m)(-n)=(-12)[m(-n)]$

774. $(-r)[(-p)(-q)]=[(-r)(-p)](-q)$

775. $[t+(-z)]+m=t+[(-z)+m]$

776. $3r+(6s+14t)=(3r+6s)+14t$

Objective 3

777. $-2+0=-2$

778. $1\cdot 0=0$

779. $-2\cdot 1=-2$

780. $0+|-8|=|-8|$

781. $18+0=18$

782. $-9.87\cdot 1=-9.87$

783. $1\cdot(-7)=-7$

784. $2.18+0=2.18$

785. $0+\left(-\dfrac{3}{4}\right)=-\dfrac{3}{4}$

786. $1\cdot\left(\dfrac{11}{7}\right)=\dfrac{11}{7}$

Objective 4

787. $-8+8=0$

788. $.25\cdot 4=1$

789. $\left(\dfrac{2}{3}\right)\cdot\left(\dfrac{3}{2}\right)=1$

790. $.25+(-.25)=0$

791. $-16+16=0$

792. $72+(-72)=0$

793. $-\dfrac{2}{5}\cdot\left(-\dfrac{5}{2}\right)=1$

794. $\dfrac{3}{25}\cdot\dfrac{25}{3}=1$

795. $-18+18=0$

796. $\left(-\dfrac{1}{8}\right)\cdot(-8)=1$

Objective 5

797. $y(6+4);\ 10y$

798. $r(10-1);\ 9r$

799. $az+a$

800. $9(x+y)$

801. $8a-8b$

802. $3(c-d)$

803. $2an-7bn+6cn$

804. $8m-12n$

805. $-10y+14z$

806. $2k+5$

807. $6(x+m)$

808. $5(7x+8z)$

Mixed Exercises

809. $x+3=3+x$ 810. $y+9=9+y$ 811. $2+7=7+2$ 812. $5(3)=3(5)$

813. $7(18)=(18)7$ 814. $-3\left(\dfrac{1}{4}\right)=\left(\dfrac{1}{4}\right)(-3)$ 815. $a(5\cdot3)=(a\cdot5)(3)$

816. $m(np)=(mn)(p)$ 817. $(pq)r=p(qr)$ 818. $(3\cdot5)(-6)=3[5(-6)]$

819. $(2+3)+7=2+(3+7)$ 820. $[-3+(-5)]+y=-3+(-5+y)$

821. $-3+3=0$; inverse property   822. $5+0=5$; inverse property

823. $\left(\dfrac{1}{2}\right)+0=\dfrac{1}{2}$; identity property   824. $\left(-\dfrac{1}{3}\right)+\dfrac{1}{3}=0$; inverse property

825. $1\cdot1=1$; either property   826. $\left(\dfrac{2}{5}\right)\cdot1=\dfrac{2}{5}$; identity property

827. $\dfrac{2}{3}\cdot\dfrac{3}{2}=1$; inverse property   828. $\left(-\dfrac{3}{5}\right)\cdot\left(-\dfrac{5}{3}\right)=1$; inverse property

829. $\left(-\dfrac{7}{9}\right)\cdot1=-\dfrac{7}{9}$; identity property   830. $4\left(\dfrac{1}{4}\right)=1$; inverse property

831. $\left(-\dfrac{14}{5}\right)\cdot\left(-\dfrac{5}{14}\right)=1$; inverse property   832. $\dfrac{4}{5}+0=\dfrac{4}{5}$; identity property

833. $2m+6$ 834. $az+a$ 835. $-3a-3b$ 836. $9(x+y)$

837. $4(m+n)$ 838. $8a-8b$

Writing/Conceptual Exercises

839. Answers will vary. One example is $12-5=7$, but $5-12=-7$.

840. Answers will vary. One example is $15\div5=3$, but $5\div15=\dfrac{1}{3}$.

841. We are not necessarily using the associative property when there are three numbers. The commutative property allows us to change the *order* of two numbers when we add or multiply, while the associative property allows us to change the *grouping*. In this example, the order of 9 and 12 has been reversed, while the grouping of the numbers is unchanged. That is why this is an example of the commutative property, not the associative property.

842. The additive inverse (or opposite) of 0 is 0 because when you add 0 to 0 you get 0. There is no multiplicative inverse (or reciprocal) for 0 because there is no number that you can multiply by 0 and get 1.

Section 1.8

Objective 1

843. $8 + 3y$ 844. $30r - 48s$ 845. $40 + 20y$ 846. $-22 + 14n$ 847. $-17 + 3b$

848. $10 - d$ 849. $11 - 6y$ 850. $7 + s$ 851. $9m - 5$ 852. $-24p - 7$

Objective 2

853. 8 854. -9 855. -6 856. 3 857. $.7$

858. -12 859. 25 860. 1 861. -1 862. $\dfrac{1}{3}$

863. $-\dfrac{5}{9}$ 864. $\dfrac{8}{7}$

Objective 3

865. Like   866. Like   867. Like   868. Unlike   869. Like   870. Unlike

871. Like   872. Unlike 873. Unlike 874. Unlike   875. Like   876. Unlike

Objective 4

877. $7x$ 878. $-2r + 12$ 879. $5.9r + 9.7$ 880. $-3a^2 + 2a^3$

881. $-\dfrac{1}{2} + \dfrac{1}{12}y$ 882. $\dfrac{3}{10}r - \dfrac{1}{2}s$ 883. $21x + 56$ 884. $-10y + 15$

885. $4r - 5$ 886. $-t - 24$ 887. $44q - 1$ 888. $10a - 14$

889. $3s - 2$ 890. $27t + 1$ 891. $-1.5y + 16$ 892. $-.2t - .2$

Objective 5

893. $9x+5; 14x$ 894. $4x-7x; -3x$ 895. $(8x+13)+3x; 11x+13$

896. $6(14+3x)+9x; 84+27x$ 897. $2x-(4x+7); -2x-7$

898. $(7x-14)+4(x+2); 11x-6$ 899. $(8-9x)-(11x+4); -20x+4$

900. $14-12(6-2x); -58+24x$ 901. $2(3x-7x)+3(x+8); -5x+24$

902. $3(-7+5x)-4[2x-(-10)]; -61+7x$

Mixed Exercises

903. $4k+1$ 904. $3m+7$ 905. $12p+8q$ 906. $10a-25b$

907. $12x+3$ 908. $6y-7$ 909. $-10n+21$ 910. $-8p+10$

911. 4   912. 3   913. 17   914. -12 915. -2 916. 3

917. 5   918. -9 919. Like   920. Like   921. Like   922. Unlike

923. Unlike   924. Like   925. $11m$ 926. $10y$ 927. r 928. x

929. $-5z$ 930. $8y+8y^2$ 931. $4a^3-4a^2$ 932. $11p+12$

933. $3q-6$ 934. $-r-2$ 935. $5k-4$ 936. $-8m+2$ 937. $-5n+17$

938. $2z+6$ 939. $7x-10x; -3x$ 940. $5x+(11x+20); 16x+20$

941. $3(2+4x)+6x; 18x+6$ 942. $12-10(x-3); 42-10x$

943. $17x-5(3x-8); 2x+40$

Writing/Conceptual Exercises

944. The sum of three times a number and eight times the number

945. Four subtracted from three times the difference between a number and one

946. The sum of a number and five, divided by seven

947. Twice the sum of a number and one subtracted from five

948. Four times a number added to twice the sum of a number and seven

CHAPTER 2   SOLVING EQUATIONS AND INEQUALITIES

Section 2.1

Objective 1

1. Yes	2. No	3. Yes	4. No	5. No	6. Yes
7. No	8. Yes	9. No	10. Yes		

Objective 2

11. 21   12. –6   13. 5   14. –3   15. $\dfrac{5}{6}$ 16. $-\dfrac{1}{8}$

17. –5   18. $\dfrac{1}{2}$ 19. 1.3   20. –2.4   21. 5   22. 3

23. –10   24. 18   25. 67   26. –65

Objective 3

27. 4   28. –6   29. 4   30. 36   31. –147   32. –45

33. –64   34. $\dfrac{1}{3}$ 35. $\dfrac{7}{9}$ 36. –3.5   37. 6.4   38. –3.8

Objective 4

39. 2   40. –2   41. 3   42. –7   43. –5   44. 14

45. –4   46. –7   47. 3   48. –2   49. –2   50. –4

Mixed Exercises

51. –4   52. –9   53. –6   54. –88   55. $-\dfrac{7}{10}$ 56. $\dfrac{1}{2}$

57. 10   58. $-\dfrac{10}{3}$ 59. –4   60. 6   61. 7.1   62. 5.8

63. 9   64. 7   65. 5   66. –2   67. 3   68. 7

69. 9.5   70. –9.5   71. 6   72. $\dfrac{1}{12}$ 73. $\dfrac{43}{6}$ 74. –13

Writing/Conceptual Exercises

75. Expression; $9z - 7$ 76. Equation; -1 77. Expression; $15r + 10$

78. (b) and (c)   79. (d)

80. Answers will vary. One example is $-\dfrac{5}{3}x = 10$.

Section 2.2

Objective 1

81. 3   82. 3   83. 6   84. –3   85. 6   86. –3

87. $-\dfrac{14}{3}$ 88. –2   89. $-\dfrac{1}{2}$ 90. –5   91. 1   92. 2

Objective 2

93. –15   94. $\dfrac{2}{11}$ 95. 3   96. $\dfrac{14}{17}$ 97. $\dfrac{83}{11}$ 98. $-\dfrac{1}{2}$

99. 325   100. 40   101. –5   102. –5   103. –3   104. 10

Objective 3

105. All real numbers   106. No solution   107. No solution

108. All real numbers   109. All real numbers   110. All real numbers

111. No solution   112. All real numbers   113. No solution

114. No solution

Objective 4

115. $17 - y$ 116. $\dfrac{18}{z}$ 117. $t + 15$ 118. $r - 10$ 119. $5x$ 120. $\dfrac{d}{20}$

121. $5x + 3y$ 122. $25r + 20s$ 123. $5n + 10d + 25q$

124. $18 - t$

Mixed Exercises

125. 1   126. 0   127. 2   128. -3 129. $\dfrac{23}{5}$ 130. $-\dfrac{9}{10}$

131. 40,000   132. 6875   133. 3   134. -5 135. 6   136. -1

137. -10 138. -5 139. 0   140. $\dfrac{15}{7}$ 141. -3.3

142. No solution   143. No solution   144. All real numbers

145. -35 146. -3 147. -4 148. All real numbers

149. 2   150. $-\dfrac{5}{33}$ 151. $t + 12$ 152. $163 - y$ 153. $5n + 10d$

154. $10x + 7.50y$

Writing/Conceptual Exercises

155. (c)   156. (d)

Section 2.3

Objective 2

157. 2   158. 8   159. 12   160. $\dfrac{16}{3}$ 161. $-\dfrac{13}{3}$ 162. 7

163. 4   164. -6 165. $\dfrac{3}{2}$ 166. 5

Objective 3

167. 15 inches   168. 176 votes   169. 52 points

170. 65 pushups   171. 76 men   172. 13 cm; 27 cm; 39 cm

173. Highest, 76; lowest, 19   174. $34   175. $21   176. $53

Objective 4

177. 150°   178. 72°   179. 20°   180. 60°   181. 66°   182. 78°

183. 43°   184. 53°   185. 138°   186. 81°

Objective 5

187. −157, −156   188. 18, 20   189. 27, 28

190. 87, 88   191. 30, 31, 32   192. 110, 111

193. 9, 11, 13   194. 29, 30   195. 80, 81, 82

196. 90, 92, 94

Mixed Exercises

197. −1   198. 11   199. 15   200. 146, 148   201. 36, 38

202. 16 oz of cranberry juice; 32 oz of orange juice; 128 oz of ginger ale

203. Lori, 5; Carolyn, 10; Becky, 8;   204. −7   205. 115, 117, 119

206. −8   207. 30, 31, 32   208. 328, 329

209. Margo, 200 boxes; Gladis, 148 boxes   210. 42°   211. 88, 90

212. Damon, 2.2 mi; Nolanda, 1.8 mi   213. 12°   214. 18, 20, 22

215. First book, 90 pages; second book, 70 pages   216. 60°   217. 82, 84

218. 186 dimes; 168 nickels   219. 74°   220. 80°

221. 110 chocolate chip cookies; 90 sugar cookies   222. 150°

223. 260 pages on the first day; 240 pages on the second day

224. 32°   225. Shirt, $24; jeans, $36   226. 50°

Section 2.4

Objective 1

227. 275   228. 22   229. 10   230. 7   231. 9   232. 153.86

233. 5500   234. 100   235. 77   236. 12   237. 14   238. 37.68

Objective 2

239. 23 ft   240. 8 in   241. 8 ft   242. 6.5 m   243. $31,400 \text{ ft}^2$

244. 10 m   245. 50 ft^2 246. $14,130 \text{ cm}^3$ 247. 4447.68 ft^2

248. 27.56 cm

Objective 3

249. $(2x)° = 40°$; $(7x)° = 140°$ 250. $(3x+5)° = 35°$; $(6x-25)° = 35°$

251. $(3x-30)° = 120°$; $(x+10)° = 60°$ 252. $(6x)° = 72°$; $(10x-48)° = 72°$

253. $(3x+8) = 32$; $(7x-24)° = 32°$ 254. $(2x+16)° = 62°$; $(7x+19)° = 118°$

255. $(4x+7)° = 59°$; $(11x-22)° = 121°$ 256. $(9x)° = 126°$; $(3x+12) = 54°$

257. $(8x+35)° = 123°$; $(4x+13)° = 57°$ 258. $(23x-13)° = 148°$; $(19x-15)° = 148°$

Objective 4

259. $L = \dfrac{V}{WH}$ 260. $t = \dfrac{A-p}{pr}$ 261. $r = \dfrac{S-a}{S}$ 262. $B = \dfrac{3V}{h}$

263. $h = \dfrac{S - 2\pi r^2}{2\pi r}$ or $h = \dfrac{S}{2\pi r} - r$ 264. $d = \dfrac{a_n - a_1}{n-1}$ 265. $k = \dfrac{mv^2}{2g}$

266. $h = \dfrac{V}{\pi r^2}$ 267. $B = \dfrac{2A}{h} - b$ or $B = \dfrac{2A - bh}{h}$ 268. $n = \dfrac{2S_n}{a_1 + a_2}$

269. $F = \dfrac{9}{5}C + 32$ 270. $v = \dfrac{d - gt^2}{t}$ 271. $r^2 = \dfrac{3V}{\pi h}$

272. $C = 180° - A - B$

Mixed Exercises

273. 48   274. 22   275. 32   276. 165   277. 9   278. 25

279. 10   280. 80   281. 8   282. 4.5   283. 2.5   284. 78.5

285. 100   286. 2   287. $c = P - a - b; 13$ 288. $W = \dfrac{V}{LH}; 6$

289. $h = \dfrac{3V}{\pi r^2}; 8.60$ 290. $r = \dfrac{C}{2\pi}; 1.50$ 291. $a = S(1 - r); 43.20$

292. $r = \dfrac{A - P}{At}; .06$ 293. 18 cm^2 294. 300 m   295. 20 cm   296. 50 ft

297. 15 m   298. 310.86 cm^3 299. $(5x - 20)° = 65°; (3x + 14)° = 65°$

300. $(3x + 42)° = 81°; (10x - 31)° = 99°$ 301. $(5x + 2)° = 82°; (3x + 34)° = 82°$

302. $(11x)° = 110°; (6x + 10)° = 70°$

Writing/Conceptual Exercises

303. Perimeter   304. Area   305. Area   306. Perimeter

Section 2.5

Objective 1

307. $\dfrac{5}{3}$ 308. $\dfrac{7}{4}$ 309. $\dfrac{5}{14}$ 310. $\dfrac{16}{1}$ or 16   311. $\dfrac{2}{3}$

312. $\dfrac{5}{48}$ 313. $\dfrac{5}{9}$ 314. $\dfrac{7}{12}$ 315. $\dfrac{56}{3}$ 316. $\dfrac{5}{63}$ 317. $\dfrac{3}{4}$

318. $\dfrac{1}{20}$ 319. $\dfrac{1}{5}$ 320. $\dfrac{3}{8}$

Objective 2

321. True   322. True   323. False   324. True   325. True   326. False

327. True   328. True   329. True   330. True   331. True   332. True

Objective 3

333. 20   334. $\dfrac{28}{3}$ 335. 12   336. 16   337. $\dfrac{25}{8}$ 338. $\dfrac{15}{2}$

339. $\dfrac{7}{15}$ 340. $-\dfrac{33}{2}$ 341. $\dfrac{25}{19}$ 342. –31   343. –2   344. 3

Objective 4

345. 105 min   346. $26.25   347. 10 in   348. $28\dfrac{4}{5}$ acres

349. 9 oz   350. 8 tanks   351. $1\dfrac{1}{2}$ bowls   352. 5-lb

353. $135   354. $250   355. 5-lb bag   356. 64-oz size

357. 29-oz can   358. 48-oz bottle   359. 48-oz jar   360. 10-oz size

361. 20-oz box   362. 64-oz carton   363. 128-oz size   364. 42-oz size

Objective 5

365. $6.75   366. $11.61   367. $12.43   368. $5.58

369. $5.22   370. $15.26   371. $487.50   372. 243 files

373. $21.00   374. $60.69

Mixed Exercises

375. $\dfrac{40}{7}$ 376. $\dfrac{1}{32}$ 377. $\dfrac{1}{16}$ 378. $\dfrac{1}{3}$ 379. $\dfrac{7}{32}$ 380. $\dfrac{5}{4}$

381. False   382. False   383. False   384. True   385. False   386. True

387. False   388. True   389. False   390. 2   391. 56   392. $\dfrac{56}{9}$

393. 8   394. 4   395. 10   396. $\dfrac{-3}{2}$ 397. $\dfrac{15}{2}$ 398. 7

399. 420 mi   400. $3\dfrac{1}{2}$ bars   401. $75

402. $28\dfrac{4}{5}$ yards   403. 24-oz size   404. 32-oz size

Section 2.6

Objective 1

405. 22.5 liters   406. 7.5 liters   407. 15.84 liters   408. $800

409. $1050   410. $936   411. $3.15   412. $9.50

413. $11.75   414. $19.50

Objective 2

415. ½ liter   416. 1.6 gal   417. 32 liters   418. 25 lb

419. 625 lb   420. 7 liters of 60%; 3 liters of 30%   421. 60 lb

422. 18 liters   423. 16.013 gal   424. 37½ lb

Objective 3

425. $28,500 at 8%; $18,500 at 6%   426. $7000 at 5%; $12,000 at 9%

427. $51,000 at 8%; $17,000 at 9½ %   428. $6000 at 12½ %; $14,000 at 14%

429. $1400   430. $7000 at 9%; $6000 at 11%   431. $15,500

432. $10,000 at 9%; $7000 at 11%   433. $4500 at 4%; $9000 at 7%

434. $18,800

Objective 4

435. 24 nickels; 46 dimes   436. 8 dimes; 5 quarters

437. 39 ten-dollar bills; 51 five-dollar bills   438. 185

439. 21   440. 195 at $13.95; 254 at $10.45

441. 15 nickels; 21 dimes; 30 quarters   442. 7 nickels; 11 dimes; 33 quarters

443. 155 adults; 123 children   444. 295

Objective 5

445. 288 mi   446. 388 mi   447. 15 hr   448. 220.5 mi

449. 408 mph   450. 7 hr   451. 411 mph   452. 60 mph

453. 2½ hr   454. 9.52 m/sec

455. Passenger train, 75 mph; freight train, 45 mph   456. 3½ hr

457. 1 mi   458. Latasha, 60 mph; Kevin, 225 mph   459. 3 hr

460. 8 mph   461. 6 mi   462. 300 mi   463. 2½ hr

464. $1\dfrac{2}{3}$ hr

Mixed Exercises

465. 38 liters   466. 20 liters   467. $480   468. $1200

469. $4.50   470. $1.85   471. $1\dfrac{7}{13}$ liters   472. 5 liters

473. $33\dfrac{3}{5}$ liters 474. 60 lb 475. 80 gal 476. 60 lb

477. $40,000 478. $3000 479. $5000 480. $8000

481. $2000 482. 22 dimes 483. 20 quarters 484. 7 nickels

485. 20 dimes 486. 100 tickets 487. 80 tickets 488. 200 tickets

489. 48 mph 490. 60 mph 491. $5\dfrac{1}{3}$ hr 492. 12 hr

493. 16 hr 494. 360 mi 495. 320 mi 496. 660 mi

497. 40 mph 498. 3 hr 499. 5 mph 500. 6 mph

501. 60 mph

Writing/Conceptual Exercises

502. (c) 503. (b)

Section 2.7

Objective 1

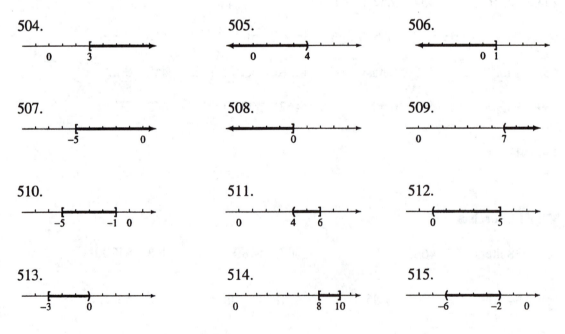

Objective 2

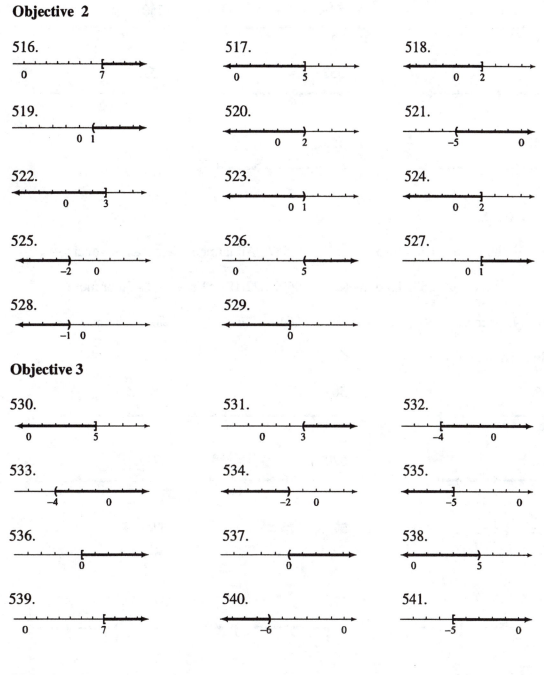

516.

0   7

517.

0   5

518.

0   2

519.

0 1

520.

0   2

521.

−5   0

522.

0   3

523.

0 1

524.

0   2

525.

−2   0

526.

0   5

527.

0 1

528.

−1 0

529.

0

Objective 3

530.

0   5

531.

0   3

532.

−4   0

533.

−4   0

534.

−2   0

535.

−5   0

536.

0

537.

0

538.

0   5

539.

0   7

540.

−6   0

541.

−5   0

Objective 4

542. $p \leq 5$ 543. $y > 5$ 544. $m \leq 44$ 545. $k \leq -2$ 546. $z > -17$ 547. $x \geq \dfrac{13}{12}$

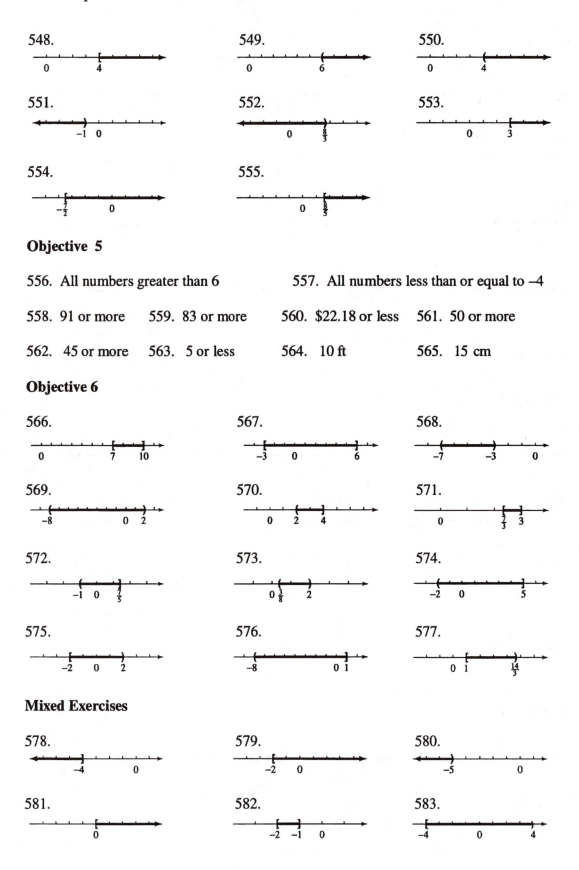

548.

549.

550.

551.

552.

553.

554.

555.

Objective 5

556. All numbers greater than 6   **557.** All numbers less than or equal to –4

558. 91 or more   **559.** 83 or more   **560.** $22.18 or less   **561.** 50 or more

562. 45 or more   **563.** 5 or less   **564.** 10 ft   **565.** 15 cm

Objective 6

566.

567.

568.

569.

570.

571.

572.

573.

574.

575.

576.

577.

Mixed Exercises

578.

579.

580.

581.

582.

583.

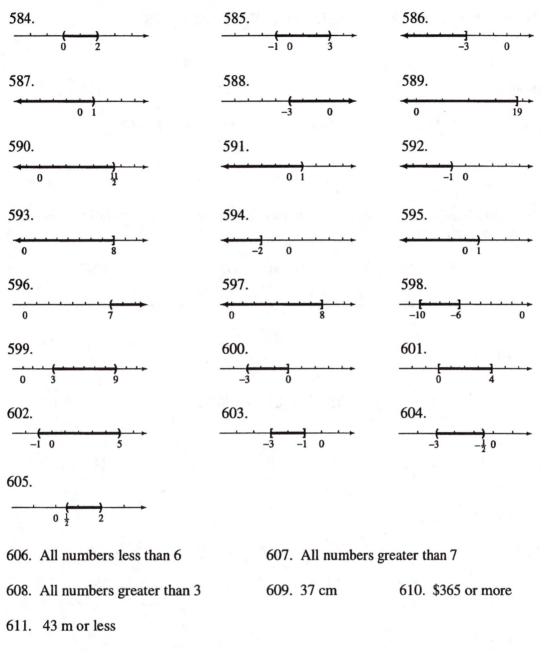

584.

585.

586.

587.

588.

589.

590.

591.

592.

593.

594.

595.

596.

597.

598.

599.

600.

601.

602.

603.

604.

605.

606. All numbers less than 6

607. All numbers greater than 7

608. All numbers greater than 3

609. 37 cm

610. $365 or more

611. 43 m or less

Writing/Conceptual Exercises

612. This would imply that
$-2 < -3$, which is false.

613. This would imply that
$0 < -7$, which is false.

614. When multiplying or dividing an inequality by a negative number, we must reverse the direction of the inequality symbol. In this case, the result will be $x > 3$.

615. We must reverse the direction of the inequality symbol when we multiply or divide by a negative number.

CHAPTER 3   LINEAR EQUATIONS IN TWO VARIABLES

Section 3.1

Objective 1

1. 45%   2. $8700   3. $6960   4. 1990   5. 1990-1991

6. 1988-1989

In Exercises 7 and 8, percents may vary slightly from those given due to difficulty in reading graph precisely.

7. 5.5%; Quarter 1, 1989   8. 1.7%; Quarter 3, 1992   9. Quarter 1, 1991

10. Miami   11. Miami   12. Dallas

Objective 2

13. $(3, 2)$ 14. $(-1, 6)$ 15. $(-5, 0)$ 16. $(2, 3)$ 17. $\left(\frac{1}{3}, -2\right)$

18. $(-4, -7)$ 19. $(8, 5)$ 20. $\left(0, \frac{1}{3}\right)$ 21. $(.2, .3)$ 22. $(4, -20)$

Objective 3

23. Yes   24. Yes   25. No   26. No   27. Yes

28. No   29. Yes   30. Yes   31. Yes   32. No

Objective 4

33. (a) $(4, 10)$ (b) $(0, -2)$ (c) $(2, 4)$ (d) $(-2, -8)$ (e) $(4, 10)$

34. (a) $(1, 2)$ (b) $(-3, 18)$ (c) $(0, 6)$ (d) $(-5, 26)$ (e) $(3, -6)$

35. (a) $(-2, 8)$ (b) $(4, -4)$ (c) $(2, 0)$ (d) $(-1, 6)$ (e) $(6, -8)$

36. (a) $(1, -1)$ (b) $(-7, 19)$ (c) $(3, -6)$ (d) $\left(0, \frac{3}{2}\right)$ (e) $\left(\frac{3}{5}, 0\right)$

37.   (a) $(-2,-2)$ (b) $(-2,0)$ (c) $(-2, 12)$ (d) $(-2, 5)$ (e) $\left(-2,-\dfrac{2}{3}\right)$

38.   (a) $(3, 4)$ (b) $(0, 4)$ (c) $(7, 4)$ (d) $(-4, 4)$ (e) $(.75, 4)$

39. $2x + y = 4$

x	2	0	1
y	0	4	2

40. $3x = y$

x	2	-3	0
y	6	-9	0

41. $3x + 2y = 6$

x	0	2	4
y	3	0	-3

42. $y = 2x - 1$

x	0	3	4
y	-1	5	7

43. $4m - n = 8$

m	0	2	-1
n	-8	0	-12

44. $3s - 4t = 12$

s	0	4	-8
t	-3	0	-9

45. $x = -5$

x	-5	-5	-5
y	0	4	-5

46. $y = 7$

x	-5	0	7
y	7	7	7

47. $x - 8 = 0$

x	8	8	8
y	-8	4	8

48. $y + 6 = 0$

x	-6	0	6
y	-6	-6	-6

Objective 5

49.   50.   51.

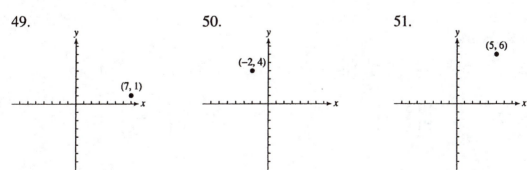

52.

53.

54.

55.

56.

57.

58.

59.

60.

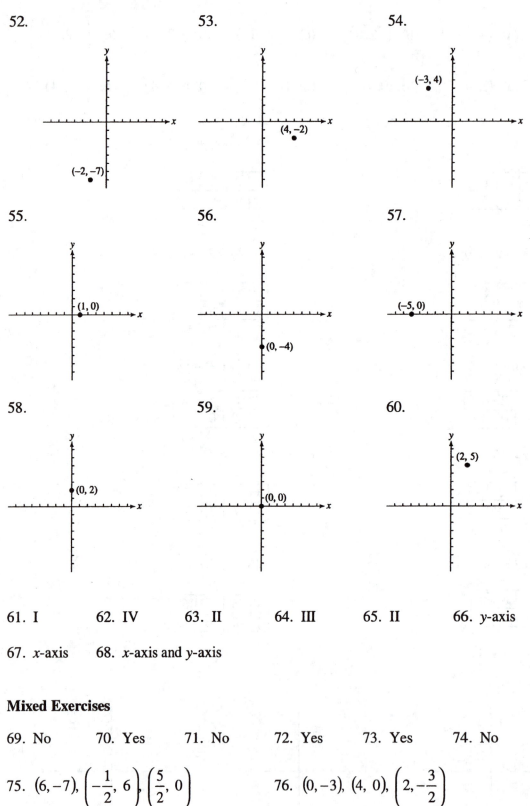

61. I   62. IV   63. II   64. III   65. II   66. y-axis

67. x-axis   68. x-axis and y-axis

Mixed Exercises

69. No   70. Yes   71. No   72. Yes   73. Yes   74. No

75. $(6, -7)$, $\left(-\frac{1}{2}, 6\right)$, $\left(\frac{5}{2}, 0\right)$ 76. $(0, -3)$, $(4, 0)$, $\left(2, -\frac{3}{2}\right)$

77. $(3, 10), (-2, 5), (2.5, 9.5)$

78. $(0, 3), (-1, 3), (100, 3)$

79. $-2x + y = 6$

x	3	-3	-4
y	12	0	-2

81. $x + 3 = -2$

x	-5	-5	-5
y	-5	0	1

80. $2r - 5s = 10$

r	0	5	-3
s	-2	0	$-\dfrac{16}{5}$

82. $5u + 7v = 35$

u	7	0	4
v	0	5	$\dfrac{15}{7}$

83. II 84. I 85. x-axis 86. III 87. IV 88. y-axis

89. U.S. 90. U.S. 91. Japan 92. Germany

Writing/Conceptual Exercises

93. No; the numbers are not in the same order. In the first pair, $x = -3$ and $y = 5$; in the second, $x = 5$ and $y = -3$.

94. No; the ordered pair $(-2, 0)$ has x-coordinate -2 and y-coordinate 0, while the point $(0, -2)$ has x-coordinate 0 and y-coordinate -2. The graph of $(-2, 0)$ is point on the x-axis, while the graph of $(0, -2)$ is a point on the y-axis.

95. 5 96. 1 97. 4 98. –2 99. y 100. x

Section 3.2

Objective 1

101. $(0, 3), (3, 0), (2, 1)$

102. $(0, -2), (-2, 0), (3, -5)$

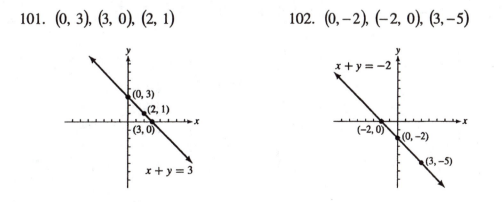

103. $(0, -4)$, $(4, 0)$, $(-2, -6)$

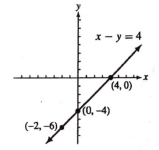

104. $(0, 1)$, $(-1, 0)$, $(4, 5)$

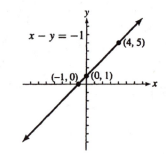

105. $(0, 7)$, $(-7, 0)$, $(-4, 3)$

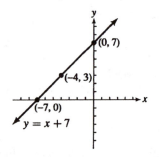

106. $(0, -4)$, $(4, 0)$, $(2, -2)$

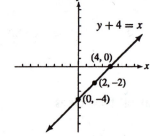

107. $(0, -2)$, $\left(\dfrac{2}{3}, 0\right)$, $(2, 4)$

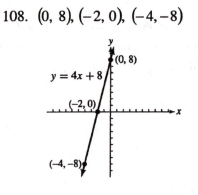

108. $(0, 8)$, $(-2, 0)$, $(-4, -8)$

109. $(0, -3)$, $(4, -3)$, $(-3, -3)$

110. $(4, 0)$, $(4, -2)$, $(4, 3)$

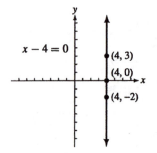
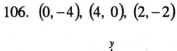

111. $(0, 1), (2, 0), (-2, 2)$

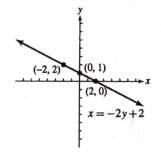

112. $(0, 2), (3, 0), (-3, 4)$

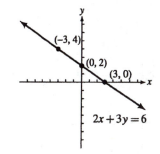

Objective 2

113. x-intercept: $(2, 0)$; y-intercept: $(0, 5)$

114. x-intercept: $(4, 0)$; y-intercept: $(0, -6)$

115. x-intercept: $(0, 0)$; y-intercept: $(0, 0)$

116. x-intercept: $(-2, 0)$; y-intercept: $\left(0, \frac{8}{7}\right)$

117. x-intercept: $(2, 0)$; y-intercept: $\left(0, -\frac{10}{3}\right)$

118. x-intercept: $\left(\frac{3}{2}, 0\right)$; y-intercept: $(0, 2)$

119. x-intercept: $\left(\frac{8}{3}, 0\right)$; y-intercept: $(0, -4)$

120. x-intercept: $(2, 0)$; y-intercept: $\left(0, \frac{14}{3}\right)$

121. x-intercept: $\left(-\frac{9}{2}, 0\right)$; y-intercept: $(0, 1)$

122. x-intercept: $(3, 0)$; y-intercept: $\left(0, \frac{9}{5}\right)$

123. x-intercept: $(2, 0)$
     y-intercept: $(0, 4)$

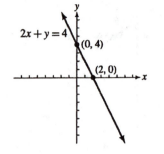

124. x-intercept: $(1, 0)$
     y-intercept: $(0, -4)$

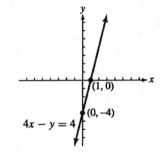

125. x-intercept: $(-4, 0)$
     y-intercept: $(0, -3)$

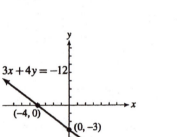

126. x-intercept: $(4, 0)$
     y-intercept: $(0, 2)$

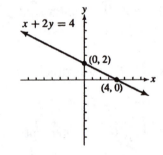

127. x-intercept: $\left(\dfrac{5}{2}, 0\right)$
     y-intercept: $(0, 3)$

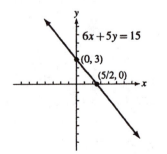

128. x-intercept: $(3, 0)$
     y-intercept: $(0, -2)$

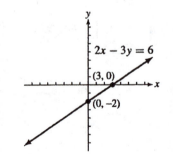

129. x-intercept: $(-2, 0)$

y-intercept: $(0, 5)$

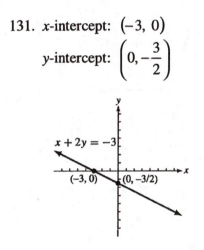

130. x-intercept: $\left(\dfrac{5}{3}, 0\right)$

y-intercept: $(0, -5)$

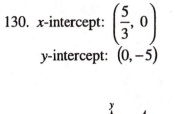
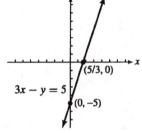

131. x-intercept: $(-3, 0)$

y-intercept: $\left(0, -\dfrac{3}{2}\right)$

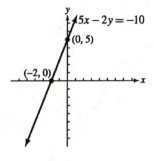

132. x-intercept: $(-6, 0)$

y-intercept: $(0, -5)$

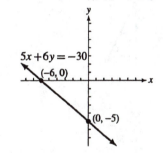

Objective 3

133.

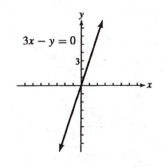

134.

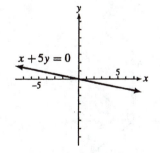

135.

136.

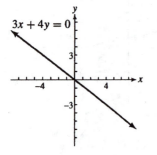

$x + y = 0$

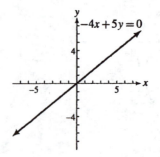

$2x + y = 0$

137.

138.

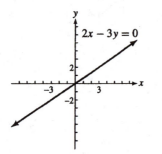

$2x + y = 0$

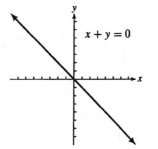

$3x + y = 0$

139.

140.

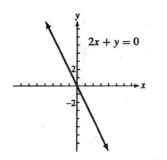

$2x - 3y = 0$

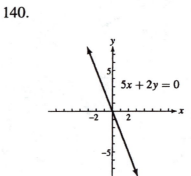

$5x + 2y = 0$

141.

142.

$3x + 4y = 0$

$-4x + 5y = 0$

Objective 4

143.

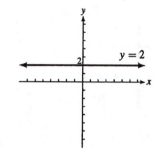

144.

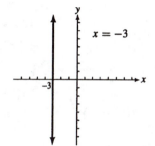

145.

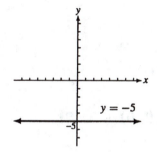

146.

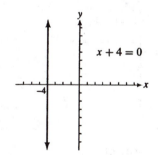

147.

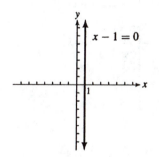

148.

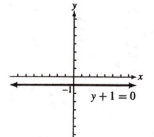

149.

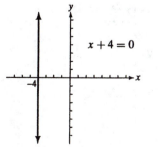

150.

151.

152.

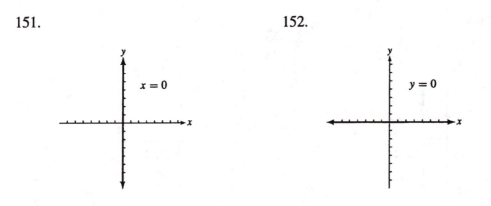

Objective 5

153. No   154. Yes   155. Yes   156. Yes   157. No   158. No

159. Yes

Mixed Exercises

160. $(0, 3), (-3, 0), (1, 4)$

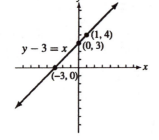

161. $\left(0, -\dfrac{1}{2}\right), (1, 0), (-3, -2)$

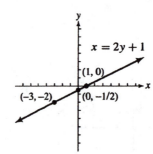

162. $(0, 2), (-6, 0), (6, 4)$

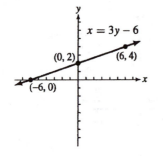

163. $(0, 4), (-2, 0), (-4, -4)$

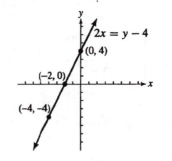

164. $(3, 1), (-1, 0), (-5, -1)$

165. $(0, 1), \left(\dfrac{5}{4}, 0\right), (5, -3)$

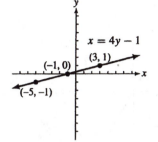

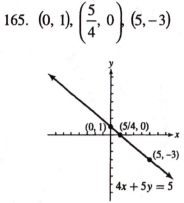

166. x-intercept: $(-1, 0)$; y-intercept: $(0, 3)$

167. x-intercept: $(-4, 0)$; y-intercept: $(0, -6)$

168. x-intercept: $\left(\dfrac{5}{2}, 0\right)$; y-intercept: $(0, -2)$

169. x-intercept: $(-2, 0)$; y-intercept: $\left(0, \dfrac{8}{7}\right)$

170. x-intercept: $(2, 0)$; y-intercept: $(0, -3)$

171. x-intercept: $(2, 0)$; y-intercept: $(0, -8)$

172. x-intercept: $(0, 0)$; y-intercept: $(0, 0)$

173. x-intercept: $\left(\dfrac{1}{2}, 0\right)$; y-intercept: $\left(0, -\dfrac{3}{7}\right)$

174. x-intercept: $(4, 0)$

   y-intercept: $(0, 6)$

175. x-intercept: $(-2, 0)$

   y-intercept: $\left(0, \dfrac{8}{7}\right)$

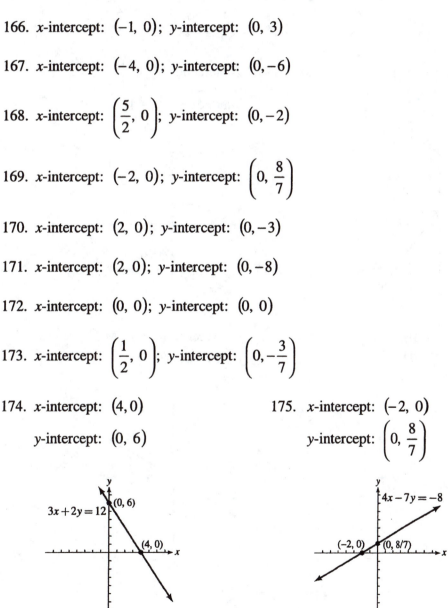

176. *x*-intercept: $(0, 0)$
    y-intercept: $(0, 0)$

177. *x*-intercept: $(0, 0)$
    y-intercept: $(0, 0)$

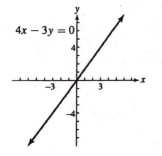

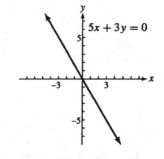

178. *x*-intercept: $\left(\dfrac{3}{2}, 0\right)$
    y-intercept: $(0, 2)$

179. *x*-intercept: $\left(\dfrac{8}{3}, 0\right)$
    y-intercept: $(0, -4)$

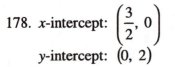

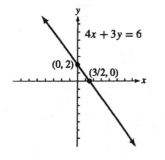

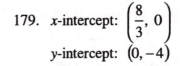

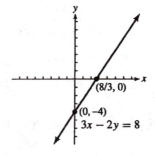

180. *x*-intercept: $(0, 0)$
    y-intercept: $(0, 0)$

181. *x*-intercept: $(0, 0)$
    y-intercept: $(0, 0)$

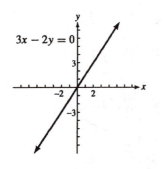

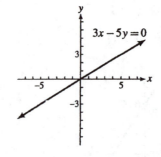

182. x-intercept: None
      y-intercept: $(0, -5)$

183. x-intercept: $(6, 0)$
      y-intercept: None

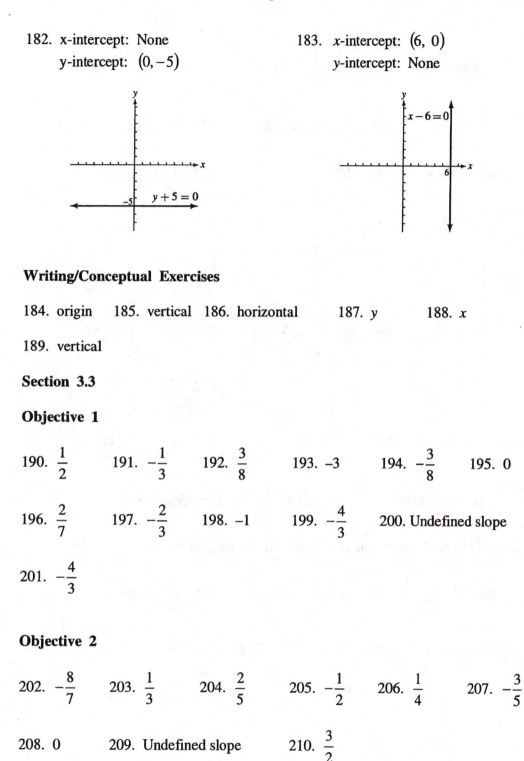

Writing/Conceptual Exercises

184. origin   185. vertical   186. horizontal   187. y 188. x

189. vertical

Section 3.3

Objective 1

190. $\dfrac{1}{2}$ 191. $-\dfrac{1}{3}$ 192. $\dfrac{3}{8}$ 193. -3 194. $-\dfrac{3}{8}$ 195. 0

196. $\dfrac{2}{7}$ 197. $-\dfrac{2}{3}$ 198. -1 199. $-\dfrac{4}{3}$ 200. Undefined slope

201. $-\dfrac{4}{3}$

Objective 2

202. $-\dfrac{8}{7}$ 203. $\dfrac{1}{3}$ 204. $\dfrac{2}{5}$ 205. $-\dfrac{1}{2}$ 206. $\dfrac{1}{4}$ 207. $-\dfrac{3}{5}$

208. 0 209. Undefined slope   210. $\dfrac{3}{2}$

Objective 3

211. 5; 5; parallel   212. 2; $-\dfrac{1}{2}$; perpendicular   213. -1; 1; perpendicular

214. -3; $\dfrac{1}{3}$; perpendicular 215. -5; $-\dfrac{1}{5}$; neither 216. -3; $\dfrac{1}{3}$; perpendicular

217. $-\dfrac{4}{5}$; $-\dfrac{5}{4}$; neither 218. -4; -4; parallel 219. 0; 0; parallel

220. 0; undefined slope; perpendicular 221. 3; 3; parallel

222. $-\dfrac{6}{5}$; $\dfrac{6}{5}$; neither

Mixed Exercises

223. $-\dfrac{2}{3}$ 224. 1 225. $\dfrac{3}{4}$ 226. $-\dfrac{2}{5}$ 227. Undefined slope

228. 0 229. $-\dfrac{1}{2}$ 230. $\dfrac{9}{4}$ 231. $\dfrac{1}{3}$ 232. $\dfrac{4}{5}$ 233. 2

234. Undefined slope 235. $\dfrac{1}{3}$ 236. -1 237. 0 238. $-\dfrac{3}{4}$

239. $-\dfrac{4}{7}$; $\dfrac{7}{4}$; perpendicular 240. $\dfrac{8}{9}$; $-\dfrac{9}{8}$; perpendicular

241. Undefined slope; 0; perpendicular 242. 0; 0; parallel

243. $\dfrac{8}{3}$; $\dfrac{8}{3}$; parallel 244. $\dfrac{7}{11}$; $-\dfrac{7}{11}$; neither 245. $\dfrac{3}{2}$; $\dfrac{3}{2}$; parallel

246. $\dfrac{3}{2}$; $-\dfrac{2}{3}$; perpendicular 247. $m=2$ 248. $m=-1$ 249. $m=\dfrac{5}{2}$

250. $m=\dfrac{1}{3}$ 251. $m=1$

Writing/Conceptual Exercises

252. False; change to "A line with negative slope falls from left or right" or "A line with positive slope rises from left to right."

253. False; change to "Horizontal lines have a slope of 0" or "Vertical lines have undefined slope."

254. False; change to "If two lines are perpendicular, their slopes are negative reciprocals."

255. (a) positive   (b) negative   256. (a) negative   (b) zero

257. (a) negative   (b) positive   258. (a) zero   (b) negative

259. (a) negative   (b) negative   260. (a) zero   (b) positive

CHAPTER 4   POLYNOMIALS AND EXPONENTS

Section 4.1

Objective 1

1. 1; 12

2. 1; −6

3. 2; 1, −1

4. 3; −2, 1, −1

5. 4; 8, 7, −5, 4

6. 2; 3, −7

7. 2; $\dfrac{1}{2}$, 1

8. 2; $-\dfrac{4}{5}, \dfrac{2}{3}$

9. 2; .75, −.25

10. 2; .10, −.20

Objective 2

11. $3s^3$

12. $-5t^6$

13. $-11x^3 + 10x^2$

14. $-y^4 - 3y^3 + y^2$

15. $-.6m^5$

16. $-4.7r^2 - 12.3r$

17. $10c^3 - 9c^2 - 5c + 6$

18. $20y^4 - 7y^3 - 5y^2 - y + 2$

19. $-3a^3 + 6a^2$

20. $-15x^3 - 2x^2 + 6x$

21. $-\dfrac{1}{4}r^3$

22. $\dfrac{9}{10}m^3 - m^2$

Objective 3

23. (a) and (b)

24. (a)

25. (c)

26. (c)

27. (c)

28. (a) and (b)

29. $5y^2 + 2y - 7$; degree 2; trinomial

30. $-m^5 + 3m^2 + 4m$; degree 5; trinomial

31. $z^4 + z^3$; degree 4; binomial

32. $p^5 + p^3$; degree 5; binomial

33. $3n^8 - n^7$; degree 8; binomial

34. $-4x^5 + 6x^2$; degree 5; binomial

35. $\left(\dfrac{5}{8}\right)x^2 - \left(\dfrac{5}{4}\right)x$; degree 2; binomial

36. $\left(\dfrac{1}{6}\right)y^2 - \left(\dfrac{7}{6}\right)y + \dfrac{1}{2}$; degree 2; trinomial

Objective 4

37. (a) 4   (b) -21 38. (a) 26   (b) -9 39. (a) 16   (b) 21

40. (a) 8   (b) 18   41. (a) -14 (b) 11   42. (a) -3 (b) -23

43. (a) 72   (b) 17   44. (a) -38 (b) -13 45. (a) 410   (b) 55

46. (a) 263   (b) 63

Objective 5

47. $8x^2 + 4x + 5$ 48. $10y^3 - 5y^2 + 16$ 49. $2m^3 - 9m^2 - 3m + 5$

50. $5w^4 + m^2 + 6m + 10$ 51. $-x^5 + 2x^2 - 2x$ 52. $4y^4 + y + 10$

53. $12p^4 + 3p^3 - 6p^2 - 6p + 15$ 54. $-6z^4 + 3z^3 - 4z^2 + 2z$

55. $11x^3 - 2x^2 - 6x - 6$ 56. $14x^3 + 2x^2 - 6x - 2$ 57. $6y^4 + 5y^3 + 8y - 6$

58. $4z^7 + z^6 - 3z^5 - 3z^4 + 2z^3 + 11$ 59. $5x^5 - 3x^4 + 7x^2 - x$

60. $m^4 - 2m^3 - m^2 + 10$ 61. $3x^2 + x$ 62. $4m^3 - 5m$

63. $-3y^2 + 5y - 13$ 64. $20k^2 + 2k - 19$ 65. $n^3 + 4n + 2$

66. $7x^2 - 3x + 7$ 67. $-4y^4 - 5y^3 - y + 1$ 68. $z^5 - 5z^4 + 3z^3 + 3z^2 - 8z$

69. $2x^3 + 4x^2 - 4x + 3$ 70. $3x^3 - x^2 + 11x + 3$ 71. $5m^3 + 2m + 15$

72. $14z^4 - 3z^3 + 5z + 1$ 73. $9a^5 - 2a^3 + 2a^2 - 2a + 15$

74. $3p^4 - 9p^3 - 3p^2 + 8p - 2$

Objective 6

75.

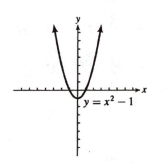

76.

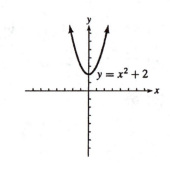

77.

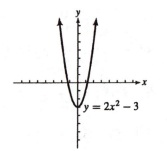

78.

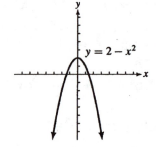

79.

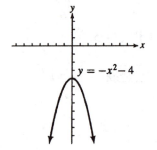

80.

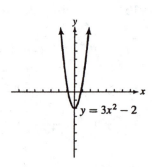

Mixed Exercises

81. $11x^3 - 2x^2 - 6x - 6$; degree 3; none of these   82. $4y^2 + 4y - 4$; degree 2; trinomial

83. $6y^4 + 5y^3 + 8y - 6$; degree 4; none of these

84. $5x^5 - 3x^4 + 7x^2 - x$; degree 5; none of these

85. $3p^4 - 9p^3 - 3p^2 + 8p - 2$; degree 4; none of these

86. $11y^2 + 2y + 6$; degree 2; trinomial   87. $2a^2 + 7$; degree 2; binomial

88. $4x^3$; degree 3; monomial   89. 3; 43; 8   90. −2; 44; 37

91. −1; 17; 17   92. 1; 61; −37   93. −2; 3; $\dfrac{27}{2}$ 94. 4; $\dfrac{26}{3}$; −17

95. −2; −205; 54   96. 7; −9; −2   97. $2x^2y - 2xy + 8y^2$

98. $4c^2 - 8cd + 8d^2$ 99. $8rs + 2r - 11s$ 100. $-2ab + ac$

101. $-x^2y + 8xy + 2xy^2$ 102. $.11a^2 - .07b^2$ 103. $x^3y - xy^3$

104. $8r^2t - 2rt - 14rt^2$ 105. $7a^3b - 6a^2b^2 + 12ab^3 + 9$

106. $-8rs + 16rt - 4st$

Writing/Conceptual Exercises

107. In the term 4^5, the base, 4, is a number rather than a variable, so the degree of the term is not 5. Because $4^5 (=1024)$ is a number and $x^0 = 1$, we can write this term as $4^5 x^0$, which shows that the degree of this term is 0.

108. Yes, the sum of polynomials in x, both of degree 4, can be of degree 3. This will happen whenever the terms of degree 4 are opposites, while the terms of degree 3 are not. For example,

$$
\begin{aligned}
&\left(2x^4 + 3x^3 - 4x^2 + x - 5\right) + \left(-2x^4 + 2x^3 - x^2 + 3x - 8\right) \\
&= 2x^4 + \left(-2x^4\right) + 3x^3 + 2x^3 + \left(-4x^2\right) + \left(-x^2\right) + x + 3x + (-5) + (-8) \\
&= 0 + 5x^3 - 5x^2 + 4x - 13 \\
&= 5x^3 - 5x^2 + 4x - 13,
\end{aligned}
$$

which is a polynomial of degree 3.

109. Yes, the sum of two polynomials in x, both of degree 3, can be of degree 1. This will happen whenever the terms of degree 3 are opposites and the terms of degree 2 are opposites, but the terms of degree 1 are not. For example,

$$\left(-5x^3+4x^2-6x+9\right)+\left(5x^3-4x^2+8x-10\right)$$
$$=-5x^3+5x^3+4x^2-4x^2+\left(-6x\right)+8x+9+\left(-10\right)$$
$$=0+0+2x-1$$
$$=2x-1,$$

which is a polynomial of degree 1.

110. Yes, it is possible to add two trinomials and obtain a sum which is a binomial. This will happen whenever one term of one of the trinomials is the opposite of the corresponding term of the other trinomial. For example,

$$\left(3x^2-5x+6\right)+\left(-2x^2+5x-8\right)$$
$$=3x^2+\left(-2x^2\right)+\left(-5x\right)+5x+6+\left(-8\right)$$
$$=x^2+0-2$$
$$=x^2-2,$$

which is a binomial.

Section 4.2

Objective 1

111. $2^5=32$ 112. $10^4=10,000$ 113. $(-1)^6=1$ 114. $12^3=1728$

115. $\dfrac{1}{3^4}=\dfrac{1}{81}$ 116. $\left(-\dfrac{2}{5}\right)^3=-\dfrac{8}{125}$ 117. r^6 118. $(-2y)^3$

119. $(ab)^4$ 120. $(.5st)^3$ 121. –64; base, -4; exponent, 3

122. –81; base, 3; exponent, 4   123. 64; base, 2; exponent, 6

124. 49; base, -7; exponent, 2   125. –64; base, 8; exponent, 2

126. –729; base, –9; exponent, 3

Objective 2

127. 4^8 128. $(-3)^6$ 129. 2^{12} 130. Cannot use product rule

131. $\left(\dfrac{1}{2}\right)^{7}$ 132. $\left(\dfrac{2}{5}\right)^{7}$ 133. $(-5)^{15}$ 134. Cannot use product rule   135. $24p^{6}$

136. $39k^{5}$ 137. $60a^{7}$ 138. $-144b^{15}$ 139. $21m^{8}$ 140. $-36x^{15}$

141. $8y^{2}; 15y^{4}$ 142. $-4x^{4}; -5x^{8}$ 143. $10a^{3}; -30a^{9}$ 144. $0; -42t^{6}$

Objective 3

145. 3^{8} 146. 7^{15} 147. 9^{12} 148. 12^{28} 149. 8^{20} 150. 11^{45}

151. $(-3)^{27}$ 152. $(-2)^{21}$ 153. -13^{24} 154. 14^{15} 155. 16^{110} 156. -21^{35}

Objective 4

157. $3^{2}b^{2}$ or $9b^{2}$ 158. $y^{2}z^{8}$ 159. $p^{10}q^{15}$ 160. $r^{15}s^{6}$ 161. $5a^{5}b^{15}$

162. $2^{3}w^{6}z^{21}$ or $8w^{6}z^{21}$ 163. $5^{3}r^{21}t^{6}$ or $125r^{21}t^{6}$ 164. $(-3)^{3}x^{3}y^{3}$ or $-27x^{3}y^{3}$

165. $(-2)^{4}r^{8}s^{4}$ or $16r^{8}s^{4}$ 166. $2 \cdot 3^{4}c^{8}d^{12}$ or $162c^{8}d^{12}$

167. $3x^{3}y^{15}$ 168. $4^{3}c^{9}d^{12}$ or $64c^{9}d^{12}$

Objective 5

169. $\dfrac{2^{3}}{3^{3}}$ or $\dfrac{8}{27}$ 170. $\dfrac{7^{2}}{8^{2}}$ or $\dfrac{49}{64}$ 171. $\dfrac{x^{5}}{y^{5}}$ 172. $\dfrac{w^{4}}{5^{4}}$ or $\dfrac{w^{4}}{625}$

173. $\dfrac{z^{3}}{10^{3}}$ or $\dfrac{z^{3}}{1000}$ 174. $\dfrac{2^{2}x^{2}}{3^{2}}$ or $\dfrac{4x^{2}}{9}$ 175. $\dfrac{x^{4}y^{4}}{z^{4}}$ 176. $\dfrac{x^{12}y^{4}}{z^{4}}$

177. $\dfrac{4^{2}x^{2}}{5^{2}}$ or $\dfrac{16x^{2}}{25}$ 178. $\dfrac{2^{4}a^{8}}{b^{12}}$ or $\dfrac{16a^{8}}{b^{12}}$ 179. $\dfrac{r^{3}}{x^{3}}$ 180. $\dfrac{1}{g^{3}}$

Mixed Exercises

181. 64; base, -2; exponent, 6 182. -64; base, 2; exponent, 6

183. -1; base, -1; exponent, 7   184. -1; base, 1; exponent, 7

185. -125; base, 5; exponent, 3   186. -125; base, -5; exponent, 3

187. 8^{20} 188. $x^6 y^3$ 189. $3^5 a^{10} b^{15}$ 190. $\dfrac{3^6 x^6}{5^6 y^6}$ 191. $7^5 x^{40} y^5$

192. $4^7 w^7 t^7$ 193. $(-5)^5 a^{14}$ 194. $5^3 a^{10} b^{21} c^{11}$ 195. z^{27}

196. $\dfrac{6^5 y^3}{7^2}$ 197. $\dfrac{9^4 w^4 x^{12}}{y^8}$ 198. $\dfrac{r^4 s^2 t^2}{2^2 n^2}$ 199. $4^5 x^5$ 200. $(-6)^9 q^9$

201. $\dfrac{4^5}{7^3}$ 202. $\dfrac{6^5 b^{10}}{11^5}$ 203. $\dfrac{3^3 \cdot 5^4 x^4}{4^3}$ 204. $\dfrac{4^3 a^3 b^6}{3^4}$ 205. z^{22}

206. $5^5 x^8 y^{17}$ 207. $\dfrac{7^5 a^{10} b^{15}}{2^5}$ 208. $\dfrac{k^6 m^{12} p^{18}}{3^6 n^{24}}$ 209. $(-2)^7 p^7 q^7$ 210. $3^2 x^{16} y^5 z^6$

Writing/Conceptual Exercises

211. The expression $(-2)^6$ means that -2 is used as a factor 6 times:

$$(-2)^6 = (-2)(-2)(-2)(-2)(-2)(-2) = 64.$$

On the other hand, -2^6 means that 2 is used as a factor 6 times and then we take the opposite of that number as the final step:

$$-2^6 = -(2 \cdot 2 \cdot 2 \cdot 2 \cdot 2 \cdot 2) = -64.$$

212. The expression $(2y)^5$ means that 2y is used as a factor 5 times:

$$(2y)^5 = (2y)(2y)(2y)(2y)(2y) = 32y^5.$$

We can obtain the same result by using the rule $(ab)^n = a^n b^n$:

$$(2y)^5 = 2^5 \cdot y^5 = 32y^5.$$

On the other hand, $2y^5$ means that we have 5 factors of y, but only one factor of 2, that is,

$$(2y)^5 = 2 \cdot y \cdot y \cdot y \cdot y \cdot y.$$

Thus,

$$(2y)^5 \neq 2y^5.$$

213. The product rule does not apply to the expression $2^5 \cdot 5^2$ because the bases are not the same. We must evaluate 2^5 and 5^2 separately and then multiply the results:

$$2^5 \cdot 5^2 = 32 \cdot 25 = 800.$$

214. The product rule states that when we multiply powers of the same base, we keep the same base and *add* the exponents, not multiply them. The correct answer is

$$(3x^4)(5x^6) = 3 \cdot 5 \cdot x^{4+6}$$
$$= 15x^{10}.$$

215. It is correct to add the exponents when multiplying x^3 and x^6, but their numerical coefficients, 8 and 9, are not exponential expressions. These numbers should be *multiplied*, not added. The correct answer is

$$(8x^3)(9x^6) = 8 \cdot 9 \cdot x^{3+6}$$
$$= 72x^9.$$

216. Combining the power rules $(ab)^m = a^m b^m$ and $\left(a^m\right)^n = a^{mn}$, we see that

$$\left(3a^3 b^4\right)^2 = 3^2\left(a^3\right)^2\left(b^4\right)^2$$
$$= 9a^6 b^8.$$

We see that each of the factors, 3, a^3, and b^4, must be squared. The error that was made in obtaining the answer, $6a^6 b^8$, was to multiply 3 by 2 rather than raising 3 to the second power.

217. Using the rule $\left(a^m\right)^n = a^{mn}$, we see that

$$\left(3^4\right)^5 = 3^{4 \cdot 5} = 3^{20}$$

and

$$\left(3^5\right)^4 = 3^{5 \cdot 4} = 3^{20}.$$

Because these results are the same, the expressions $\left(3^4\right)^5$ and $\left(3^5\right)^4$ are equivalent.

218. In the first expression, we have c^5 in the denominator. If $c = 0$, the value of the denominator will be 0. Because division by 0 is undefined, the expression will be undefined. To avoid this possibility, we have the restriction $c \neq 0$.

In the second expression, on the other hand, we can allow $c = 0$. If $c = 0$, the value of the expression will be $\left(3 \cdot a^3 \cdot b^4 \cdot 0\right)^2 = 0^2 = 0$. Unlike division by 0, multiplication by 0 is not a problem, and there is no need to restrict the variable.

Section 4.3

Objective 1

219. $28y^5$ 220. $72y^8$ 221. $15z^3 + 21z$

222. $-32p^5 + 28p^4 + 20p^3$ 223. $2k^6 + 8k^5 - 4k^4 + 12k^3 - 18k$

224. $-12y^5 - 6y^4 + 24y^3 - 33y^2$ 225. $6k + 4k^2 + 10k^5$

226. $21m - 35m^3 + 14m^4$ 227. $-12a + 8a^2 - 28a^6$

228. $-6r^6 + 21r^5 - 15r^4$

229. $32m^3n - 40m^2n^2 + 48mn^3$

230. $-20r^4s^3 + 6r^3s^2 - 10r^3s^3$

Objective 2

231. $x^2 + 8x + 15$

232. $x^2 - 8x + 12$

233. $2p^2 + 11p + 12$

234. $56 + 5a - 6a^2$

235. $10n^2 - 33n - 7$

236. $6x^2 - 17x + 5$

237. $27r^2 + 24r - 35$

238. $x^3 + 8$

239. $y^3 - 125$

240. $2x^4 - x^3 - 3x^2 + 4x + 4$

241. $6y^4 - 7y^3 + 4y^2 - 15y + 7$

242. $10m^5 - 15m^4 - 23m^3 - 3m^2 - 5m$

243. $2z^6 - 2z^5 - z^4 + z^3 - z^2 + 3z - 3$

244. $10x^4 - 9x^3 + 18x^2 + 12x + 5$

Objective 3

245. $x^2 - x - 12$

246. $y^2 + 8y + 12$

247. $r^2 - r - 56$

248. $z^2 - 16z + 55$

249. $4k^2 + 4k - 3$

250. $4m^2 - 27m - 40$

251. $16y^2 + 18y + 5$

252. $6x^2 + xy - 2y^2$

253. $12m^2 - 17mn - 7n^2$

254. $6p^2 + 11pq - 10q^2$

255. $8 + 14m - 15m^2$

256. $121k^2 - 100$

257. $3 - 7x - 6x^2$

258. $3 + 26a + 48a^2$

259. $5 + 6y - 8y^2$

260. $21p^2 - p - 2$

261. $15r^2 + rs - 6s^2$

262. $-36m^2 + 9mn + 10n^2$

263. $6x^4 - 11x^2 - 10$

264. $5v^4 - 14v^2w^2 - 3w^4$

Mixed Exercises

265. $20x^5$

266. $12m^2 + 28m$

267. $q^2 - 13q + 42$

268. $54p^2 - 21pr - 20r^2$ 269. $15t^2 - 4tu - 4u^2$ 270. $8y^5 - 4y^4 - 24y^3 + 4y^2$

271. $12x^4 + 38x^3 + 22x^2 - 19x - 14$ 272. $3x^3 - 14x^2 + 10x + 12$

273. $6x^5 - 3x^4 - 32x^3 + 20x^2 - 10x + 3$ 274. $6x^5 - 7x^4 + 14x^3 - 5x^2 + 5x + 2$

275. $4x^4 + 4x^3 - 2x^2 + 4x + 6$ 276. $12y^4 + 2y^3 - 5y^2 + 16y - 5$

277. $8a^5 - 4a^4 + 2a^3 - 2a^2 - a$ 278. $4b^6 - 12b^5 + 2b^4 + 2b^3 + 2b^2 + 2b - 2$

279. $8x^5 - 12x^4 + 10x^3 + 9x^2 - 16x + 12$ 280. $-9y^6 + 3y^5 + 6y^4 - 5y^3 - y^2 + 4y + 2$

281. $-4m^6 - 14m^4 + 10m^3 - 12m^2 + 19m - 4$

282. $16x^4 + 24x^3 - 11x^2 - 4x$

Section 4.4

Objective 1

283. $z^2 + 12z + 36$ 284. $t^2 - 20t + 100$ 285. $4x^2 + 12x + 9$

286. $25y^2 - 10y + 1$ 287. $4m^2 - 28m + 49$ 288. $9p^2 + 48p + 64$

289. $9 - 6x + x^2$ 290. $25 + 60y + 36y^2$ 291. $16p^2 - 24pq + 9q^2$

292. $4m^2 - 28mp + 49p^2$ 293. $25z^2 - 5z + \dfrac{1}{4}$ 294. $4x^2 - xy + \dfrac{1}{16}y^2$

295. $9x^2 + 2xy + \dfrac{1}{9}y^2$ 296. $49a^2 - 7ab + \dfrac{1}{4}b^2$

Objective 2

297. $z^2 - 49$ 298. $k^2 - 144$ 299. $16b^2 - 49$ 300. $64k^2 - 9p^2$

301. $16p^2 - 9q^2$ 302. $25 - 9x^2$ 303. $81 - 4y^2$ 304. $x^2 - \dfrac{1}{4}$

305. $y^2 - \dfrac{4}{9}$ 306. $16m^2 - \dfrac{1}{16}$ 307. $49a^2 - \dfrac{4}{9}b^2$ 308. $\dfrac{9}{16}s^2 - \dfrac{4}{25}t^2$

309. $y^4 - 9$ 310. $25m^4 - 4n^6$

Objective 3

311. $x^3 - 3x^2 + 3x - 1$ 312. $r^3 + 6r^2 + 12r + 8$ 313. $y^3 + 15y^2 + 75y + 125$

314. $8x^3 - 12x^2 + 6x - 1$ 315. $8x^3 + 60x^2 + 150x + 125$

316. $k^4 + 4k^3 + 6k^2 + 4k + 1$ 317. $t^4 - 8t^3 + 24t^2 - 32t + 16$

318. $81x^4 + 216x^3 y + 216x^2 y^2 + 96xy^3 + 16y^4$

319. $z^4 - 20z^3 + 150z^2 - 500z + 625$

320. $256s^4 + 768s^3 t + 864s^2 t^2 + 432st^3 + 81t^4$

321. $8x^3 - 36x^2 + 54x - 27$ 322. $j^4 - 4j^3 + 6j^2 - 4j + 1$

Mixed Exercises

323. $16y^2 - 24y + 9$ 324. $36b^2 - 121$ 325. $64b^2 - 48b + 9$

326. $49t^2 - 16u^2$ 327. $81k^2 + 36km + 4m^2$ 328. $625 - 256w^2$

329. $9x^2 - \dfrac{25}{16}$ 330. $36y^2 - 4y + \dfrac{1}{9}$ 331. $25j^2 + 5jk + \dfrac{1}{4}k^2$

332. $\dfrac{16}{49}t^2 - 81u^2$ 333. $\dfrac{1}{4}x^2 - \dfrac{1}{3}xy + \dfrac{1}{9}y^2$ 334. $\dfrac{4}{9}x^2 - \dfrac{9}{16}y^2$

335. $8x^3 - 36x^2 y + 54xy^2 - 27y^3$ 336. $81a^4 - 432a^3 b + 864a^2 b^2 - 768ab^3 + 256b^4$

337. $\dfrac{1}{16}x^4 + \dfrac{1}{2}x^3 y + \dfrac{3}{2}x^2 y^2 + 2xy^3 + y^4$ 338. $8x^3 - 6x^2 y + \dfrac{3}{2}xy^2 - \dfrac{1}{8}y^3$

339. $36x^2 - 4y^2$ 340. $x^4 + 8x^3 y + 24x^2 y^2 + 32xy^3 + 16y^4$

341. $25a^2 - 2ab + \dfrac{1}{25}b^2$ 342. $16x^4 - 32x^3 y + 24x^2 y^2 - 8xy^3 + y^4$

343. $\dfrac{1}{8}x^3 - \dfrac{1}{4}x^2y + \dfrac{1}{6}xy^2 - \dfrac{1}{27}y^3$ **344.** $9x^2 - \dfrac{1}{36}y^2$

Writing/Conceptual Exercises

345. In the expression $x^2 - y^2$, we square two numbers and then subtract the results. In the expression $(x-y)^2$, we subtract first and then square the result. For example, if $x=5$ and $y=4$,

$$x^2 - y^2 = 5^2 - 4^2 = 25 - 16 = 9,$$

while

$$(x-y)^2 = (5-4)^2 = 1.$$

In general,

$$(x-y)^2 = x^2 - xy + y^2 \neq x^2 - y^2.$$

346. Let $x=5$. Then

$$(x-3)^2 = (5-3)^2 = 2^2 = 4,$$

while

$$x^2 - 9 = 5^2 - 9 = 25 - 9 = 16.$$

This example shows that

$$(x-3)^2 \neq x^2 - 9.$$

347. When we square a binomial, the middle term of the resulting trinomial is *twice* the product of the two terms. This student remembered to use the product of the two terms to find the middle term of the trinomial, but forgot to multiply this product by 2. The correct answer is

$$(x-3y)^2 = x^2 - 2(x)(3y) + (3y)^2$$
$$= x^2 - 6xy + 9y^2.$$

348. We have seen that $(a+b)^2 = a^2 + 2ab + b^2$ has 3 terms, $(a+b)^3$ will have 4 terms, and $(a+b)^4$ will have 5 terms. The number of terms in the simplified product is always one more than the power to which we are raising the binomial. We expect the same pattern to continue, so that the simplified answer for $(a+b)^5$ will have 6 terms.

Section 4.5

Objective 1

349. 1   350. 1   351. -1 352. 1   353. 2   354. 2

355. 0   356. 2   357. 0   358. -2 359. -1 360. 0

Objective 2

361. $\dfrac{1}{49}$ 362. $\dfrac{1}{12}$ 363. $\dfrac{1}{25}$ 364. $-\dfrac{1}{32}$ 365. $\dfrac{16}{9}$ 366. $\dfrac{9}{8}$

367. $\dfrac{8}{15}$ 368. $\dfrac{1}{20}$ 369. $-\dfrac{3}{8}$ 370. $\dfrac{1}{r^4}$ 371. $\dfrac{1}{y^8}$ 372. $2r^4$

373. $\dfrac{1}{6r^5}$ 374. $\dfrac{2y^2}{3x^3}$ 375. $\dfrac{25}{16}$

Objective 3

376. 4^4 377. 5^5 378. $(-2)^7$ 379. $\dfrac{1}{(-4)^7}$ 380. $\dfrac{1}{7x^5}$ 381. $\dfrac{k^4}{2m^4}$

382. $\dfrac{y}{12^3}$ 383. 9   384. $\dfrac{1}{12}$ 385. x^7 386. $\dfrac{p^7}{3^5 m}$ 387. $8^5 b^4 c^{10}$

388. z 389. $a^6 b^8$ 390. $\dfrac{xy^2}{5^5}$

Objective 4

391. 7^2 392. 9   393. 8^{10} 394. a^7 395. $\dfrac{1}{2^{12}}$ 396. $\dfrac{3^4 y^2}{x^6}$

397. $\dfrac{4^2}{5^2 w^2 y^{10}}$ 398. $\dfrac{q^5}{p^{10}}$ 399. $\dfrac{3^2}{2^5 y^3}$ 400. $\dfrac{1}{9xy}$ 401. $\dfrac{1}{t^3}$

402. q^{23} 403. $\dfrac{x^4}{y^8}$ 404. $\dfrac{1}{k^4 t^{10}}$

Mixed Exercises

405. 2   406. $\dfrac{1}{25}$ 407. 0   408. 1   409. 1   410. 1

411. $\dfrac{27}{8}$ 412. $\dfrac{256}{81}$ 413. $\dfrac{7}{36}$ 414. 7   415. $\dfrac{1}{a^{10}}$ 416. $\dfrac{1}{2}$

417. $\dfrac{1}{5^5}$ 418. $\dfrac{64}{27}$ 419. $\dfrac{5}{k^5}$ 420. $\dfrac{x^6}{y^4}$ 421. x 422. $\dfrac{2}{5x^5}$

423. z^{18} 424. y^5 425. $\dfrac{4n^2 p^4}{m}$

Writing/Conceptual Exercises

426. Negative   427. Positive   428. Positive   429. Negative

430. Positive   431. Positive   432. Zero   433. Zero

434. Positive   435. Negative   436. $9x^{10}$ sq cm   437. $8y^{12}$ cu ft

438. We keep the same base and add the exponents when we are *multiplying* powers of the same base, not adding them. We cannot apply any of the laws of exponents here. Instead, evaluate the two terms separately and then add the results:

$$2^{-2} + 2^{-3} = \frac{1}{2^2} + \frac{1}{2^3}$$
$$= \frac{1}{4} + \frac{1}{8} = \frac{3}{8}.$$

Section 4.6

Objective 1

439. $2m + 3$ 440. $4m^2 - 3m + 2$ 441. $5m^3 - 3m$ 442. $10m - 6 + \dfrac{1}{m}$

443. $9m+6-\dfrac{2}{m}$ 444. $\dfrac{4}{3}m-\dfrac{2}{3}$ 445. $-18m+10+\dfrac{2}{m}$ 446. $1-\dfrac{1}{m^2}$

447. $2-\dfrac{1}{m}+\dfrac{3}{m^2}$ 448. $5m^3-\dfrac{5}{3}m^2+1-\dfrac{2}{3m}+\dfrac{1}{m^2}$ 449. $2p+6p^4$

450. $2x^4+3x^3+5x$ 451. $2y^6-\dfrac{9}{4}y$ 452. $3z^4+9z^2-1+\dfrac{10}{3z}$

453. $30x^3-10x$ 454. $\dfrac{3}{2}m^4-m^2+3$ 455. $\dfrac{1}{2}m+\dfrac{3}{2}-\dfrac{6}{m}$ 456. $4p^2-5p-\dfrac{1}{p}$

457. $5q^2-3+\dfrac{2}{q}$ 458. $8y^4+9y^3-\dfrac{11}{y^2}-\dfrac{12}{y^3}$ 459. $4z^2+\dfrac{8}{3}z-2+\dfrac{5}{3z^2}$

460. $9+6x^3+\dfrac{3}{4}x^7$ 461. $-2y^4+y^3-3y-\dfrac{9}{y}$ 462. $-2+\dfrac{2}{y}-\dfrac{10}{y^2}$

Objective 2

463. $x-3$ 464. $y+4$ 465. $x+7$ 466. $p+10$ 467. $x+2$

468. $r+3+\dfrac{-5}{r-5}$ 469. $a-7+\dfrac{37}{2a+3}$ 470. $3w+1$ 471. $5w-2+\dfrac{-4}{w-4}$

472. $6b-5+\dfrac{42}{b+7}$ 473. $5m-4$ 474. $y-1$ 475. $2x+3$

476. $3y^2-5y+6$ 477. $z^2-2z+5+\dfrac{1}{2z+1}$ 478. $2m^2+5m-3\dfrac{-2}{3m-5}$

479. $9p^3-2p+6$ 480. $4x-3+\dfrac{4x+5}{3x^2-2x+5}$ 481. $2x^2-5x+4+\dfrac{-5}{3x^2+1}$

482. $y^3-2y^2+4+\dfrac{-3}{4y^2-3}$ 483. $y^2+4y-4+\dfrac{-1}{y^2+1}$ 484. $3x^2+2x+1+\dfrac{-1}{x^2-1}$

485. y^2+y+1 486. b^2-1 487. $9x^4+3x^3+2x$

488. $16x^4+24x^3+36x^2+54x+81$

Writing/Conceptual

489. It is a polynomial in the variable y having degree 4.

490. Mara's answer is incorrect. The last term of the answer should be $\dfrac{3z}{3z}=1$. The problem should be worked as follows:

$$\frac{9z^3-12x^2+3z}{3z}=\frac{9z^3}{3z}-\frac{12z^2}{3z}+\frac{3z}{3z}$$
$$=3z^2-4z+1.$$

491. Yes, Marc was correct. When a polynomial is divided by a monomial, the quotient should have the same number of terms as the dividend. In this case, the dividend is a trinomial, so the quotient should be a trinomial, not a binomial.

492. Checking Mara's answer gives

$$3z\!\left(3z^2-4z\right)=9z^3-12z^2.$$

This result is missing the last term of the dividend. Checking the correct answer gives

$$3z\!\left(3z^2-4z+1\right)=9z^3-12z^2+3z,$$

the original dividend.

493. Josh did not leave space for the missing terms in the dividend, so he will not have room to put the x^4-terms, x^3-terms, x^2-terms, and x –terms that will appear in the division process. He should have set up the problem as follows:

$$x-1\overline{)x^5+0x^4+0x^3+0x^2+0x-1}.$$

494. (c)

495. This remainder cannot be correct because the remainder must be of smaller degree than the divisor. In this problem, the divisor, $x-4$, is of degree 1, so the remainder must be of degree 0. This means that the remainder must be a number rather than a variable expression.

496. The two answers are equivalent, so both can be correct.

Section 4.7

Objective 1

497. 6.75×10^2 498. 3.295×10^3 499. 3.5744×10^4 500. 6.09906×10^5

501. 8.5×10^0 502. 1.3205×10^7 503. 2×10^{-2} 504. 1.76×10^{-1}

505. 5×10^{-5} 506. 7.36×10^{-3} 507. -9.3572×10^4 508. -4.8×10^{-4}

Objective 2

509. 350   510. 63,700   511. −275,000   512. 7.045

513. 40,000,000   514. 150,000,000   515. .062   516. .000755

517. .2001   518. .00002753   519. −8.02   520. − .00999

Objective 3

521. 3,000,000   522. 253   523. .00000713   524. 24,000

525. 4,000,000   526. .0144   527. 50,000   528. .02

529. 6000   530. 200   531. 9   532. 2.1

Mixed Exercises

533. 4.905×10^6 534. 3.2×10^1 535. 1×10^{-1} 536. 7.5×10^{-3}

537. -5.32×10^0 538. -6.09×10^{-5} 539. 342,000,000   540. .00000271

541. −203,000   542. −.000000924   543. 1,236,000,000   544. −.043762

545. 600,000   546. −.00007   547. 544   548. .014

549. 10   550. 300,000   551. 3000   552. 15,000

553. 6000   554. 200   555. 10   556. 100   557. 1600   558. 350

Writing/Conceptual Exercises

559. In scientific notation

560. Not in scientific notation; 2.0×10^1

561. Not in scientific notation; 3.852×10^3

562. Not in scientific notation; 5×10^{-4}

563. Not in scientific notation; 9.2×10^4

564. In scientific notation

565. Not in scientific notation; 3×10^3

566. Not in scientific notation; 7.5×10^0

567. Answers will vary. Some reasons that may be given are the following:

1. Scientists often use numbers that are very large or very small. Without scientific notation, it requires lots of zeros to write these numbers, which is inconvenient.

2. If we write numbers with lots of zeros, it is easy to lose count and write the wrong number of zeros.

3. It is much easier to calculate with numbers written in scientific notation, using the laws of exponents, than with numbers written without exponents.

4. Calculators cannot display numbers containing large numbers of zeros but can display numbers written in scientific notation.

568. We use 10 as the base in scientific notation because 10 is the base of our number system. We can easily multiply or divide a number by a power of 10 by simply moving the decimal point.

CHAPTER 5   FACTORING AND APPLICATIONS

Section 5.1

Objective 1

1. 6 2. 18 3. 5 4. 14 5. 10 6. 1

7. 6 8. 14 9. 32 10. 28 11. $5x$ 12. $7y^2$

13. $2ab$ 14. m^3 15. $3x^2$ 16. z^2 17. km^2n^4 18. w^2x^2y

19. $15a^2y$ 20. $9xy$

Objective 2

21. 7 22. $3x^2$ 23. $3y^2$ 24. $-3a^2y$ 25. $3(3r+8t)$

26. $19(3m+2n)$ 27. $9q(2q-5)$ 28. $9x(3y+2+4x)$

29. $15w(w+6x)$ 30. $8a(3b+1+2c)$ 31. $14t(3w+2+5t)$

32. $11r(3s+2+5r)$ 33. $5a^2(a^2+5a-4)$ 34. No common factor (except 1)

35. $13x^8(2+x^4-6x^2)$ 36. $5b^2(20b^2+3)$ 37. $8xy(2x+3+5y)$

38. $(a+b)(5+x)$ 39. $(a-b)(c-d)$ 40. $(r-2s)(x^2+y^2)$

Objective 3

41. $(y+4)(y+6)$ 42. $(a+7)(a+3)$ 43. $(d-5)(d+2)$ 44. $(y-7)(y+3)$

45. $2(m+3)(m+n)$ 46. $(3z+2q)(6z-q)$ 47. $(1-m)(1-n)$ 48. $(w+3)(4-z)$

49. $(7-x)(y-9)$ 50. $(2a+b^2)(3a^3-b^3)$ 51. $(2a-3b)(a^2+b^2)$

52. $(4r^2-s)(2r-s^2)$ 53. $(2x+y)(x^3+2y^3)$ 54. $(5-d)(4-c)$

Mixed Exercises

55. 14

56. 9

57. $13y^3$

58. $17pq^2$

59. No common factor (except 1)

60. $-21luv$

61. $7a(3a-5)$

62. $8q(p+3q)$

63. $5ab(a+2+3b)$

64. $(x-y)(2a-5b)$

65. $(m-15)(m+4)$

66. $(14-q)(2-q)$

67. $(5m+9)(2m-1)$

68. $(3r-4s)(r^2+s^2)$

69. $9(m^2n-3mn+1)$

70. $(1+p)(1-q)$

71. $12x^7(4x^5-3x^3+2)$

72. $4y^3m^2(5y^4-9m^2)$

73. $13y^2(y+3)$

74. $9q^2p^3(5q^2p-4+9p)$

75. $(2-3p)(4-3p^3)$

76. $(2c+3)(6c-5)$

77. $4mn(n-7)$

78. $4a^2(24a^3+12a+1)$

79. $(7m+5n)(4m-1)$

80. $(3x+4y)(x^2+y^2)$

Writing/Conceptual Exercises

81. Answers will vary. One example is 7, 14, 21, 28.

82. Answers will vary. One example is $8x^2yz^3$, $4x^3y^2z^3$, $-12x^4y^4z^4$.

83. In factored form

84. Not in factored form

85. Not in factored form

86. In factored form

87. The given expression cannot be factored because the two terms, $2x(3y-4)$ and $5(3y+4)$, do not have a common factor.

88. The given expression can be factored because the two terms have a common factor, $2u+v$.
$$r^2(2u+v)+s^2(2u+v)=(2u+v)(r^2+s^2)$$

Section 5.2

Objective 1

89. 1 and 28, -1 and -28, 2 and 14, -2 and -14, 4 and 7, -4 and -7; the pair with a sum of 11 is 4 and 7.

90. 1 and 40, -1 and -40, 2 and 20, -2 and -20, 4 and 10, -4 and -10, 5 and 8, -5 and -8; the pair with a sum of -13 is -5 and -8.

91. 1 and -16, -1 and 16, 2 and -8, -2 and 8, 4 and -4, -4 and 4; the pair with a sum of 6 is -2 and 8.

92. 1 and -54, -1 and 54, 2 and -27, -2 and 27, 3 and -18, -3 and 18, 6 and -9, -6 and 9; the pair with a sum of -3 is 6 and -9.

93. $x+2$ 94. $t-9$ 95. $m+12$ 96. $p-5$ 97. $(y+5)(y+6)$

98. $(x-9)(x-3)$ 99. $(b-1)(b+2)$ 100. $(z-5)^2$ or $(z-5)(z-5)$

101. $(x-15)(x+3)$ 102. $(n-11)(n+3)$ 103. $(r+14)(r+1)$

104. $(x-7y)(x-8y)$ 105. $(x-7y)(x+3y)$ 106. $(m-3n)(m+n)$

Objective 2

107. $3(a+5)(a+4)$ 108. $2p(p+5)(p-1)$ 109. $3hk(h+12)(h-1)$

110. $8(b+5)(b-4)$ 111. $7(a-4)(a-2)$ 112. $2p^5(p-7)(p+2)$

113. $ab(a+3b)(a-2b)$ 114. $2(x-8)(x-4)$ 115. $10(n-5)(n-6)$

116. $3j(h-4)(h-2)$ 117. $10k^4(k+4)(k+8)$ 118. $16x^5(x-1)(x-2)$

119. $6xy^2(x+2y)(x+3y)$ 120. $4a^2b(a-3b)(a-5b)$

Mixed Exercises

121. $(a-7)(a-4)$ 122. Prime   123. $x(x-7)(x-1)$

124. $5r(r-6)(r-5)$ 125. $(y-8)(y+5)$ 126. $2m(m-5)(m+1)$

127. $3n^3(n-5)(n-1)$ 128. $(q-13)(q+2)$ 129. $(b-5c)^2$ or $(b-5c)(b-5c)$

130. $r^8(r-7q)(r+3q)$ 131. $(c-7d)(c-5d)$ 132. $3(m-1)^2$ or $3(m-1)(m-1)$

133. $2t(s+7)(s-5)$ 134. $2x(x-5y)(x-2y)$ 135. $2(m+5)(m+1)$

136. $3(n+3)(n+1)$ 137. $(x-5y)(x+3y)$ 138. $(p+4)(p+8)$

139. $4(r-2)(r-9)$ 140. $10(z-5)(z+7)$ 141. $5(a+b)(a-2b)$

142. $(f-5g)(f+2g)$ 143. $2(s-9)(s-11)$ 144. $x(x+3y)(x-4y)$

145. $2r(r+2)(r-5)$ 146. $3a(a+9)(a-2)$

Writing/Conceptual Exercises

147. (d)   148. (a)   149. Factor out the greatest common factor, 3.

150. Factor out the greatest common factor, $4k$.

151. The answer $(p-2)(4p+12)$ is not correct. The given polynomial is not completely factored because $4p+12$ has a common factor of 4 and can be factored as $4(p+3)$. To avoid this error, the greatest common factor of 4 should be factored out of the given polynomial as the first step.

152. There is no pair of integers whose product is 7 and whose sum is -4, so s^2-4s+7 cannot be factored out.

Section 5.3

Objective 1

153. $(4b+1)(2b+1)$ 154. $(x+3)(2x+7)$ 155. $(5a-2)(3a+1)$

156. $(2m+3)(m+2)$ 157. $(7b-2)(b+2)$ 158. $(3m+2)(m+3)$

159. $p(3p+1)(p+5)$ 160. $2(4m+n)(m+3n)$ 161. $b(7a-4)(a+2)$

162. $(2s-t)(s+10t)$ 163. $2(2y+z)(y-z)$ 164. $(5a+2b)(5a+3b)$

Objective 2

165. $a+4$ 166. $4k+5$ 167. $3t-4$ 168. $8y-5$

169. $(5x+2)(2x+1)$ 170. $(2z+1)(2z-3)$ 171. $(a-6)(2a+1)$

172. $(w-2)(15w+1)$ 173. $(2q+1)(5q+3)$ 174. $(7m+1)(2m-5)$

175. $(m+3)(3m-2)$ 176. $(5q+6)(3q+1)$ 177. $(3a+b)(a-2b)$

178. $(5w+2z)^2$ or $(5w+2z)(5w+2z)$ 179. $(2s-3w)(4s+3w)$

180. $(2x+3y)(3x-2y)$ 181. $(6x-y)(3x-4y)$ 182. $(4y-3)(4y+5)$

Mixed Exercises

183. $(2a+3)(a+3)$ 184. $(2p+1)(3p+5)$ 185. $(6y-1)(y+1)$

186. $(2z+1)(4z+5)$ 187. $(p-5)(3p+1)$ 188. $(3r-1)(4r+5)$

189. $(x-4)(7x+1)$ 190. $2(3x+7y)(x-y)$ 191. $x^2(2x-3)(x-4)$

192. $2a(3a-2b)(2a+3b)$ 193. $(9r-8t)(3r+2t)$ 194. $y^3z^2(2y+z)(y-3z)$

195. $3x^2y(4x+5y)(6x-7y)$ 196. $2x^2y^3z^2(6y+5z)(9y-8z)$

197. $4x^5(8x-9y)(9x+8y)$ 198. $5(3x-2yz)(6x+5yz)$ 199. $(7x+12y)(9x-7y)$

200. $(3-z)(4+z)$ 201. $(5-3x)(3-2x)$ 202. $-2x(4x-3)(2x+5)$

203. $y(2y-7)(3y+1)$ 204. $-1(x+5)(x-4)$ 205. $-1(2a+b)(4a-b)$

206. $-1(3p-2q)(2p+q)$

Writing/Conceptual Exercises

207. (b) 208. (c)

209. The binomial $3x-6$ has a common factor of 3. If $3x-6$ were a factor of $6x^2-11x+3$, then 3 would have to be a factor of $6x^2-11x+3$, but it is not. Therefore, $3x-6$ cannot be a factor of $6x^2-11x+6$.

210. Both signs will be negative.

Section 5.4

Objective 1

211. $(t+10)(t-10)$ 212. Prime   213. $(8+a)(8-a)$ 214. $(y+15)(y-15)$

215. $(2a+9)(2a-9)$ 216. $\left(6j+\dfrac{5}{7}\right)\left(6j-\dfrac{5}{7}\right)$ 217. $(12b+7)(12b-7)$

218. $(11m+6n)(11m-6n)$ 219. $(x^2+9)(x+3)(x-3)$ 220. $(z^2+12)(z^2-12)$

221. $(a^2+100)(a+10)(a-10)$ 222. $(16y^2+1)(4y+1)(4y-1)$

223. Prime   224. $m^2(mn+1)(mn-1)$

Objective 2

225. $(y-3)^2$ 226. $(q+9)^2$ 227. $(t-6)^2$ 228. $(c-12)^2$

229. $\left(a-\dfrac{1}{2}\right)^2$ 230. $(3w+1)^2$ 231. $(6q-5)^2$ 232. $(7j+2)^2$

233. $(8p^2+3q^2)^2$ 234. $\left(10p-\dfrac{5}{8}r\right)^2$ 235. $(r-1.3)^2$ 236. $-4(x-1)^2$

237. $-3(2a+5b)^2$ 238. $2(3x-4y)^2$

Objective 3

239. $(a-1)(a^2+a+1)$ 240. $(b-3)(b^2+3b+9)$ 241. $(z-2)(z^2+2z+4)$

242. $(c-6)(c^2+6c+36)$ 243. $(d-5)(d^2+5d+25)$ 244. $(m-4)(m^2+4m+16)$

245. $(2y-1)(4y^2+2y+1)$ 246. $(5z-2)(25z^2+10z+4)$

247. $(3r-1)(9r^2+3r+1)$ 248. $(7t-1)(49t^2+7t+1)$

249. $(c^3-d^2)(c^6+c^3d^2+d^4)$ 250. $(t^2-r)(t^4+t^2r+r^2)$

251. $(q-2p)(q^2+2qp+4p^2)$

252. $(5m-2p)(25m^2+10mp+4p^2)$

253. $(2z-3y)(4z^2+6zy+9y^2)$

254. $(4x-3y)(16x^2+12xy+9y^2)$

255. $\left(2m-\dfrac{1}{3}\right)\left(4m^2+\dfrac{2}{3}m+\dfrac{1}{9}\right)$

256. $\left(p-\dfrac{1}{2}\right)\left(p^2+\dfrac{1}{2}p+\dfrac{1}{4}\right)$

257. $(5c-6d)(25c^2+30cd+36d^2)$

258. $(10a-3b)(100a^2+30ab+9b^2)$

Objective 4

259. $(x+1)(x^2-x+1)$

260. $(y+3)(y^2-3y+9)$

261. $(z+2)(z^2-2z+4)$

262. $(m+4)(m^2-4m+16)$

263. $(n+6)(n^2-6n+36)$

264. $(p+5)(p^2-5p+25)$

265. $(3a+1)(9a^2-3a+1)$

266. $(2b+1)(4b^2-2b+1)$

267. $(5c+2)(25c^2-10c+4)$

268. $(7d+3)(49d^2-21d+9)$

269. $(t^2+1)(t^4-t^2+1)$

270. $(s^2+5t)(s^4-5s^2t+25t^2)$

271. $(w^2+2z)(w^4-2w^2z+4z^2)$

272. $(4x+3y)(16x^2-12xy+9y^2)$

273. $(3a+2b)(9a^2-6ab+4b^2)$

274. $(5c^2+1)(25c^4-5c^2+1)$

275. $(6m+5p)(36m^2-30mp+25p^2)$

276. $(y^2+z^3)(y^4-y^2z^3+z^6)$

277. $\left(z+\dfrac{1}{2}\right)\left(z^2-\dfrac{1}{2}z+\dfrac{1}{4}\right)$

278. $\left(3t+\dfrac{1}{4}\right)\left(9t^2-\dfrac{3}{4}t+\dfrac{1}{16}\right)$

Mixed Exercises

279. $(a-8)^2$

280. $(10k+m)(10k-m)$

281. $(3p+11)(3p-11)$

282. $(2f+1)^2$

283. Prime

284. $(5h+2)^2$

285. $(2x+5)(4x^2-10x+25)$

286. $(2y+3z)(4y^2-6yz+9z^2)$

287. $(a^2-9b)^2$

288. $(z-.8)^2$

289. $4x(x-2)(x^2+2x+4)$

290. $\left(r^2+25\right)\left(r+5\right)\left(r-5\right)$ 291. $\left(j+\dfrac{1}{7}\right)^2$ 292. $\left(\dfrac{1}{2}x+3\right)\left(\dfrac{1}{4}x^2-\dfrac{3}{2}x+9\right)$

293. $\left(x^2-\dfrac{1}{5}\right)\left(x^4+\dfrac{1}{5}x^2+\dfrac{1}{25}\right)$ 294. $\left(\dfrac{1}{3}x+6y\right)^2$

295. $\left(x^3-4y\right)\left(x^6+4x^2y+16y^2\right)$ 296. $3\left(4m^2+25\right)$

297. $\left(x-1\right)\left(x^2+x+1\right)\left(x^6+x^3+1\right)$ 298. $4\left(4s-5\right)^2$

299. $-2\left(x+5\right)^2$ 300. Prime

301. $\left(5x+4y^2\right)\left(25x^2-20xy^2+16y^4\right)$ 302. $12\left(z-1\right)\left(z^2+z+1\right)$

303. $\left(11t^2-12v^3\right)\left(11t^2+12v^3\right)$ 304. $\left(3p+2q\right)\left(3p-2q\right)\left(9p^2+4q^2\right)$

Writing/Conceptual Exercises

305. A perfect square   306. A perfect cube   307. Neither of these

308. Both of these   309. Neither of these   310. Both of these

311. Answers will vary. One example is $4x^2+36$. To find other examples of this type of binomial, start with the sum of any two squares and use the distributive property to multiply this expression by any number other than 1. We can obtain

$$4\left(x^2+9\right)=4x^2+36$$
$$10\left(x^2+9\right)=10x^2+90$$
$$-5\left(x^2+9\right)=-5x^2-45,$$

and so on.

312. Ask the student to check the factorization by multiplying. Using FOIL will assure that the middle term is not omitted.

$$\left(x+5\right)\left(x+5\right)=x^2+5x+5x+25$$
$$=x^2+10x+25$$
$$\neq x^2+25.$$

The binomial x^2+125 is a sum of two squares with no common factor, so it is prime.

Section 5.5

Objective 1

313. $-6, \dfrac{5}{2}$

314. $-\dfrac{3}{4}, \dfrac{1}{2}$

315. $-3, -2$

316. $-9, 9$

317. $-8, -3$

318. $-4, 8$

319. $0, 9$

320. $\dfrac{5}{2}, 3$

321. $-3, \dfrac{2}{3}$

322. $-5, \dfrac{5}{4}$

323. $\dfrac{2}{5}$

324. $-10, 10$

325. $-\dfrac{5}{3}, \dfrac{3}{2}$

326. $-\dfrac{2}{5}, 2$

327. $-4, \dfrac{3}{5}$

328. $-3, \dfrac{1}{2}$

Objective 2

329. $-6, -4, 0$

330. $-7, 0, \dfrac{1}{2}$

331. $-\dfrac{9}{2}, 0, \dfrac{9}{2}$

332. $-6, 0, 6$

333. $-11, 0, 11$

334. $0, 1, 4$

335. $-2, 0, \dfrac{3}{2}$

336. $-7, 2, 7$

337. $-10, 0, 1$

338. $-4, 0, \dfrac{1}{3}$

339. $-3, 2, 5$

340. $-7, 2, 3, 7$

341. $-5, 0, 4$

342. $-2, \dfrac{1}{2}, 3$

343. $-\dfrac{5}{2}, \dfrac{1}{2}, 3$

344. $-\dfrac{1}{2}, \dfrac{1}{2}, \dfrac{2}{3}$

Mixed Exercises

345. $-8, -\dfrac{9}{4}$

346. $-5, 0, 7$

347. $-5, 2$

348. $0, 5$

349. $-5, 0, 3$

350. $-\dfrac{2}{3}, 6$

351. $-\dfrac{5}{2}, \dfrac{5}{2}$

352. $-13, 0, 13$

353. $-5, 1, 5$

354. $-5, \dfrac{7}{3}$

355. $0, 2, 4$

356. $-4, \dfrac{1}{3}$

357. $6, 8$

358. $-1, 0, 1, 2$

359. $-\dfrac{3}{2}, 5$

360. 0

361. –4, 9   362. –6, 5   363. –2, 0, 2   364. $-\dfrac{9}{2}, \dfrac{9}{2}$

365. –4, 0, 9   366. 6, 7   367. –6, –5, 3   368. –3, –2, 2, 3

369. –9, –1, 0, 1, 9   370. $-\dfrac{1}{2}$, 0, 3

Writing/Conceptual Exercises

371. $x^2 + 4x - 21 = 0$ 372. $3x^2 - 20x + 12 = 0$ 373. $10x^2 + 11x + 3 = 0$

374. $8x^2 - 2x - 15 = 0$ 375. $x^2 + 5x = 0$ 376. $x^2 - 14x + 49 = 0$

377. The zero-factor property only applies when the product of two factors is *zero*. There is no comparable property that applies when the product is 8. To solve this equation, it is necessary to multiply the binomial factors on the left, then rewrite the equation in standard form and solve by factoring.

378. Dividing both sides of an equation by x (or any variable expression) is not permitted because the value of x might be 0, and division by 0 is undefined. Although the method shown in this exercise yields a correct answer, $\dfrac{1}{3}$, the other solution of $6x^2 = 2x$, which is 0, has been "lost". A correct solution follows.

$$6x^2 = 2x$$
$$6x^2 - 2x = 0$$
$$2x(3x - 1) = 0$$
$$2x = 0 \;\; or \;\; 3x - 1 = 0$$
$$x = 0 \;\; or \;\;\;\; x = \dfrac{1}{3}$$

Section 5.6

Objective 1

379. Length, 12 cm; width, 9 cm   380. Length, 14 in; width, 7 in

381. Length, 11 ft; width, 8 ft

382. First rectangle: length, 12 m; width, 4 m; second rectangle: length, 8 m; width, 6 m

383. First square, 5 m; second square, 3 m   384. Base, 15 cm; height, 6 cm

385. Length, 6 cm; width, 2 cm   386. Length, 14 cm; width, 7 cm

387. 8 ft   388. 3 m

Objective 2

389. 12 in   390. 13 m   391. 7 m, 24 m, 25 m   392. Train, 80 mi; car, 60 mi

393. 13 ft   394. 75 ft   395. 9 ft   396. 60 m, 80 m, 100 m

397. 26 mi   398. 16 ft

Objective 3

399. 33.4 mpg   400. 36.5 mpg   401. −54.0 mpg

402. No. Negative car mileage means the car can't run.

Mixed Exercises

403. Length, 11 cm; width, 6 cm   404. Length, 8 in; width, 4 in

405. 7 in   406. Length, 6 ft width, 5 ft   407. 8 ft

408. Judy, 9 km; Sheri, 12 km   409. Length, 12 cm; width, 4 cm

410. Length, 8 ft; width, 7 ft   411. 25 mi   412. 4 m and 6 m

413. (a) $2\frac{1}{2}$ $\left(or\ \frac{5}{2}\ or\ 2.5\right)$ sec   (b) 5 sec

414. (a) $1\frac{1}{2}$ $\left(or\ \frac{3}{2}\ or\ 1.5\right)$ sec   (b) 1 sec and 2 sec

415. 2, 3 or −3, −2   416. 5, 6 or −2, −1   417. 3, 5 or −3, −1

418. 4, 6 or −6, −4   419. 8, 10 or −8, −6   420. 11, 13, 15 or −1, 1, 3

Section 4.7

421. $x < -1$ *or* $x > 4$

422. $-1 \leq m \leq 2$

423. $g \leq -5$ *or* $g \geq -4$

424. $-1 < a < 6$

425. $-5 < r < -2$

426. $q \leq 1$ *or* $q \geq 12$

427. $-\dfrac{1}{7} \leq r \leq 1$

428. $x < -6$ *or* $x > 0$

429. $b < \dfrac{1}{2}$ *or* $b > 5$

430. $p < -\dfrac{3}{2}$ *or* $p > \dfrac{2}{3}$

431. $-2 \leq r \leq 4$

432. $-\dfrac{3}{2} < k < \dfrac{3}{2}$

433. $m \leq -4$ *or* $m \geq \dfrac{3}{2}$

434. $t < -9$ *or* $t > 9$

435. $-6 \leq x \leq 6$

436. $3 < k < 4$

437. $g < -\dfrac{5}{2}$ *or* $g > 3$

438. $-2 \leq z \leq -1$

439. $r < -5$ *or* $r > 2$

440. $-3 \leq x \leq 1$

CHAPTER 6   RATIONAL EXPRESSIONS

Section 6.1

Objective 1

1. 0   2. 2   3. –3   4. 0, 7   5. 3, 5

6. Never undefined   7. –6, 6   8. Never undefined

9. 2   10. –3, –2   11. –5, 0, 2   12. –2, 2

Objective 2

13. (a) $\dfrac{14}{3}$ (b) $-\dfrac{11}{3}$ 14. (a) -2 (b) -7 15. (a) $\dfrac{32}{5}$ (b) $-\dfrac{9}{10}$

16. (a) 7 (b) $\dfrac{1}{3}$ 17. (a) $\dfrac{7}{11}$ (b) $-\dfrac{13}{6}$ 18. (a) $\dfrac{6}{5}$ (b) $-\dfrac{8}{15}$

19. (a) $\dfrac{3}{32}$ (b) $-\dfrac{4}{49}$ 20. (a) 3 (b) $-\dfrac{2}{21}$

21. (a) Undefined (b) Undefined   22. (a) $\dfrac{27}{8}$ (b) $-\dfrac{8}{3}$

23. (a) $-\dfrac{9}{4}$ (b) $\dfrac{11}{4}$ 24. (a) -2 (b) $-\dfrac{16}{13}$

Objective 3

25. $\dfrac{1}{3}k$ 26. $-\dfrac{1}{3}xy$ 27. $\dfrac{7b^2}{6c^2}$ 28. $\dfrac{3}{2}$ 29. $z-2$ 30. $\dfrac{b-3}{b-1}$

31. $\dfrac{2a+1}{2a-1}$ 32. $\dfrac{x+2}{x+4}$ 33. $-(y+1)$ *or* $-y-1$ 34. $\dfrac{-1}{p+1}$

35. $4(k-1)$ *or* $4k-4$ 36. $\dfrac{3q-z}{q-z}$ 37. $\dfrac{6r-5s}{r+s}$ 38. $\dfrac{5x-2y}{x+y}$ 39. $\dfrac{v+3}{v-2}$

40. $\dfrac{2r+3s}{4r-s}$

Objective 4

Answers will vary. One possible answer is given

41. $-\dfrac{2x+3}{2-3x}$

42. $-\dfrac{5-4a}{4-5a}$

43. $\dfrac{p-6}{d+9}$

44. $\dfrac{9y+2}{2-4y}$

45. $-\dfrac{11y-6}{6+11z}$

46. $\dfrac{2v-8u}{3-2z}$

47. $\dfrac{3z-5y}{3y-5z}$

48. $\dfrac{7p-y}{4y+8p}$

49. 1

Mixed Exercises

50. -5

51. $-\dfrac{5}{3}, 2$

52. $-\dfrac{1}{3}$

53. $-5, 5$

54. $-6, 0, 6$

55. $-2, 3$

56. $-5, 3$

57. $-\dfrac{5}{2}, \dfrac{5}{2}$

58. (a) $\dfrac{11}{2}$ (b) $-\dfrac{9}{2}$

59. (a) 2   (b) Undefined

60. (a) $\dfrac{11}{20}$ (b) $-\dfrac{9}{5}$

61. (a) $-\dfrac{11}{5}$ (b) $\dfrac{4}{5}$

62. (a) $\dfrac{7}{3}$ (b) $-\dfrac{2}{7}$

63. (a) $-\dfrac{27}{26}$ (b) $\dfrac{4}{3}$

64. (a) 5   (b) 0

65. (a) $-\dfrac{3}{5}$ (b) $\dfrac{4}{5}$

66. $\dfrac{3}{4bc}$

67. $\dfrac{3qt^3}{2r^2}$

68. $\dfrac{2}{r}$

69. $\dfrac{9(x+3)}{2}$

70. $\dfrac{x+y}{x-4y}$

71. -1

72. $\dfrac{2a-b}{a+b}$

73. $\dfrac{z^2+3z+9}{z+3}$

74. $\dfrac{2t+3}{4t+3}$

75. $\dfrac{b^2+1}{d}$

76. $\dfrac{2x-7y}{3x+2y}$

77. $\dfrac{m+1}{m}$

Writing/Conceptual Exercises

78. (c)   79. (b)

80. A ration expression is undefined whenever its denominator is 0, since division by 0 is undefined. In this expression ,when $x = 3$, the value of the denominator is 0, so the expression is undefined.

81. The instructor should explain to the student that she made the very common error of trying to write a rational expression in lowest terms *before factoring*. Notice that in the given expression, the numerator and denominator are not in factored form. The fundamental property is applied only *after* the numerator and denominator are expressed in factored form:

$$\frac{5x+10}{10x+20} = \frac{5(x+2)}{10(x+2)} = \frac{5}{10} = \frac{1}{2}.$$

Section 6.2

Objective 1

82. $\dfrac{56m}{n}$ 83. $\dfrac{4}{3b}$ 84. -2 85. $\dfrac{x+6}{x+3}$ 86. $\dfrac{(a+1)^2}{(a+2)(a-2)}$

87. $\dfrac{2m-3}{2m+3}$ 88. $\dfrac{x-3}{2x-3}$ 89. $\dfrac{y-3}{y+2}$ 90. $\dfrac{5}{7}$ 91. $\dfrac{18z^2}{r}$

Objective 2

92. $\dfrac{3}{32}$ 93. $8z$ 94. $-\dfrac{1}{2}$ 95. $\dfrac{18}{x^2 z(x+y)}$ 96. $\dfrac{2(m+5)}{m-3}$

97. $\dfrac{4(2a-1)}{a}$ 98. $\dfrac{m(m-n)}{(m-1)(m+n)}$ 99. $-\dfrac{a(a+1)}{a-1}$ 100. $\dfrac{4k-1}{3k-2}$ 101. $\dfrac{z-3}{2z-3}$

Mixed Exercises

102. $\dfrac{36q^2}{5}$ 103. $\dfrac{q}{p}$ 104. $\dfrac{x^2}{6}$ 105. $\dfrac{p^3}{q}$ 106. $-\dfrac{5}{7}$ 107. $\dfrac{5}{21}$

108. $\dfrac{5(r+1)}{r-3}$ 109. $\dfrac{x-5}{x+9}$ 110. $\dfrac{2y+3}{2y-3}$ 111. $\dfrac{k+4}{k+2}$ 112. $\dfrac{2(x+4)}{x-3}$

113. $(m+4)(m-3)$ 114. $\dfrac{-3}{10}$ 115. $\dfrac{x-2}{2(x+1)}$ 116. $\dfrac{y+6z}{y+z}$

117. $\dfrac{4(b+2)^2}{(b+1)(b-3)}$ 118. $\dfrac{a+2}{a+3}$ 119. $\dfrac{y+4}{y+1}$ 120. $20rs^5$ 121. $\dfrac{3n}{2m}$

122. $\dfrac{4(y-1)}{3(y-3)}$ 123. $\dfrac{4}{5}$ 124. $\dfrac{7}{p+q}$ 125. $\dfrac{1}{6(n+5)}$ 126. $\dfrac{18}{(r-1)(r+2)}$

127. $\dfrac{x-3}{x-5}$ 128. $\dfrac{8(z+1)}{5z(z+2)}$ 129. $\dfrac{b+5}{b-2}$ 130. $\dfrac{3x+2}{4x-1}$ 131. $\dfrac{1}{b-2}$

Section 6.3

Objective 1

132. 90   133. 24   134. 120   135. 350   136. $80y$ 137. $96b^4$

138. $21r^3(r-5)$ 139. $6(2t-9)$ 140. $x-y$ or $y-x$

141. a^2-b^2 or b^2-a^2 142. $2a(a-5)(a-1)$ 143. $3v^3(v-6)(v+3)$

Objective 2

144. $\dfrac{35}{42}$ 145. $\dfrac{36}{4r}$ 146. $\dfrac{20mn^3}{32n^4}$ 147. $\dfrac{-4k}{18k-36}$

148. $\dfrac{-5y(y+3)}{6(y+3)^2}$ or $\dfrac{-5y^2-15y}{6(y+3)^2}$ 149. $\dfrac{2z(z+2)}{5z^2+5z-10}$ or $\dfrac{2z^2+4z}{5z^2+5z-10}$

150. $\dfrac{35(r+1)}{5r(r+7)(r+1)}$ or $\dfrac{35r+35}{5r(r+7)(r+1)}$ 151. $\dfrac{7(y+3)}{(y+3)^2(y-3)}$ or $\dfrac{7y+21}{(y+3)^2(y-3)}$

152. $\dfrac{27p^2}{21p^3-126p^2}$ 153. $\dfrac{(t+3)(t+1)}{t^3+27}$ or $\dfrac{t^2+4t+3}{t^3+27}$

Mixed Exercises

154. 42   155. 600   156. $15x$ 157. $18a^3$ 158. $w(w+5)(w-5)(w-1)$

159. $(t+1)(t-4)(t+2)$ 160. $(2z-1)(z+4)(z-3)$ 161. $m^2(m-5)(m+3)$

162. $z^2(z-2)(z+3)(z-3)$ 163. $(p+6)(p-3)^2$ 164. $(2q-5)(q+2)(q-4)$

165. $(3r-4)(3r+2)(3r+1)$ 166. $\dfrac{36}{96}$ 167. $\dfrac{135w}{72w^2}$ 168. $\dfrac{42}{18c-90}$

169. $\dfrac{77q}{42q^2-252q}$ 170. $\dfrac{2p(p+4)}{p^2-16}$ or $\dfrac{2p^2+8p}{p^2-16}$

171. $\dfrac{(r-3)(r+1)}{r^3+r^2-12r}$ or $\dfrac{r^2-2r-3}{r^3+r^2-12r}$ 172. $\dfrac{12(k+4)}{k^3+7k^2+12k}$ or $\dfrac{12k+48}{k^3+7k^2+12k}$

173. $\dfrac{2x(3x+2)}{2x^3-8x}$ or $\dfrac{6x^2+4x}{2x^3-8x}$ 174. $\dfrac{-5}{7-r}$

175. $\dfrac{3p(5p-6)}{10p^2+23p-42}$ or $\dfrac{15p^2-18p}{10p^2+23p-42}$ 176. $\dfrac{18(w-4)}{w^3+5w^2-36w}$ or $\dfrac{18w-72}{w^3+5w^2-36w}$

177. $\dfrac{xy(x+y)}{x^3y+2x^2y^2+xy^3}$ or $\dfrac{x^2y+xy^2}{x^3y+2x^2y^2+xy^3}$

178. $\dfrac{2k(k-3)}{(k+7)(k-3)(k-5)}$ or $\dfrac{2k^2-6k}{(k+7)(k-3)(k-5)}$

179. $\dfrac{36z(z+1)}{(z-3)(z+2)(z+1)}$ or $\dfrac{36z^2+36z}{(z-3)(z+2)(z+1)}$

180. $\dfrac{4x(x+2)}{(x-5)(x-3)(x+2)}$ or $\dfrac{4x^2+8x}{(x-5)(x-3)(x+2)}$ 181. $\dfrac{x^2-y^2}{x^3-y^3}$

Writing/Conceptual Exercises

182. For this problem, 108 is a common denominator, although it is not the *least* common denominator. You will get the same answer with either denominator, as long as you make sure to write the answer in lowest terms.

183. Both are correct. Either $r-s$ or it's opposite, $s-r$, may be chosen as the LCD.

184. Yes, $(5-3r)(1-2r)$ is also acceptable because

$$(3r-5)(2r-1)=6r^2-13r+5$$

and $\quad (5-3r)(1-2r)=5-3r+6r^2$
$$=6r^2-13r+5.$$

185. Since $x^2-x-20=(x-5)(x+4)$, the original fraction must be multiplied by $x+4$ in both the numerator and denominator.

Section 6.4

Objective 1

186. $\dfrac{6x}{x-1}$ 187. $\dfrac{3+k}{k-7}$ 188. $\dfrac{5x}{x^2-9}$ 189. $\dfrac{9t+2}{2t+1}$ 190. $\dfrac{1}{b-1}$ 191. $\dfrac{1}{x+1}$

192. $\dfrac{1}{y-2}$ 193. $\dfrac{4}{m+1}$ 194. $\dfrac{6}{x+1}$ 195. $\dfrac{1}{y-3}$ 196. $\dfrac{1}{2x-3y}$ 197. $\dfrac{2}{2v+3w}$

Objective 2

198. $\dfrac{3x+8}{12}$ 199. $\dfrac{7m+6}{21p}$ 200. $\dfrac{2(3a-1)}{a^2-1}$ 201. $\dfrac{10h-9}{1-h^2}$

202. $\dfrac{3y^2+y+2}{(y+3)(y+2)(y-1)}$ 203. $\dfrac{2m^2+5m-12}{(m-4)(m-1)(m+1)}$ 204. $\dfrac{5m+11}{4m}$

205. $\dfrac{8x+20}{(x-5)(x+5)}$ 206. $\dfrac{3n-4}{2(n-2)(n+2)}$ 207. $\dfrac{13s^2+5s-13}{(3x-2)(s+4)(2s-3)}$

208. $\dfrac{13p^2+11p+7}{(2p-1)(p+3)(3p+2)}$ 209. $\dfrac{2x^2-z+1}{(z-1)(z+1)^2}$

210. $\dfrac{9x+3}{(2x+1)(2x-1)(x+2)}$

211. $\dfrac{1}{(x+3)(x+2)}$

Objective 3

212. 6

213. $z+y$

214. $\dfrac{1}{x-5}$

215. $\dfrac{1}{3x+5y}$

216. $\dfrac{28-5x}{20}$

217. $\dfrac{3k+6}{10}$

218. $\dfrac{7}{24}$

219. $\dfrac{4-m}{m^2-9}$

220. $\dfrac{8+y}{x-y}$

221. $\dfrac{-3}{2(x-2)}$

222. $\dfrac{2q-21}{(2q+3)(q+5)(2q-1)}$

223. $\dfrac{2m^2-m+1}{(m-1)(m+1)^2}$

224. $\dfrac{4b^2-10b+5}{(b+4)(b-2)(b-1)}$

225. $\dfrac{3}{(n-1)(n+1)(n+2)}$

Mixed Exercises

226. $\dfrac{1}{x+2}$

227. $\dfrac{-1}{m+5}$

228. $\dfrac{19z}{6(z+1)}$

229. $\dfrac{a^2}{a^2-1}$

230. $\dfrac{5p+8}{6}$

231. $\dfrac{6z^2+23z-2}{(z+2)(z+1)(z+5)}$

232. $\dfrac{-11}{m-2}$ or $\dfrac{11}{2-m}$

233. $\dfrac{-2z+9}{(z-2)(z+2)(z+3)}$

234. $\dfrac{9b+14}{(3b+4)(3b-4)(b+2)}$

235. $\dfrac{6m-4}{(2m-1)(m+4)(m-3)}$

236. $\dfrac{5r+3s}{r+2s}$

237. $\dfrac{2c-35}{(2c+1)(c+5)(c-4)}$

238. $\dfrac{3p^2-7p-4}{(p+3)(p-4)(p-2)}$

239. $\dfrac{4z}{(z-1)(z+1)}$ or $\dfrac{4z}{z^2-1}$

240. $\dfrac{11x^2-x-11}{(2x-1)(x3)(3x+2)}$

241. $\dfrac{6z^2+4z+2}{(z-1)^2(z+1)}$

242. $\dfrac{x}{6}$

243. $\dfrac{9t+13}{24}$

244. $\dfrac{9}{a-2}$

245. $\dfrac{7x+2}{(x-2)(x+2)^2}$

246. $\dfrac{4b+3}{2(b+6)}$

247. $\dfrac{11x-50y}{(x+2y)(x-2y)}$

248. $\dfrac{2(c^2+3cd+4d^2)}{(d+c)^2(3d+c)}$ or $\dfrac{2c^2+6cd+8d^2}{(d+c)^2(3d+c)}$

249. $\dfrac{8z}{(z+2)(z-2)}$ or $\dfrac{8z}{z^2-4}$

Writing/Conceptual Exercises

250. $r-5$ and $5-r$

251. $\dfrac{-5y}{y-2}$ or $-\dfrac{5y}{y-2}$

252. The error occurs in line (2). Because the quantity $2(x-3)$ is being subtracted, we should have $-2x+6$ rather than $-2x-6$. Starting with line (2), the correct solution is given below.

$$\frac{3(x-2)-2(x-3)}{(x-3)(x-2)} = \frac{3x-6-2x+6}{(x-3)(x-2)}$$
$$= \frac{x}{(x-3)(x-2)}$$

253. (a)

Section 6.5

Objective 1

254. $-\dfrac{8}{7}$

255. $\dfrac{5}{9}$

256. $\dfrac{a^2}{3bc}$

257. $\dfrac{2}{r-s}$

258. $\dfrac{3x+1}{3x-2}$

259. $\dfrac{5m-4}{2+m}$

260. $\dfrac{6p-3}{4p+6}$

261. $\dfrac{2}{z}$

262. $\dfrac{6(3s+1)}{s(3s+2)}$

263. $\dfrac{9}{p}$

264. $\dfrac{s}{r}$

265. $\dfrac{7(5k-m)}{4}$ or $\dfrac{35k-7m}{4}$

Objective 2

266. $-\dfrac{21}{22}$

267. $-\dfrac{35}{9}$

268. $\dfrac{1}{a}$

269. $\dfrac{2x}{1-x}$

270. $\dfrac{x}{x+y}$

271. $\dfrac{r^2+1}{5+rt}$

272. $\dfrac{(x-5)^2}{x(x+5)}$ 273. $\dfrac{s^2+1}{1-s^2}$ 274. $\dfrac{9}{10}$ 275. $\dfrac{2(1-3h)}{h(1+2h)}$

276. $\dfrac{2y+3x}{2(4x-y)}$ 277. $\dfrac{a+ab-b^2}{a(a+b)}$

Mixed Exercises

278. xy 279. $\dfrac{24m^2}{1+m}$ 280. 4 281. $\dfrac{5(a+3)}{a+5}$ 282. $\dfrac{z}{2}$ 283. -1

284. $\dfrac{4}{y}$ 285. $\dfrac{5w}{1-w}$ 286. $\dfrac{-5-3v}{7+3v}$ 287. $\dfrac{-n-4}{n+6}$ 288. $\dfrac{9t+6}{2t+7}$ 289. $\dfrac{2s-18}{5}$

290. $(a+1)^2$ 291. $\dfrac{7+2z}{-5-2z}$ 292. $\dfrac{4}{w-1}$ 293. $\dfrac{2s(2r-1)}{r(3-s)}$ 294. $\dfrac{4-s}{4+3s}$

295. $\dfrac{2s-5}{-2s+7}$ 296. $\dfrac{x+4}{3}$ 297. $\dfrac{(k-23)(k+2)}{5k(k+1)}$ 298. $\dfrac{5(x+3)}{(x+4)}$ 299. $\dfrac{7}{10}$

300. $\dfrac{3x^2-3y^2+4x-4y}{6x}$ 301. $8xy(2x-3y)$

Writing/Conceptual Exercises

302 Choice (c) is correct because the numerator is the opposite of the original numerator and the denominator is the opposite of the original denominator. This complex fraction may be obtained by multiplying the given complex fraction by $\dfrac{-1}{-1}=1$.

303. Choice (b) is correct because it is the only negative number among the choices. Because the complex fraction has a negative numerator and a positive denominator, it must be a negative number.

304. For this complex fraction, x^3y^3 is a common denominator, although it is not the *least* common denominator. The LCD of all the denominators is x^2y^2. Although it is less efficient to use x^3y^3 than x^2y^2, the student will obtain the correct answer as long as the final answer is written in lowest terms.

305. The student has not made any errors so far, but the solution is incomplete because the answer is not written in lowest terms. The work should be continued as follows:

$$\frac{100z+40}{5z^2+2z}=\frac{20(5z+2)}{z(5z+2)}=\frac{20}{z}.$$

Section 6.6

Objective 1

306. Expression; $\dfrac{13}{15}x$ 307. Equation; $\dfrac{15}{13}$ 308. Expression; $\dfrac{7}{10}y$

309. Equation; $\dfrac{20}{7}$ 310. Equation; -24 311. Expression; $-\dfrac{1}{6}z$

312. Expression; $\dfrac{31}{24}r$ 313. Equation; $\dfrac{21}{31}$ 314. Equation; -1

315. Expression; $-\dfrac{1}{4}s$

Objective 2

316. 5   317. 4   318. 2   319. 27   320. $-\dfrac{1}{3}$, 4   321. -4, 16

322. 5   323. 4   324. 25   325. $-\dfrac{3}{5}$, 3   326. No solution   327. 2

Objective 3

328. $d=\dfrac{kF}{N}$ 329. $D=\dfrac{Fd-k}{F}$ or $D=d-\dfrac{k}{f}$ 330. $r=\dfrac{S-a}{S}$ or $r=1-\dfrac{a}{S}$

331. $b=\dfrac{2A-hB}{h}$ or $b=\dfrac{2A}{h}-B$ 332. $M=\dfrac{Fd^2}{Gm}$ 333. $R_2=\dfrac{RR_1}{R_1-R}$

334. $T_2 = \dfrac{T_1 V_2 P_2}{V_1 P_1}$ 335. $d_0 = \dfrac{fd_1}{d_1 - f}$ 336. $m_2 = \dfrac{Fd^2}{Gm_1}$

337. $a_n = \dfrac{2S_n - na_1}{n}$ or $a_n = \dfrac{2S_n}{n} - a_1$

Mixed Exercises

338. 2 339. 8 340. $-\dfrac{27}{5}$ 341. No solution 342. No solution

343. $-\dfrac{4}{3}, 1$ 344. 4 345. 1 346. -3 347. -2 348. $-24, 1$

349. $-6, \dfrac{1}{2}$ 350. $-24, 1$ 351. $-6, \dfrac{1}{2}$ 352. $R = \dfrac{nE - Inr}{I}$ or $R = \dfrac{nE}{I} - nr$

353. $r = \dfrac{eR}{E - e}$ 354. $V = \dfrac{Bh}{3}$ 355. $V_1 = \dfrac{T_1 V_2 P_2}{P_1 T_2}$ 356. $v_s = \dfrac{Fv - fv - fv_0}{F}$

357. $q = \dfrac{2pf - Ab}{Ab}$ or $q = \dfrac{2pf}{Ab} - 1$ 358. $b_1 = \dfrac{2A - hB_2}{h}$ or $b_1 = \dfrac{2A}{h} - b_2$

359. $R_1 = \dfrac{AR_r}{R_2 - A}$ 360. $t = \dfrac{s + r}{s - r}$ 361. $F = \dfrac{9}{5}C + 32$

362. $B = \dfrac{2A}{h} - b$ or $\dfrac{2A - bh}{h}$ 363. $b = \dfrac{ac}{a - c}$

364. Expression; $\dfrac{5x + 1}{(x + 1)(x - 1)}$ or $\dfrac{5x + 1}{x^2 - 1}$ 365. Equation; -7 366. Expression; $\dfrac{1}{6}x$

367. Expression; $\dfrac{81}{28x}$ 368. Equation; $\dfrac{7}{11}$ 369. Equation; $-\dfrac{17}{7}$

Writing/Conceptual Exercises

370. (d) 371. -3 and 0

372. This is an expression to be simplified, not an equation to be solved.

373. If $x = 3$, the rational expression $\dfrac{8}{x-3}$ is undefined. The given equation is true for all real numbers except 3.

374. Put all terms with t on one side, and all other terms on the other side.

375. Factor out z on the left side.

Section 6.7

Objective 1

376. 24   377. -2 378. -5 379. 3   380. $\dfrac{7}{10}$ 381. $\dfrac{1}{4}$ *or* 4

382. $-\dfrac{1}{2}$ *or* 2   383. $-\dfrac{2}{3}$ *or* 1   384. Bjorn, \$26,000; Wei-Li, \$32,500

385. 32

Objective 2

386. 3 mph   387. 900 mi   388. 150 mi   389. 240 mph   390. 426 mph

391. 8 mph   392. 380 mph   393. 25 mph   394. 24 mph   395. 50 mph

Objective 3

396. $\dfrac{12}{7}$ *or* $1\dfrac{5}{7}$ hr   397. $\dfrac{6}{5}$ *or* $1\dfrac{1}{5}$ hr   398. $\dfrac{15}{8}$ *or* $1\dfrac{7}{8}$ hr   399. $\dfrac{45}{14}$ *or* $3\dfrac{3}{14}$ hr

400. $\dfrac{3}{10}$ hr   401. $\dfrac{15}{11}$ *or* $1\dfrac{4}{11}$ hr   402. 3 hr   403. $\dfrac{15}{2}$ *or* $7\dfrac{1}{2}$ hr

404. $\dfrac{36}{5}$ *or* $7\dfrac{1}{5}$ hr   405. $\dfrac{11}{4}$ *or* $2\dfrac{3}{4}$ hr

Objective 4

406. 50　　　　　407. $\dfrac{5}{2}$　　　　　408. $162.50　　　　　409. 24 min

410. 10.83 pounds per square inch　　　　411. 7.2 pounds per square inch

412. 69.08 cm　　　413. 100 amps　　　414. 15.6 cm^2　　　415. 20 lb

Mixed Exercises

416. $-\dfrac{2}{3}$　　417. 3 mph　　418. 7　　419. $\dfrac{28}{5}$ or $5\dfrac{3}{5}$ days　　420. 153.86 cm^2

421. 3.5 mph　　　422. 780 mi　　　423. $-\dfrac{1}{4}$ or 2　　　424. $\dfrac{84}{19}$ or $4\dfrac{8}{19}$ hr

425. 40 mph　　　426. $32,000　　　427. $\dfrac{100}{11}$ or $9\dfrac{1}{11}$ min　　428. $\dfrac{1}{2}$ or 4

429. $\dfrac{3}{4}$ or $\dfrac{1}{3}$　　430. 48 mph　　431. $\dfrac{1}{3}$　　　432. 15 foot-candles

433. 24,000 watts　　434. 2.5 ohms　　435. 9 min　　436. 30 mi

437. 240 mph　　438. James, $22,000; Hazel, $38,500　　439. 60 hr

Writing/Conceptual Exercises

440. With the wind, $(525 + y)$ miles per hour; against the wind, $(525 - y)$ miles per hour.

441. $\dfrac{3}{5}$ of the apartment　　442. Inverse　　　443. Direct

CHAPTER 7   EQUATIONS OF LINES

Section 7.1

Objective 1

1. $y = 3x - 2$

2. $y = -x$

3. $y = \dfrac{1}{2}x - \dfrac{1}{2}$

4. $y = \dfrac{2}{3}x + 4$

5. $y = x - 1$

6. $y = 5$

7. $y = 2x + 3$

8. $y = \dfrac{3}{4}x - \dfrac{1}{2}$

9. $y = -5x$

10. $y = -x + 1$

11. $y = -2$

12. $y = x + \dfrac{4}{3}$

Objective 2

13.

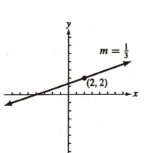

14.

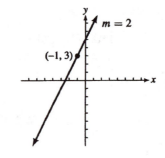

15.

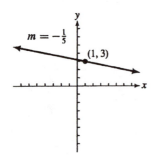

16.

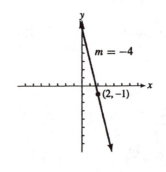

17.

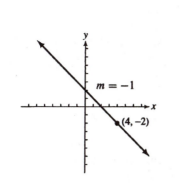

18.

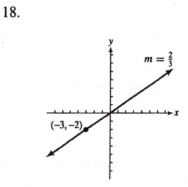

19.

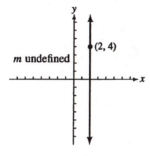

20.

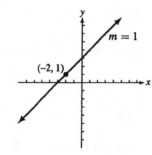

21.

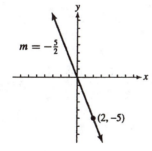

22.

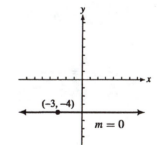

Objective 3

23. $3x - y = 2$

24. $x - y = -3$

25. $x + 3y = -7$

26. $5x + y = 3$

27. $4x + y = -1$

28. $x - 4y = 0$

29. $3x + 4y = -7$

30. $y = 0$

31. $2x + 5y = 4$

32. $x = -6$

33. $5x - 2y = 23$

34. $x + y = -4$

35. $2x + 4y = -17$

36. $5x - 4y = 12$

Objective 4

37. $x-y=-2$ 38. $3x-y=14$ 39. $2x+5y=4$ 40. $9x-y=-22$

41. $5x+y=0$ 42. $x=-3$ 43. $2x-y=-5$ 44. $x-y=-5$

45. $9x-5y=-3$ 46. $11x-10y=-36$ 47. $10x-3y=-11$

48. $11x+6y=-8$ 49. $2x+12y=9$ 50. $5x-8y=-5$

Objective 5

51. $m=19$; $y=19x+82$

52. The approximation of y is 158. The actual value of y is 160.

53. $m=\dfrac{58}{3}$; $y=\dfrac{58}{3}x+\dfrac{253}{3}$

54. The approximation of y is $142.\overline{3}$. The actual value of y is 136.

Mixed Exercises

55. $y=-2x+3$ 56. $y=-4x+5$ 57. $y=\dfrac{1}{2}x-\dfrac{2}{3}$ 58. $y=-3$

59. $y=x$ 60. $y=\dfrac{3}{4}x-\dfrac{4}{3}$ 61. $y=-\dfrac{2}{3}x+5$ 62. $y=-3x+\dfrac{3}{5}$

63.

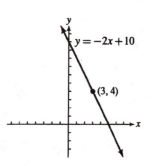

64.

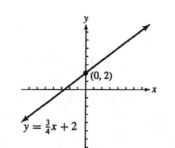

65.

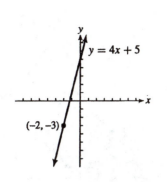

66.

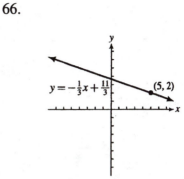

67.

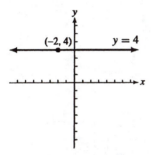

68.

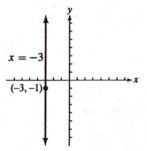

69. $x - 2y = 7$ 70. $2x - y = 9$ 71. $y = -5$ 72. $8x - y = -4$

73. $3x + y = 0$ 74. $x = 5$ 75. $x + 2y = -8$ 76. $2x + 3y = -13$

77. $3x + 4y = 44$ 78. $x - 3y = -18$ 79. $y = \frac{1}{2}x - 3$ 80. $y = 2x - 4$

81. $y = -3x + 1$ 82. $y = -.1x + 1$ 83. $y = 3x - 2$

Writing/Conceptual Exercises

84. H   85. C   86. I   87. B   88. G   89. A   90. F   91. D   92. E

Section 7.2

Objective 1

93.

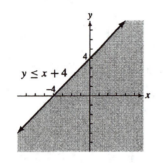

94.

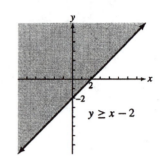

95.

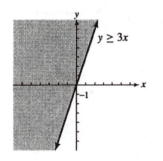

96.

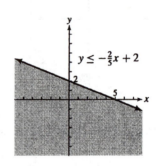

97.

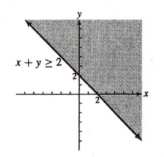

98.

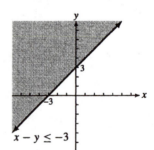

99.

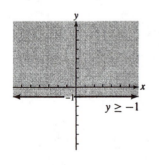

100.

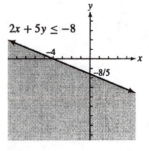

101.

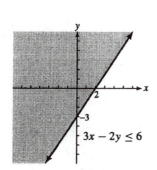

$3x - 2y \le 6$

102.

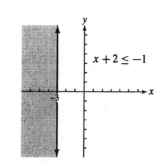

$x + 2 \le -1$

Objective 2

103.

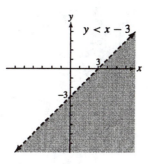

$y < x - 3$

104.

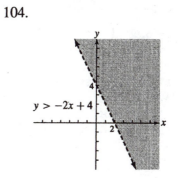

$y > -2x + 4$

105.

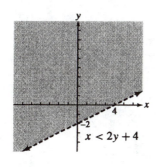

$x < 2y + 4$

106.

$x + 3y < 3$

107.

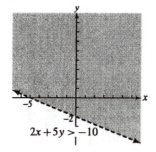

$2x + 5y > -10$

108.

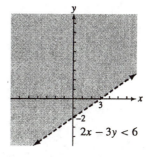

$2x - 3y < 6$

109.

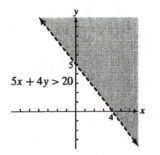

110.

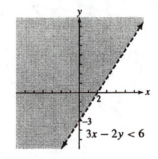

111.

112.

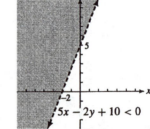

Objective 3

113.

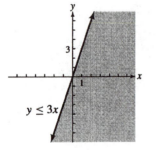

114.

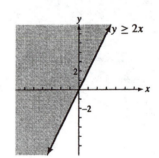

115.

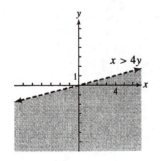

116.

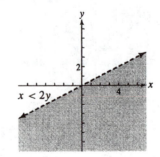

117.

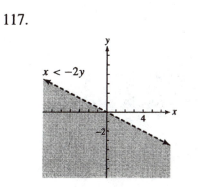

118.

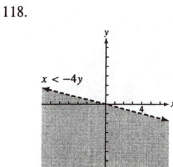

119.

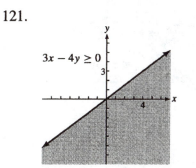

120.

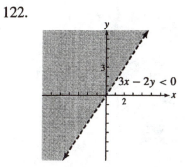

121.

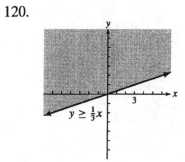

122.

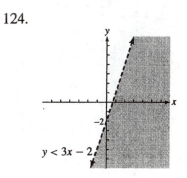

Mixed Exercises

123.

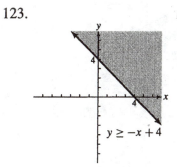

124.

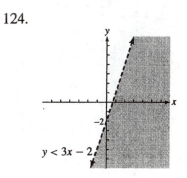

125.

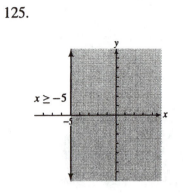

$x \geq -5$

126.

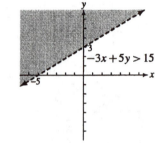

$y < 3$

127.

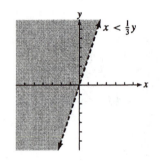

$3x + 2y \leq -6$

128.

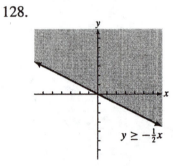

$y \geq -\frac{1}{2}x$

129.

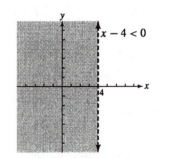

$x < \frac{1}{3}y$

130.

$-3x + 5y > 15$

131.

$x - 4 < 0$

132.

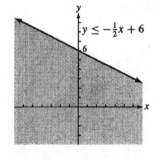

$y \leq -\frac{1}{2}x + 6$

133.

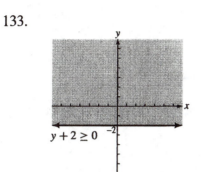

$y + 2 \geq 0$

134.

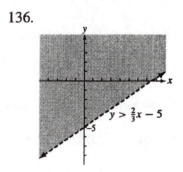

$3x - 4y - 12 > 0$

135.

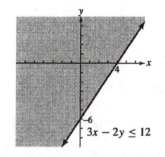

$2x \leq 3y$

136.

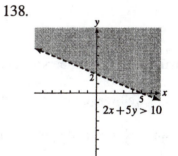

$y > \frac{2}{3}x - 5$

137.

$3x - 2y \leq 12$

138.

$2x + 5y > 10$

139.

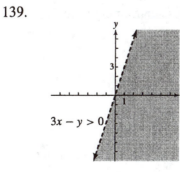

$3x - y > 0$

140.

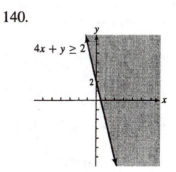

$4x + y \geq 2$

141. B   143. D   143. A   144. C   145. C   146. B   147. D   148. A

Section 7.3

Objectives 1 and 2

149. Function	150. Function	151. Not a function	152. Function
153. Function	154. Function	155. Not a function	156. Not a function
157. Function	158. Not a function		

Objective 3

159. Function	160. Function	161. Function	162. Function
163. Function	164. Function	165. Not a function	166. Function
167. Not a function	168. Not a function	169. Function	170. Function
171. Function	172. Function	173. Function	174. Not a function
175. Not a function	176. Not a function	177. Not a function	178. Function

Objective 4

179. Domain: {all real numbers}; range: {all real numbers}

180. Domain: {all real numbers}; range: {all real numbers}

181. Domain: {all real numbers}; range: {all real numbers greater than or equal to 4}

182. Domain: {all real numbers}; range: {all real numbers greater than or equal to –7}

183. Domain: {all real numbers}; range: {all real numbers less than or equal to 5}

184. Domain: {all real numbers}; range: {2}

185. Domain: {all real numbers}; range: {all nonnegative real numbers}

186. Domain: {all real numbers}; range: {all real numbers}

187. Domain: {all real numbers}; range: {all real numbers greater than or equal to 6}

188. Domain: {all real numbers}; range: {all nonnegative real numbers}

Objective 5

189. (a) 10   (b) −11   190. (a) 29   (b) 2

191. (a) 0   (b) 9   192. (a) −2   (b) 13

193. (a) 8   (b) 5   194. (a) 5   (b) −4

195. (a) 10   (b) −11   196. (a) −6   (b) 0

197. (a) 36   (b) 9   198. (a) 25   (b) 16

199. (a) −9   (b) −9   200. (a) 0   (b) −9

201. (a) 3   (b) 6   202. (a) −1   (b) −4

203. (a) 3   (b) 2   (c) 27   204. (a) −6   (b) 0   (c) 0

205. (a) 0   (b) −8   (c) 27   206. (a) −1   (b) −13   (c) 32

207. (a) −16   (b) 0   (c) 65 208. (a) 1   (b) −7   (c) 58

Objective 6

209. $\{(22.0,9.9),(21.0,11.4),(18.0,8.8),(17.0,9.2),(14.0,8.6)\}$

210. Domain: $\{22.0,21.0,18.0,17.0,14.0\}$

211. Range: $\{9.9,11.4,8.8,9.2,8.6\}$

212. $f(21.0)=11.4$, $f(14.0)=8.6$

213. $x=17.0$

214. $x=18.0$

215. 11,400,000 viewers

Mixed Exercises

216. Not a function; domain: {1, 2, 3}; range: {−1, 3, 4, 7}

217. Function; domain: {−1, 0, 1}; range: {2, 5, 8}

218. Function; domain: $\{2, 3, 4\}$; range: $\{-4, -3, -2\}$

219. Function; domain: $\{0, 2, 4, 6\}$; range: $\{-3, -2, -1, 0\}$

220. Not a function; domain: $\{0, 3\}$; range: $\{0, 2, 4, 5\}$

221. Not a function; domain: $\{1\}$; range: $\{0, 1, 2, 7\}$

222. Function; domain: [all real numbers}; range: {all real numbers}

223. Not a function   224. Not a function

225. Function; domain: {all real numbers except 0}; range: {all real numbers except 0}

226. Function; domain: {all real numbers}; range: {all real numbers}

227. Function; domain: {all real numbers}; range: {all real numbers greater than or equal to 1}

228. Function; domain: {all real numbers}; range: {all nonnegative numbers}

229. Function; domain: {all real numbers}; range: $\{3\}$

230. (a) 1   (b) 13   231. (a) 10   (b) -2 232. (a) 12   (b) 48

233. (a) 8   (b) 8   234. (a) $\dfrac{4}{5}$ (b) $\dfrac{4}{17}$ 235. (a) $-\dfrac{3}{5}$ (b) $\dfrac{9}{5}$

236. Function   237. Not a function 238. Not a function   239. Function

240. Function   241. Function   242. Not a function   243. Function

244. (a) 7   (b) 1   (c) $y = 3x + 4$ 245. (a) −3   (b) 2   (c) $y = -5x + 7$

246. (a) 3   (b) 0   (c) $y = -.5x + 6$ 247. (a) −7   (b) 0   (c) $y = \dfrac{1}{7}x - 8$

248. (a) 33   (b) −3   (c) $y = \dfrac{7}{3}x + 19$

Writing/Conceptual Exercises

249. A relation is *any* set of ordered pairs. A function is a special kind of relation. In a function, each first component must go with exactly one second component.

250. A function is a relationship between two sets, called the domain and the range. In a function, each element in the domain must correspond to exactly one element in the range.

251. A relation is a function if every vertical line cuts the graph in no more than one point.

CHAPTER 8   LINEAR SYSTEMS

Section 8.1

Objective 1

1. No   2. No   3. Yes   4. Yes   5. No   6. Yes

7. No   8. Yes   9. No   10. Yes   11. Yes   12. Yes

13. No   14. Yes   15. Yes

Objective 2

16. $(2, 3)$ 17. $(-2, 2)$ 18. $(-3, -6)$ 19. $(5, -3)$ 20. $(2, -1)$

21. $(4, 4)$ 22. $(0, 8)$ 23. $(-5, -5)$ 24. $(-4, 2)$ 25. $(0, 0)$

26. $(-1, 1)$ 27. $(-6, 4)$ 28. $(2, 4)$ 29. $(0, 3)$ 30. $(-5, 4)$

Objective 3

31. Infinite number of solutions   32. Infinite number of solutions

33. No solution   34. No solution   35. No solution

36. Infinite numbers of solution   37. Infinite number of solutions

38. No solution   39. No solution   40. No solution

41. Infinite number of solutions   42. No solution

Objective 4

43. (a) Inconsistent   (b) Parallel lines   (c) No solution

44. (a) Dependent   (b) One line   (c) Infinite number of solutions

45. (a) Inconsistent   (b) Parallel lines   (c) No solution

46. (a) Neither   (b) Intersecting lines   (c) One solution

47. (a) Dependent   (b) One line   (c) Infinite number of solutions

48. (a) Inconsistent (b) Parallel lines (c) No solution

49. (a) Neither (b) Intersecting lines (c) One solution

50. (a) Inconsistent (b) Parallel lines (c) No solution

51. (a) Neither (b) Intersecting lines (c) One solution

52. (a) Dependent (b) One line (c) Infinite number of solutions

53. (a) Neither (b) Intersecting lines (c) One solution

54. (a) Inconsistent (b) Parallel lines (c) No solution

Objective 5

55. $(1, 2)$ 56. $(4, -8)$ 57. $(3, -5)$ 58. $(7, 4)$ 59. $(6, 7)$ 60. $(-12, 4)$

61. $(15, 6)$ 62. $(4, 3)$ 63. Infinite solutions   64. No solution

65. $(2, 3)$ 66. $\left(\dfrac{1}{5}, 4\right)$

Mixed Exercises

67. Yes   68. No   69. Yes   70. No   71. Yes   72. Yes

73. $(2, 0)$ 74. Infinite number of solutions (dependent equations)   75. $(6, -2)$

76. No solution (inconsistent system)   77. $(-6, 4)$ 78. $(-5, -6)$

79. No solution   80. One solution   81. One solution   82. One solution

83. Infinite number of solutions   84. One solution   85. One solution

86. Infinite number of solutions   87. No solution   88. One solution

89. No solution   90. One solution   91. One solution

92. Infinite number of solutions   93. Infinite number of solutions

94. No solution   95. One solution   96. No solution

Writing/Conceptual Exercises

97. There cannot be exactly two solutions. There will be one solution, no solution of an infinite number of solutions.

98. We cannot always read the exact coordinates of a point, especially if the coordinates are fractions.

99. There are an infinite number of solutions.

100. Answers will vary, but the lines must intersect at $(4,-2)$. One example is

$$x+2y=0$$
$$x-\ y=6.$$

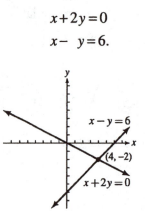

Section 8.2

Objective 1

101. $(2, 6)$ 102. $(1, 3)$ 103. $(0, 0)$ 104. $(8,-1)$

105. $(1,-4)$ 106. $(4,-3)$ 107. $(2, 3)$ 108. $(-8,-5)$

109. $(-4,-9)$ 110. $(4,-10)$ 111. $(3, 1)$ 112. $(5, 1)$

Objective 2

113. No solution   114. Infinite number of solutions

115. Infinite number of solutions   116. No solution   117. No solution

118. Infinite number of solutions   119. No solution

120. Infinite number of solutions   121. No solution

122. Infinite number of solutions   123. Infinite number of solutions

124. No solution

Objective 3

125. No solution   126. $(3, -2)$ 127. $(-9, -11)$ 128. $(1, 6)$

129. $\left(\dfrac{22}{9}, \dfrac{8}{9}\right)$ 130. $(2, -4)$ 131. Infinite number of solutions

132. $(-6, 2)$ 133. $(-12, 0)$ 134. $(0, -6)$

Mixed Exercises

135. $(-6, 2)$ 136. $(5, 5)$ 137. $(4, -2)$ 138. $(-7, 10)$

139. No solution   140. Infinite number of solutions   141. $(4, -2)$

142. $(-3, -2)$ 143. $(1, 4)$ 144. Infinite number of solutions

145. Infinite number of solutions   146. No solution

147. Infinite number of solutions   148. No solution   149. No solution

150. Infinite number of solutions   151. Infinite number of solutions

152. Infinite number of solutions   153. $(5, 6)$ 154. $(8, -9)$

155. Infinite number of solutions   156. $\left(\dfrac{3}{5}, \dfrac{1}{5}\right)$ 157. $(6, -12)$

158. $(-8, 4)$ 159. $(1, 5)$ 160. $(-2, 4)$ 161. $\left(7, \dfrac{1}{3}\right)$

162. $(-1, 6)$

Writing/Conceptual Exercises

163. To avoid fractions, solve the first equation for y.

164. To avoid fractions, solve the first equation for y.

165. To avoid fractions, solve the second equation for x.

166. To avoid fractions, solve the second equation for x.

Section 8.3

Objective 1

| 167. $(3, 1)$ | 168. $(2, -1)$ | 169. $(8, 2)$ | 170. $(3, 2)$ |

171. $\left(-\dfrac{1}{2}, 5\right)$ 172. $\left(\dfrac{3}{2}, -\dfrac{7}{2}\right)$ 173. $(0, 3)$ 174. $(-4, -1)$

| 175. $(5, 2)$ | 176. $(-2, 2)$ | 177. $(2, -1)$ | 178. $(-7, 10)$ |

Objective 2

179. $(3, -2)$ 180. $\left(\dfrac{1}{2}, 3\right)$ 181. $(2, 0)$ 182. $(-3, -2)$

| 183. $(4, -2)$ | 184. $(1, -8)$ | 185. $(5, -7)$ | 186. $(-3, 1)$ |

| 187. $(1, 1)$ | 188. $(2, -5)$ | 189. $(1, 1)$ | 190. $(5, 3)$ |

Objective 3

191. $(4, 2)$ 192. $4, -3$ 193. $(-8, -2)$ 194. $\left(-\dfrac{1}{3}, 4\right)$

195. $\left(\dfrac{19}{8}, \dfrac{9}{8}\right)$ 196. $\left(\dfrac{9}{2}, -2\right)$ 197. $\left(-\dfrac{2}{13}, \dfrac{23}{13}\right)$ 198. $\left(\dfrac{5}{37}, -\dfrac{3}{37}\right)$

199. $\left(-\dfrac{3}{29}, -\dfrac{7}{29}\right)$ 200. $\left(-\dfrac{3}{13}, -\dfrac{15}{26}\right)$ 201. $(5, 4)$ 202. $(-1, 1)$

Objective 4

203. No solution   204. Infinite number of solutions

205. Infinite number of solutions   206. No solution   207. No solution

208. Infinite number of solutions   209. No solution

210. Infinite number of solutions   211. No solution

212. Infinite number of solutions   213. Infinite number of solutions

214. No solution

Mixed Exercises

215. $(6, 3)$ 216. $(-1, 11)$ 217. $(2, 5)$ 218. $(-3, 0)$

219. Infinite number of solutions   220. No solution   221. $(-2, -6)$

222. $(-6, 3)$ 223. $\left(-\dfrac{1}{3}, 4\right)$ 224. $(-3, 5)$

225. Infinite number of solutions   226. $\left(-\dfrac{5}{4}, \dfrac{3}{4}\right)$ 227. $(4, 6)$

228. $(6, 1)$ 229. $(4, 2)$ 230. $\left(2, \dfrac{1}{6}\right)$ 231. $\left(-\dfrac{1}{3}, 3\right)$

232. $\left(\dfrac{1}{2}, -\dfrac{1}{6}\right)$ 233. $\left(-\dfrac{1}{4}, \dfrac{1}{2}\right)$ 234. $\left(\dfrac{5}{6}, \dfrac{1}{4}\right)$ 235. No solution

236. $(-3, 2)$ 237. $(0, -2)$ 238. $(0, 5)$ 239. $(2, 2)$

240. $(3, 0)$ 241. $(2, -1)$ 242. $(-6, 0)$ 243. $(-2, 3)$

244. Infinite number of solutions

Writing/Conceptual Exercises

245. (b)   246. (a) 4   (b) -2

247. It is impossible to have two numbers such that the difference between the larger and the smaller is both 12 and 10.

248. Neither is correct. Dividing both sides of the equation $5x = 0$ by 5 gives us $x = 0$. Then substitute 0 for x in either of the original equations to get $y = 5$. The system has one solution. The ordered pair $(0, 5)$.

Section 8.4

Objective 1

249. 45 and 29   250. −25 and −78   251. 13 and 4   252. 9 and 3

253. 22 and 11   254. Rockland: 4664; Brookdale: 3421

255. Ms. Ferrari: 28; Mrs. Bayliess: 25   256. 43.5 cm; 14.5 cm

257. Length: 14 ft; width: 6 ft   258. 32 cm, 32 cm, 52 cm

Objective 2

259. 5000 adults; 3000 children   260. 300 regular; 100 student

261. 25 tens; 52 twenties   262. 1000 25¢ stamps; 250 15¢ stamps

263. 300 student; 111 non-student   264. 30 five; 60 twenties;

265. 296 children's tickets; 127 adult tickets

266. 1500 general admission; 750 reserved   267. $3000 at 5%; $7000 at 8%

268. $4000 at 7%; $8000 at 9%

Objective 3

269. 15 pounds at $6 per pound; 30 pounds at $3 per pound

270. 20 pounds at $1.60 per pound; 10 pounds at $2.50 per pound

271. 24 barrels at $40 a barrel; 26 barrels at $65 a barrel

272. 8 pounds at $4.20 per pound; 6 pounds at $3.50 per pound

273. .6 pound of caramels; .4 pound of creams

274. 24 liters of 35% solution; 16 liters of 55% solution

275. 10 liters of 90% solution; 20 liters of 75% solution

276. 10 liters of 15% solution; 5 liters of 30% solution

277. 70 liters of 10% solution; 30 liters of 50% solution

278. 40 liters of 25% solution; 20 liters of 40% solution

Objective 4

279. 35 mph; 65 mph

280. 90 mph; 110 mph

281. Barry: 6 mph; Dave: 4 mph

282. 55 mph; 40 mph

283. Gila: 55 mph; Gareth: 50 mph

284. Mr. Burton: 54 mph; Mr. Quigley: 52 mph

285. Boat, 9 mph; current, 7 mph

286. Wind, 35 mph; plane, 265 mph

287. Current, 5mph; kayak, 11 mph

288. Plane A: 400 mph; plane B: 360 mph

Mixed Exercises

289. Length: 25 cm; width: 9 cm

290. 6 at $12 each; 4 at $15 each

291. 9 fives; 5 tens

292. 560 km/hr; 500 km/hr

293. 75 milliliters

294. $15,000 at 6%, $35,0000 at 7%

295. Current: 4 mph; boat: 12 mph

296. 50 pounds at $6 per pound; 100 pounds at $3 per pound

297. Andy: 76 km/hr; Ralph: 36 km/hr

298. 6 liters

299. −13, −4

300. 5, −8

301. 65°, 25°

302. 118°, 62°

303. 50 ones; 33 fives

304. 23 reserved; 80 general admission

305. $4000 at 14%; $2000 at 11%

306. $5000 at 9% ; $10,000 at 12%

307. 45 mph; 50 mph

308. 6 mi   309. 60 mph   310. 78 pennies; 78 nickels; 276 dimes

311. 7 mph   312. 55 mph

Section 8.5

Objective 1

313.

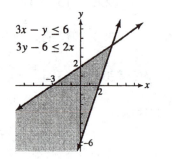

$4x + 5y \leq 20$

$y \leq x + 3$

314.

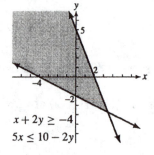

$3x + 5y \geq 15$

$y \geq x - 2$

315.

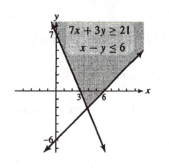

$x + y \leq 3$

$5x - y \geq 5$

316.

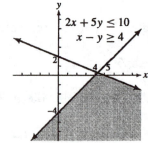

$x + y \geq 3$

$x - 2y \leq 4$

317.

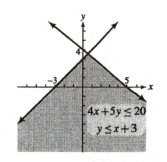

$7x + 3y \geq 21$

$x - y \leq 6$

318.

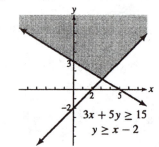

$2x + 5y \leq 10$

$x - y \geq 4$

319.

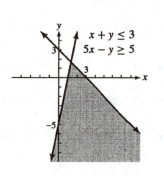

$3x - y \leq 6$

$3y - 6 \leq 2x$

320.

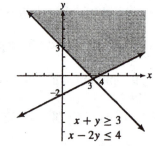

$x + 2y \geq -4$

$5x \leq 10 - 2y$

321.

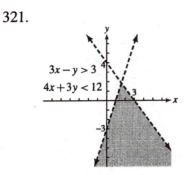

$3x - y > 3$
$4x + 3y < 12$

322.

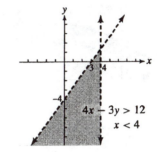

$6x - y > 6$
$2x + 5y < 10$

323.

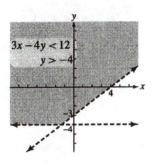

$3x - 4y < 12$
$y > -4$

324.

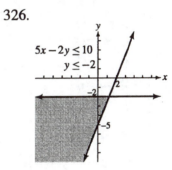

$4x - 3y > 12$
$x < 4$

325.

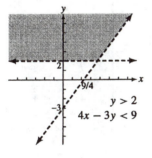

$y > 2$
$4x - 3y < 9$

326.

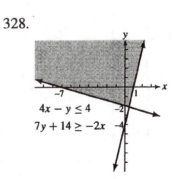

$5x - 2y \le 10$
$y \le -2$

327.

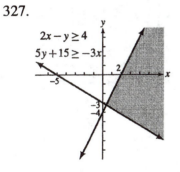

$2x - y \ge 4$
$5y + 15 \ge -3x$

328.

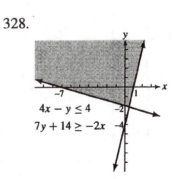

$4x - y \le 4$
$7y + 14 \ge -2x$

329.

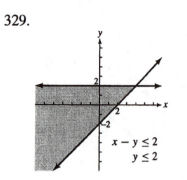

$$x - y \leq 2$$
$$y \leq 2$$

330.

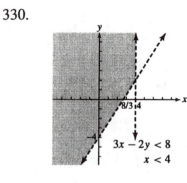

$$3x - 2y < 8$$
$$x < 4$$

331.

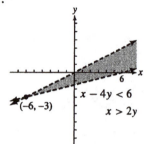

$$x - 4y < 6$$
$$x > 2y$$

332.

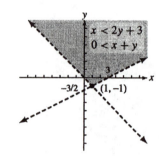

$$x < 2y + 3$$
$$0 < x + y$$

333.

$$2x + 3y \leq 6$$
$$y \leq x$$
$$x \leq 2$$
$$y \geq 0$$

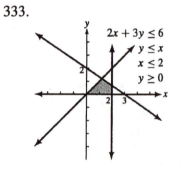

334.

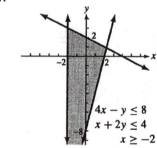

$$4x - y \leq 8$$
$$x + 2y \leq 4$$
$$x \geq -2$$

335.

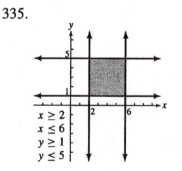

$$x \geq 2$$
$$x \leq 6$$
$$y \geq 1$$
$$y \leq 5$$

336.

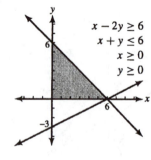

$$x - 2y \geq 6$$
$$x + y \leq 6$$
$$x \geq 0$$
$$y \geq 0$$

Writing/ Conceptual Exercises

337. There are an infinite number of solutions in a region which is a solution of a system of linear inequalities which is not necessary the entire *xy*-plane.

338. If an inequality is a "≤" or a "≥" inequality, we draw a solid line for the boundary. If an equality is a "<" or a ">" inequality, the boundary is a dashed line.

339. The graph below represents solutions in all quadrants.

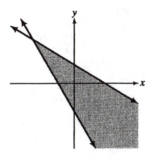

340. The graph below represents solutions only in one quadrant.

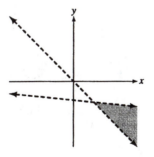

CHAPTER 9   ROOTS AND RADICALS

Section 9.1

Objective 1

1. $9, -9$ 2. $13, -13$ 3. $16, -16$ 4. $20, -20$ 5. $18, -18$

6. $24, -24$ 7. $19, -19$ 8. $35, -35$ 9. $\dfrac{7}{8}, -\dfrac{7}{8}$ 10. $\dfrac{11}{12}, -\dfrac{11}{12}$

11. $\dfrac{13}{25}, -\dfrac{13}{25}$ 12. $\dfrac{14}{17}, -\dfrac{14}{17}$ 13. 10 14. -12 15. -1

16. 6 17. 15 18. -30 19. 100 20. -23

21. $\dfrac{8}{15}$ 22. $-\dfrac{13}{21}$ 23. $\dfrac{16}{19}$ 24. $-\dfrac{5}{8}$

Objective 2

25. Rational   26. Irrational   27. Rational   28. Not a real number

29. Not a real number   30. Irrational   31. Rational   32. Irrational

33. Rational   34. Rational   35. Irrational   36. Not a real number

Objective 3

37. 6.708 38. 5.568 39. -3.873 40. -9.487 41. -12.247

42. 17.321 43. -11.446 44. 15.652 45. 28.844 46. -33.166

47. 111.803 48. -162.351

Objective 4

49. $c = 17$ 50. $c = 35$ 51. $b = 24.495$ 52. $a = 7$ 53. $b = 5.745$

54. $c = 8.544$ 55. 9.165 cm   56. 13.038 cm   57. 20 inches   58. $2\,$m

59. 24 ft   60. 48 ft   61. 35 mi   62. 17 ft   63. 5.657 ft

64. 12 mi

Objective 5

65. $\sqrt{13}$ 66. $\sqrt{10}$ 67. $\sqrt{45}$ or $3\sqrt{5}$ 68. $\sqrt{244}$ or $2\sqrt{61}$

69. $\sqrt{117}$ or $3\sqrt{13}$ 70. 5   71. 12   72. $\sqrt{32}$ or $4\sqrt{2}$

73. $\sqrt{73}$ 74. $\sqrt{130}$ 75. $\sqrt{26}$ 76. $\sqrt{173}$

Objective 6

77. 3   78. −5   79. −1   80. −10   81. −6

82. 2   83. −3   84. 4   85. Not a real number

86. −7   87. −5   88. 2   89. Not a real number

90. −10   91. Not a real number   92. 8

Mixed Exercises

93. 14   94. −1   95. Not a real number   96. 4   97. −2

98. Not a real number   99. $-\dfrac{15}{17}$ 100. $\dfrac{3}{5}$ 101. Rational; 15

102. Irrational; 2.646   103. Rational; −7   104. Irrational; 8.660

105. Rational; $-\dfrac{9}{13}$ 106. Rational; $\dfrac{19}{21}$ 107. Irrational; 2.915

108. Rational; −.3   109. $c = 26$ 110. $a = 7.483$ 111. $b = 11.958$

112. $b = 9$ 113. $c = 9.434$ 114. $a = 7.550$ 115. $b = 14.422$ 116. $a = 9.220$

Writing/Conceptual Exercises

117. 1   118. 2   119. 0   120. (b) and (d) 121. (b) and (c)

122. b must be negative   123. b must be positive   124. b must be negative

125. b must be negative   126. b must be 0

Section 9.2

Objective 1

127. $\sqrt{15}$ 128. 6   129. 7   130. 8   131. $\sqrt{42}$

132. 12   133. 23   134. 10   135. $\sqrt{143}$ 136. $\sqrt{6r}$

137. $\sqrt{14x}$ 138. $6\sqrt{ab}$

Objective 2

139. $2\sqrt{6}$ 140. $6\sqrt{2}$ 141. $10\sqrt{5}$ 142. $-5\sqrt{6}$ 143. $-2\sqrt{14}$

144. $12\sqrt{2}$ 145. $-6\sqrt{6}$ 146. $30\sqrt{10}$ 147. $y\sqrt{y}$ 148. pq^2

149. $4a^2\sqrt{5a}$ 150. $5r^3s^3\sqrt{5r}$ 151. $5\sqrt{3}$ 152. $2\sqrt{15}$ 153. $6\sqrt{2}$

154. $10\sqrt{3}$ 155. $11\sqrt{2}$ 156. $12\sqrt{3}$

Objective 3

157. $\dfrac{4}{5}$ 158. $\dfrac{13}{11}$ 159. $\dfrac{15}{2}$ 160. $\dfrac{\sqrt{13}}{7}$ 161. 3

162. $\dfrac{\sqrt{3}}{9}$ 163. $\dfrac{3}{2}$ 164. 30   165. 1   166. 6

167. $\dfrac{\sqrt{11}}{6}$ 168. $\dfrac{1}{25}$ 169. $\dfrac{\sqrt{15}}{4}$ 170. $\dfrac{4\sqrt{5}}{9}$ 171. $\dfrac{\sqrt{65}}{6}$

Objective 4

172. $2\sqrt[3]{3}$ 173. $3\sqrt[3]{3}$ 174. $4\sqrt[3]{2}$ 175. $3\sqrt[4]{3}$ 176. $2\sqrt[5]{2}$

177. $2\sqrt[4]{20}$ 178. $-\dfrac{4}{3}$ 179. $\dfrac{1}{6}$ 180. $\dfrac{2}{5}$ 181. $\dfrac{6}{5}$

182. $-\dfrac{2}{3}$ 183. $\dfrac{5}{4}$ 184. 3   185. 5   186. 3

Mixed Exercises

187. 11   188. $\sqrt{195}$ 189. 15   190. $10\sqrt{5}$ 191. $13\sqrt{2}$

192. 79   193. $3\sqrt{11}$ 194. $7\sqrt{5}$ 195. $-5\sqrt{7}$ 196. $15\sqrt{3}$

197. 42   198. $3x\sqrt{3}$ 199. $-\dfrac{12}{15}$ 200. $4\sqrt{3}$ 201. $-\dfrac{3}{2}$

202. $\dfrac{5\sqrt{2}}{a}$ 203. $\dfrac{3}{5}$ 204. $\dfrac{2}{3}$ 205. $5a\sqrt{3ab}$ 206. $2\sqrt[3]{5}$

207. $-2\sqrt[5]{2}$ 208. 2   209. $7x^3 y^3 \sqrt{2y}$ 210. $3\sqrt[4]{3}$

Writing/Conceptual Exercises

211. True   212. False   213. False   214. True   215. (b) and (c)

216. (a) and (d)

Section 9.3

Objective 1

217. $5\sqrt{5}$ 218. $2\sqrt{11}$ 219. $2\sqrt[3]{10}$ 220. $\sqrt{7}$

221. Cannot be simplified further   222. $10\sqrt{3}$ 223. $-\sqrt{6}$

224. $4\sqrt{5}$ 225. $13\sqrt[4]{12}$ 226. $5\sqrt[5]{4}$

Objective 2

227. $5\sqrt{2}$ 228. $9\sqrt{2}$ 229. $12\sqrt{2}$ 230. $-7\sqrt{11}$ 231. $4\sqrt{6}+6\sqrt{5}$

232. $-\sqrt{7}$ 233. $\sqrt{2}-3\sqrt{3}$ 234. $-4\sqrt{2}$ 235. $5\sqrt{2}-3\sqrt{3}$ 236. 0

237. $3\sqrt{2z}$ 238. $10\sqrt{6w}$ 239. $-4\sqrt{5x}$ 240. $\sqrt[4]{2}$ 241. 0

242. $13r\sqrt[3]{r^2}$

Objective 3

243. $4\sqrt{15}$ 244. $2\sqrt{42}$ 245. $5\sqrt{14}$ 246. $12\sqrt{35}$ 247. $8\sqrt{x}$

248. $18\sqrt{y}-5\sqrt{6y}$ 249. $x\sqrt{5}$ 250. $39w\sqrt{6}$ 251. $4k\sqrt{2k}$ 252. $13\sqrt{5y}$

Writing/Conceptual Exercises

253. $\sqrt{19}$ 254. $\sqrt[3]{11}$ 255. $\sqrt{3}$ 256. $10\sqrt{5}$ 257. $9\sqrt{7m}$

258. $-6\sqrt[3]{2}$ 259. $\sqrt{77}-3\sqrt{7m}$ 260. $8r\sqrt{2}$ 261. $2\sqrt{3}$

262. $4\sqrt{2}$ 263. $18\sqrt{2}+8\sqrt{3}-12\sqrt{7}$ 264. $22\sqrt[3]{2}$ 265. $12\sqrt{13}$

266. $49\sqrt{2}$ 267. $24\sqrt[3]{5}$ 268. $-13\sqrt[4]{3}$ 269. $-15\sqrt{6}$ 270. $7\sqrt{2}$

271. $6\sqrt[3]{5}$ 272. $-4\sqrt[3]{3}$ 273. $13\sqrt{2z}$ 274. $9\sqrt[3]{3}$ 275. $9\sqrt{3y}$

276. $5\sqrt[3]{2r}$ 277. $27\sqrt{3}$ 278. $12\sqrt{x}$

Writing/Conceptual Exercises

279. $4\sqrt{7}+8\sqrt{7}=(4+8)\sqrt{7}=12\sqrt{7}$ 280. $11\sqrt[3]{3}-4\sqrt[3]{3}=(11-4)\sqrt[3]{3}=7\sqrt[3]{3}$

281. Answers will vary. One example is $\sqrt{81}+\sqrt[3]{27}=9+3=12$.

282. Answers will vary. One example is $\sqrt{100}-\sqrt[3]{8}=10-2=8$.

Section 9.4

Objective 1

283. $\dfrac{2\sqrt{5}}{5}$ 284. $-\dfrac{4\sqrt{13}}{13}$ 285. $-\dfrac{\sqrt{6}}{3}$ 286. $\dfrac{\sqrt{15}}{3}$ 287. $\sqrt{2}$

288. $\dfrac{5\sqrt{10}}{2}$ 289. $-2\sqrt{11}$ 290. $-\dfrac{4\sqrt{3}}{3}$ 291. $\dfrac{\sqrt{21}}{7}$ 292. $\dfrac{\sqrt{10}}{4}$

293. $\dfrac{\sqrt{2}}{2}$ 294. $\dfrac{2\sqrt{7}}{7}$ 295. $\dfrac{1}{5}$ 296. $\dfrac{1}{5}$ 297. $\dfrac{5\sqrt{2}}{2}$

298. $\dfrac{3\sqrt{2}}{4}$

Objective 2

299. $\dfrac{\sqrt{5}}{5}$ 300. $\dfrac{\sqrt{42}}{7}$ 301. $\dfrac{\sqrt{6}}{4}$ 302. $\dfrac{4\sqrt{6}}{9}$ 303. $\dfrac{3\sqrt{14}}{10}$

304. $\dfrac{\sqrt{5}}{2}$ 305. $\dfrac{2\sqrt{6}}{3}$ 306. $\dfrac{\sqrt{55}}{5}$ 307. $\dfrac{\sqrt{42}}{35}$ 308. $\dfrac{\sqrt{6xy}}{x}$

309. $\dfrac{\sqrt{7rs}}{s}$ 310. $\dfrac{a\sqrt{15b}}{5}$ 311. $\dfrac{\sqrt{mk}}{k^2}$ 312. $\dfrac{f\sqrt{fd}}{d^2}$ 313. $\dfrac{a\sqrt{30b}}{3}$

Objective 3

314. $\dfrac{\sqrt[3]{18}}{3}$ 315. $\dfrac{\sqrt[3]{18}}{2}$ 316. $\dfrac{\sqrt[3]{4}}{4}$ 317. $\dfrac{\sqrt[3]{20}}{2}$ 318. $\dfrac{\sqrt[3]{175}}{5}$

319. $\dfrac{\sqrt[3]{45}}{3}$ 320. $\dfrac{2\sqrt[3]{5}}{5}$ 321. $\sqrt[3]{6}$ 322. $\dfrac{\sqrt[3]{9}}{6}$ 323. $\dfrac{\sqrt[3]{300}}{20}$

324. $\dfrac{\sqrt[3]{36}}{18}$ 325. $\dfrac{\sqrt[3]{100r}}{5r}$ 326. $\dfrac{\sqrt[3]{6r^2s}}{3r}$ 327. $\dfrac{\sqrt[3]{6vw}}{6w}$ 328. $\dfrac{2\sqrt[3]{5u^2t}}{5u}$

Mixed Exercises

329. $9\sqrt{7}$

330. $-4\sqrt{6}$

331. $\sqrt{6}$

332. $\dfrac{\sqrt{15}}{3}$

333. $\dfrac{\sqrt{55}}{10}$

334. $3\sqrt{2}$

335. $\dfrac{\sqrt[3]{98}}{7}$

336. $\dfrac{\sqrt{33}}{11}$

337. $\dfrac{\sqrt[3]{180x^2}}{6x}$

338. $\dfrac{3\sqrt{2}}{14}$

339. $\dfrac{2\sqrt{6qt}}{t}$

340. $\dfrac{\sqrt[3]{63x^2y}}{9y}$

341. $\dfrac{-14\sqrt{3}}{9}$

342. $\dfrac{\sqrt[3]{28r}}{14}$

343. $\dfrac{3}{4}$

344. $t^5\dfrac{\sqrt[3]{x^2}}{x}$

345. $\dfrac{\sqrt[3]{42}}{6}$

346. $\dfrac{3\sqrt{2ty}}{t^2}$

347. $\dfrac{9x^2\sqrt{2t}}{t^3}$

348. $\dfrac{7\sqrt{3}}{15}$

349. $\dfrac{\sqrt{10}}{4x}$

350. $\dfrac{\sqrt{38}}{8}$

351. $\dfrac{c\sqrt[3]{d}}{d}$

352. $\dfrac{3\sqrt{2t}}{t}$

Writing/Conceptual Exercises

353. (c)

354. (d)

355. (a)

356. The first step is to rewrite $\sqrt[3]{\dfrac{7}{11}}$ as $\dfrac{\sqrt[3]{7}}{\sqrt[3]{11}}$.

This step uses the quotient rule for radicals.

Section 9.5

Objective 1

357. $\sqrt{35}-\sqrt{21}$

358. $5\sqrt{2}+2\sqrt{15}$

359. $5\sqrt{3}+2\sqrt{15}$

360. $3\sqrt{77}-14$

361. $2\sqrt{10}+2\sqrt{35}+\sqrt{6}+\sqrt{21}$

362. $\sqrt{15}+\sqrt{10}-\sqrt{33}-\sqrt{22}$

363. $12+2\sqrt{15}-2\sqrt{30}-5\sqrt{2}$

364. $-28+13\sqrt{6}$

365. -5

366. 5

367. $17-4\sqrt{15}$

368. $21-2\sqrt{110}$

369. $89+24\sqrt{5}$

370. $30+12\sqrt{6}$

Objective 2

371. $\dfrac{2\left(3+\sqrt{2}\right)}{7}$ or $\dfrac{6+2\sqrt{2}}{7}$

372. $-5\left(2-\sqrt{5}\right)$ or $-10++5\sqrt{5}$

373. $\dfrac{5\left(4+\sqrt{3}\right)}{13}$ or $\dfrac{20+5\sqrt{3}}{13}$

374. $\dfrac{4\sqrt{3}+3}{3}$

375/ $\dfrac{\sqrt{5}\left(\sqrt{3}+1\right)}{2}$ or $\dfrac{\sqrt{15}+\sqrt{5}}{2}$

376. $\sqrt{5}-\sqrt{3}$

377. $\dfrac{-9\left(\sqrt{5}-\sqrt{7}\right)}{2}$ or $\dfrac{-9\sqrt{5}+9\sqrt{7}}{2}$

378. $\dfrac{-6\left(\sqrt{3}-\sqrt{10}\right)}{7}$ or $\dfrac{-6\sqrt{3}+6\sqrt{10}}{7}$

379. $\sqrt{21}+\sqrt{14}+\sqrt{6}+2$

380. $\dfrac{-2\sqrt{3}-4\sqrt{6}-\sqrt{2}-4}{14}$ or $-\dfrac{2\sqrt{3}+4\sqrt{6}+\sqrt{2}+4}{14}$

381. $\dfrac{\sqrt{14}+2+\sqrt{21}+6}{5}$

382. $19+6\sqrt{10}$

Objective 3

383. $4-\sqrt{3}$

384. $\dfrac{3+\sqrt{7}}{6}$

385. $\sqrt{3}+3$

386. $\dfrac{2\sqrt{2}+4}{7}$

387. $\dfrac{3+2\sqrt{2}}{4}$

388. $\dfrac{2\sqrt{7}-3}{5}$

389. $2+\sqrt{2}$

390. $\dfrac{1+\sqrt{3}}{2}$

391. $\dfrac{2\sqrt{6}-3}{4}$

392. $\dfrac{2+\sqrt{3}}{4}$

393. $9\sqrt{3}+3$

394. $\dfrac{9\sqrt{5}-2\sqrt{2}}{3}$

Mixed Exercises

395. $\sqrt{30}-\sqrt{66}$

396. $8\sqrt{5}+4-2\sqrt{30}-\sqrt{6}$

397. $13+2\sqrt{22}$

398. 36

399. $\dfrac{4+\sqrt{11}}{5}$

400. $\dfrac{3\sqrt{5}+\sqrt{10}+6+2\sqrt{2}}{7}$

401. $\dfrac{\sqrt{35}-\sqrt{5}-\sqrt{7}+1}{6}$

402. $\dfrac{15\sqrt{5}+25\sqrt{3}}{6}$

403. $\dfrac{\sqrt{14}-\sqrt{6}}{2}$

404. $\dfrac{5+\sqrt{21}}{2}$

405. $215+5\sqrt{3}$

406. $70+20\sqrt{10}$

407. $14-4\sqrt{6}$

408. $2\sqrt{15}-\sqrt{110}+3\sqrt{2}-\sqrt{33}$

409. 23

410. $\dfrac{26\left(\sqrt{11}-\sqrt{2}\right)}{9}$

411. $\dfrac{2-\sqrt{7}}{4}$

412. $\dfrac{10-5\sqrt{10}}{2}$

413. $2\sqrt{10}-3$

414. $\dfrac{\sqrt{11}-3}{2}$

415. $5\sqrt{2}-4$

416. $\dfrac{\sqrt{6}-7\sqrt{3}}{5}$

417. $\dfrac{1-\sqrt{2}}{2}$

418. $\dfrac{1-\sqrt{3}}{3}$

419. $\dfrac{2-9\sqrt{2}}{3}$

420. $\dfrac{3+2\sqrt{15}}{4}$

Section 9.6

Objective 1

421. 81   422. 1   423. 4   424. 20   425. 5   426. $\dfrac{5}{2}$

427. 2   428. 5   429. $\dfrac{1}{2}$ 430. 19   431. 7   432. -3

Objective 2

433. No solution   434. 64   435. No solution   436. 25

437. No solution   438. No solution   439. No solution   440. -2

441. -4 442. No solution   443. 2   444. No solution

445. 5   446. -3

Objective 3

447. 0, 3   448. −1   449. 5, 6   450. 3   451. −1

452. No solution   453. 4, 9   454. 1, 4   455. 9

456. 4   457. 6   458. −1, 8   459. −1   460. 4, 16

461. 3   462. 8   463. 7   464. 5   465. −1

466. 4   467. 11   468. −7, −2

Mixed Exercises

469. 33   470. $\dfrac{13}{2}$ 471. $-\dfrac{7}{3}$ 472. No solution

473. 16   474. 6   475. 5   476. 12   477. 4

478. No solution   479. 5   480. 2   481. 7

482. 11   483. No solution   484. 2   485. 12

486. 6   487. 3   488. −1, 3   489. 2   490. 3, 4

491. $\dfrac{3}{4}$ 492. 5   493. No solution   494. 6

495. 3, 4   496. −9   497. 3   498. 8

Writing/ Conceptual Exercises

499. Since $\sqrt{2x+1}$ must be greater than or equal to zero for any replacement for x, it cannot equal −9, a negative number.

500. In the equation $x^2 = 81$, x can have either a positive or a negative value, since the square of either a positive or negative number is positive. In fact, it has the two solutions, −9 and 9, since $9^2 = 81$ and $(-9)^2 = 81$. On the other hand, $\sqrt{x} = 9$ has only one solution, 81, because $\sqrt{x}$ represents only the nonnegative square root.

501. To solve equations with radicals such as this one, we use the squaring property of equality. Using this property can give a new equation with *more* solutions than the original one. You must check all potential solutions from the squared equation in the original equation. For the equation given here, there are two potential solutions, 3 and 6 but only 6 checks in the original equation.

502. The expression $\sqrt{z^2+3z+4}$ represents a nonnegative square root. Since the value of this expression can never be negative and this expression is equal to z, we see that z can never be negative. Any solution of the equation is a value of z that makes the equation true, so we know that the equation cannot have a negative solution.

Section 9.7

Objective 1

503. 4	504. 5	505. 13	506. 15	507. 2
508. 3	509. 4	510. 5	511. 6	512. 8
513. 9	514. 10	515. 2	516. 3	517. 5
518. 6	519. 2	520. 3	521. 2	522. 2

Objective 2

523. 64	524. 125	525. 216	526. 512	527. 4
528. 16	529. 9	530. 81	531. 16	532. 25
533. 36	534. 64	535. $\dfrac{1}{4}$	536. $\dfrac{1}{125}$	537. $\dfrac{1}{216}$
538. $\dfrac{1}{512}$	539. $\dfrac{1}{8}$	540. $\dfrac{1}{9}$	541. $\dfrac{1}{81}$	542. $\dfrac{1}{100}$

Objective 3

543. 5^2	544. 6^2	545. 7	546. 4^2	547. $\dfrac{1}{8}$
548. $\dfrac{1}{9^2}$	549. $25^{\frac{1}{2}}$	550. $\dfrac{1}{8}$	551. $6^{\frac{7}{5}}$	552. r^6s
553. 9^2	554. 4^6	555. $\dfrac{1}{12^{\frac{9}{8}}}$	556. $\dfrac{2^3}{3^3}$	557. $\dfrac{3^2}{4^2}$

558. $\dfrac{a^8}{b^3}$ 559. $r^{\frac{1}{3}}s^{\frac{1}{3}}$ 560. $\dfrac{1}{y^3}$ 561. $\dfrac{1}{r^2}$ 562. z

563. $\dfrac{c^4}{x^2}$ 564. 8   565. x^2 566. a

Objective 4

567. $\sqrt{7}$ 568. $\sqrt{3}$ 569. 2   570. 3   571. 5

572. 3   573. $\sqrt[3]{p}$ 574. $\sqrt{z}$ 575. $\sqrt[3]{r}$ 576. $\sqrt[4]{a}$

577. 2   578. 10

Mixed Exercises

579. 5   580. $\dfrac{1}{64}$ 581. 3   582. 81   583. $\dfrac{16}{625}$

584. 81   585. 2   586. $\dfrac{1}{512}$ 587. $\dfrac{1}{2}$ 588. 8

589. 9   590. 216   591. 10   592. 16   593. 7

594. $\dfrac{1}{1000}$ 595. 81   596. $\dfrac{1}{216}$ 597. $\dfrac{1}{y^3}$ 598. $\dfrac{1}{x^3}$

599. $\dfrac{x^{18}}{y^{16}}$ 600. $\dfrac{1}{12^{\frac{7}{8}}}$ 601. $r^{\frac{1}{3}}s^{\frac{3}{8}}$ 602. $\dfrac{a^{\frac{1}{12}}}{b^{\frac{1}{8}}}$

Writing/Conceptual Exercises

603. (c)   604. (d)   605. (a)   606. (c)

CHAPTER 10   QUADRATIC EQUATIONS

Section 10.1

Objective 1

1. -10, 10 2. -15, 15 3. -20, 20 4. -3.5, 3.5 5. $-\sqrt{15}$, $\sqrt{15}$

6. $-3\sqrt{7}$, $3\sqrt{7}$ 7. $-4\sqrt{2}$, $4\sqrt{2}$ 8. $-2\sqrt{7}$, $2\sqrt{7}$ 9. $-6\sqrt{2}$, $6\sqrt{2}$

10. No real number solution   11. $-5\sqrt{6}$, $5\sqrt{6}$ 12. $-\sqrt{6}$, $\sqrt{6}$

Objective 2

13. -5, -1 14. -2, 12 15. -8, 4 16. 2, 12 17. 5 18. -3

19. -6, -4 20. $-4+2\sqrt{2}$, $-4-2\sqrt{2}$ 21. $3+5\sqrt{2}$, $3-5\sqrt{2}$

22. $7+3\sqrt{5}$, $7-3\sqrt{5}$ 23. $-5+\sqrt{15}$, $-5-\sqrt{15}$ 24. $\dfrac{\left(2+3\sqrt{3}\right)}{3}$, $\dfrac{\left(2-3\sqrt{3}\right)}{3}$

Mixed Exercises

25. -8, 8 26. -1, 5 27. No real number solution

28. 0, 10 29. $-3\sqrt{6}$, $3\sqrt{6}$ 30. $3+\sqrt{5}$, $3-\sqrt{5}$

31. No real number solution   32. -7 33. -3, $\dfrac{3}{5}$

34. $\dfrac{\left(-4+\sqrt{14}\right)}{3}$, $\dfrac{\left(-4-\sqrt{14}\right)}{3}$ 35. No real number solution

36. $-4\sqrt{5}$, $4\sqrt{5}$ 37. -1.8, 1.8 38. -5.5, 5.5

39. $\dfrac{\left(-7+3\sqrt{3}\right)}{3}$, $\dfrac{\left(-7-3\sqrt{3}\right)}{3}$ 40. $\dfrac{-5\sqrt{2}}{13}$, $\dfrac{5\sqrt{2}}{13}$ 41. -18, 10

42. $12+6\sqrt{5}$, $12-6\sqrt{5}$ 43. $-\dfrac{2}{5}$, $\dfrac{2}{5}$ 44. $-\dfrac{11}{8}$, $\dfrac{11}{8}$ 45. $\dfrac{-2\sqrt{10}}{17}$, $\dfrac{2\sqrt{10}}{17}$

46. No real number solution   47. -1.4, 1.4 48. -1.8, 1.8

49. $\dfrac{\left(-4+5\sqrt{3}\right)}{5}$, $\dfrac{\left(-4-5\sqrt{3}\right)}{5}$ 50. $\dfrac{\left(7+3\sqrt{2}\right)}{3}$, $\dfrac{\left(7-3\sqrt{2}\right)}{3}$ 51. -1, 0

52. No real number solution   53. -24, 8 54. $20+10\sqrt{5}$, $20-10\sqrt{5}$

Writing/Conceptual Exercises

55. True

56. False; there will be either two rational solutions or two irrational solutions.

57. False; the solutions will be irrational because a prime number cannot be a perfect square.

58. False; the equation will have no real number solutions. (Irrational numbers are real numbers.)

59. False; the equation is equivalent to $x^2 = 9$ and has two real numbers solutions, 3 and -3.

60. (b)   61. (b)   62. (c)

Section 10.2

Objective 1

63. -9, 1 64. 3, 9 65. -10, 0

66. -4, 2 67. 1, 3 68. $-6+2\sqrt{3}$, $-6-2\sqrt{3}$

69. 3, 5 70. $1+\sqrt{6}$, $1-\sqrt{6}$ 71. $-2+2\sqrt{3}$, $-2-2\sqrt{3}$

72. $\dfrac{\left(3+\sqrt{5}\right)}{2}$, $\dfrac{\left(3-\sqrt{5}\right)}{2}$ 73. $\dfrac{\left(5+3\sqrt{5}\right)}{2}$, $\dfrac{\left(5-3\sqrt{5}\right)}{2}$ 74. -4, -2

Objective 2

75. $-2, 0$ 76. $-7, 3$ 77. $\dfrac{(3+3\sqrt{7})}{2}, \dfrac{(3-3\sqrt{7})}{2}$ 78. $\dfrac{(1+\sqrt{6})}{3}, \dfrac{1-\sqrt{6}}{3}$

79. $\dfrac{(3+\sqrt{15})}{3}, \dfrac{(3-\sqrt{15})}{3}$ 80. $\dfrac{(2+\sqrt{14})}{2}, \dfrac{(2-\sqrt{14})}{2}$ 81. No real number solution

82. $-\dfrac{1}{3}, 1$ 83. $-\dfrac{1}{2}, 3$ 84. $-\dfrac{5}{2}, -1$ 85. $1+\sqrt{2}, 1-\sqrt{2}$ 86. $-3, 7$

Objective 3

87. $\dfrac{(1+2\sqrt{2})}{2}, \dfrac{1-2\sqrt{2}}{2}$ 88. No real number solution   89. No real number solution

90. $\dfrac{(-2+\sqrt{10})}{2}, \dfrac{(-2-\sqrt{10})}{2}$ 91. $-2+\sqrt{3}, -2-2\sqrt{3}$ 92. $-8, 2$ 93. $-8, -2$

94. $1+\sqrt{5}, 1-\sqrt{5}$ 95. $-6, 4$ 96. $\dfrac{(-5+\sqrt{13})}{2}, \dfrac{(-5-\sqrt{13})}{2}$

Mixed Exercises

97. $0, 6$ 98. $2, 12$ 99. No real number solution   100. $-\dfrac{3}{2}, \dfrac{1}{2}$

101. $\dfrac{(-5+\sqrt{37})}{2}, \dfrac{(-5-\sqrt{37})}{2}$ 102. $-1+\sqrt{5}, -1-\sqrt{5}$ 103. $\dfrac{(4+\sqrt{46})}{6}, \dfrac{(4-\sqrt{46})}{6}$

104. No real number solution   105. $3+\sqrt{5}, 3-\sqrt{5}$ 106. $5+\sqrt{17}, 5-\sqrt{17}$

107. 4 108. $-21, 1$ 109. $-\dfrac{5}{2}, -\dfrac{1}{2}$ 110. $\dfrac{(1+\sqrt{7})}{2}, \dfrac{(1-\sqrt{7})}{2}$

111. $\dfrac{(2+3\sqrt{2})}{2}, \dfrac{(2-3\sqrt{2})}{2}$ 112. $-\dfrac{4}{3}, 1$ 113. No real number solution

114. $\dfrac{(-1+\sqrt{7})}{4}, \dfrac{(-1-\sqrt{7})}{4}$ 115. No real number solution   116. $-2, 1$

117. $\dfrac{(3+\sqrt{5})}{2}, \dfrac{(3-\sqrt{5})}{2}$ 118. $\dfrac{(-3+\sqrt{13})}{2}, \dfrac{(-3-\sqrt{13})}{2}$ 119. $\dfrac{3}{2}$ 120. $-5,\ 7$

Writing/Conceptual Exercises

121. Divide both sides by 3 so that the coefficient of the square term will be 1.

122. It is not a quadratic equation because of the x^3 term.

123. Yes, any quadratic equation with rational solutions (including integer solutions) can be solved by factoring. Working backwards from the solutions to the equation, we obtain

$$(x+4)(x-5)=0$$
$$x^2 - x - 20 = 0$$

124. No

Section 10.3

Objective 1

125. $a = 3,\ b = 2,\ c = 1$ 126. $a = 4,\ b = -3,\ c = 2$ 127. $a = 1,\ b = 5, c = 0$

128. $a = 5,\ b = 0,\ c = -3$ 129. $a = 7,\ b = 3,\ c = -2$ 130. $a = 3,\ b = -1,\ c = 1$

131. $a = 1,\ b = -2,\ c = -7$ 132. $a = 3,\ b = -5,\ c = -4$ 133. $a = 7, b = 0,\ c = -5$

134. $a = -4,\ b = -2,\ c = 3$ 135. $a = 1,\ b = 1,\ c = -5$ 136. $a = 1,\ b = 5,\ c = 7$

137. $a = 1,\ b = -4,\ c = -4$ 138. $a = -1,\ b = -1,\ c = 9$

Objective 2

139. $\dfrac{(-3+\sqrt{5})}{2}, \dfrac{(-3-\sqrt{5})}{2}$ 140. $-10,\ 3$ 141. $\dfrac{(-5+\sqrt{37})}{6}, \dfrac{(-5-\sqrt{37})}{6}$

142. $\dfrac{(1+\sqrt{7})}{2}, \dfrac{(1-\sqrt{7})}{2}$ 143. $\dfrac{(-1+\sqrt{21})}{2}, \dfrac{(-1-\sqrt{21})}{2}$ 144. $-1,\ 0$

145. No real number solution   146. $-\sqrt{7}, \sqrt{7}$ 147. No real number solution

148. No real number solution   149. $\dfrac{(1+\sqrt{3})}{2}, \dfrac{(1-\sqrt{3})}{2}$ 150. $-\dfrac{1}{2}, \dfrac{3}{2}$

151. $4+3\sqrt{2}, \ 4-3\sqrt{2}$ 152. No real number solution

153. $\dfrac{(3+\sqrt{2})}{2}, \dfrac{(3-\sqrt{2})}{2}$ 154. $\dfrac{2}{3}, 6$ 155. $\dfrac{(5+\sqrt{34})}{3}, \dfrac{(5-\sqrt{34})}{3}$ 156. $-7, \ -\dfrac{1}{2}$

Objective 3

157. -3 158. 6   159. $-\dfrac{1}{4}$ 160. 5   161. 8   162. $\dfrac{3}{2}$

163. 7   164. 10   165. $\dfrac{2}{5}$ 166. $-\dfrac{3}{10}$ 167. $\dfrac{5}{8}$ 168. $-\dfrac{7}{12}$

Objective 4

169. $-1+\sqrt{6}, \ -1-\sqrt{6}$ 170. $\dfrac{(3+\sqrt{17})}{2}, \dfrac{(-3-\sqrt{17})}{2}$ 171. $\dfrac{(4+\sqrt{34})}{6}, \dfrac{(4-\sqrt{34})}{6}$

172. $-1+\sqrt{5}, \ -1-\sqrt{5}$ 173. No real number solution

174. $\dfrac{(-2)+\sqrt{10}}{3}, \dfrac{(-2-\sqrt{10})}{3}$ 175. $\dfrac{(-1+\sqrt{7})}{3}, \dfrac{(-1-\sqrt{7})}{3}$

176. $\dfrac{(3+\sqrt{57})}{4}, \dfrac{(3-\sqrt{57})}{4}$ 177. $\dfrac{(-1+2\sqrt{7})}{3}, \dfrac{(-1-2\sqrt{7})}{3}$

178. $\dfrac{(-4+\sqrt{19})}{3}, \dfrac{(-4-\sqrt{19})}{3}$ 179. $-\dfrac{3}{5}, 2$ 180. $-\dfrac{7}{3}, 1$

Objective 5

181. 3 sec   182. 1 sec and 3 sec; both are reasonable   183. 10%

184. About .5 sec   185. 2 sec   186. 2.5 sec   187. .4 sec, 2.6 sec   188. 21.9 ft

189. 2 months, 4 months   190. Ethan, 9.1 hr; Mario, 11.1 hr

Mixed Exercises

191. $a = 4$, $b = 5$, $c = -6$ 192. $a = 1$, $b = -12$, $c = 0$ 193. $a = 4$, $b = 0$, $c = -4$

194. $a = 1$, $b = -1$, $c = -6$ 195. $a = 6$, $b = -8$, $c = -5$ 196. $a = 10$, $b = -5$, $c = -3$

197. $\dfrac{\left(-5+\sqrt{5}\right)}{2}$, $\dfrac{\left(-5-\sqrt{5}\right)}{2}$ 198. $\dfrac{\left(-5+\sqrt{65}\right)}{10}$, $\dfrac{\left(-5-\sqrt{65}\right)}{10}$ 199. -13, 1

200. $\dfrac{\left(-6+\sqrt{26}\right)}{2}$, $\dfrac{\left(-6-\sqrt{26}\right)}{2}$ 201. $\dfrac{\left(-1+\sqrt{5}\right)}{4}$, $\dfrac{\left(-1-\sqrt{5}\right)}{4}$ 202. $-3\sqrt{2}$, $3\sqrt{2}$

203. 9   204. $-\dfrac{8}{3}$ 205. -3, -1 206. -4, -2

207. No real number solution   208. $-\dfrac{3}{5}$, 1 209. -2, $\dfrac{5}{2}$ 210. $\dfrac{9}{7}$

211. $\dfrac{\left(-1+\sqrt{13}\right)}{4}$, $\dfrac{\left(-1-\sqrt{13}\right)}{4}$ 212. $-\dfrac{4}{3}$ 213. $\dfrac{\left(2+\sqrt{14}\right)}{2}$, $\dfrac{\left(2-\sqrt{14}\right)}{2}$

214. -3, $\dfrac{5}{2}$ 215. 2 sec   216. 1 sec and 3 sec

217. 3 units or 7 units   218. 5.9 mph

Writing/Conceptual Exercises

219. The equation has two real number solutions.

220 This is not correct. An applied problem may have more than one answer. We only reject the solutions that do not make sense in the original problem.

221. (d); since $\sqrt{10} > 3$, we know that $3 - \sqrt{10} < 0$. this cannot possibly be an answer because the height of a triangle cannot be negative. All of the other choices are positive numbers.

222. The answer should be simplified.

Since
$$\frac{3+5}{2} = \frac{8}{2} = 4$$

and
$$\frac{3-5}{2} = \frac{-2}{2} = -1,$$

the solutions are -1 and 4.

Section 10.4

Objective 1

223. $5i$ 224. $9i$ 225. $10i$ 226. $13i$ 227. $i\sqrt{7}$ 228. $i\sqrt{11}$

229. $4i\sqrt{3}$ 230. $6i\sqrt{2}$ 231. $12i\sqrt{2}$ 232. $10i\sqrt{10}$ 233. $5i\sqrt{3}$ 234. $8i\sqrt{2}$

Objective 2

235. $6+16i$ 236. $-2+9i$ 237. $-3-5i$ 238. $-18+7i$ 239. $-6+13i$ 240. $4-12i$

241. $14+9i$ 242. $-11+4i$ 243. $8-6i$ 244. $-12+i$

Objective 3

245. $10-35i$ 246. $2+10i$ 247. $23+i$ 248. $13-i$ 249. 53 250. 82

251. $57-i$ 252. $36+22i$ 253. 125 254. $-24+10i$

Objective 4

255. $1+i$

256. $\frac{6}{5} - \frac{3}{5}i$

257. $\frac{3}{2} - \frac{5}{2}i$

258. $\frac{19}{10} - \frac{13}{10}i$

259. $\frac{19}{29} - \frac{4}{29}i$

260. $-\frac{6}{13} - \frac{17}{13}i$

261. $7-2i$

262. $-2-3i$

263. $-\frac{1}{4} - \frac{1}{2}i$

264. $-\frac{5}{2} + \frac{3}{2}i$

265. $\frac{5}{41} + \frac{4}{41}i$

266. $-\frac{1}{2} + \frac{1}{2}i$

Objective 5

267. $-4+i$, $-4-i$ 268. $2+5i$, $2-5i$ 269. $\dfrac{1}{2}+2i$, $\dfrac{1}{2}-2i$

270. $\dfrac{2}{3}+2i\sqrt{2}$, $\dfrac{2}{3}-2i\sqrt{2}$ 271. $-\dfrac{6}{5}+i\sqrt{2}$, $-\dfrac{6}{5}-i\sqrt{2}$ 272. $\dfrac{5}{2}+i\sqrt{2}$, $\dfrac{5}{2}-i\sqrt{2}$

273. $1+i$, $1-i$ 274. $5+4i$, $5-4i$ 275. $-1+i\sqrt{2}$, $-1-i\sqrt{2}$

276. $\dfrac{1}{3}+\dfrac{\sqrt{11}}{3}i$, $\dfrac{1}{3}-\dfrac{\sqrt{11}}{3}i$ 277. $-\dfrac{3}{4}+\dfrac{\sqrt{7}}{4}i$, $-\dfrac{3}{4}-\dfrac{\sqrt{7}}{4}i$ 278. $\dfrac{3}{4}+\dfrac{\sqrt{31}}{4}i$, $\dfrac{3}{4}-\dfrac{\sqrt{31}}{4}i$

Mixed Exercises

279. $6i$ 280. $20i$ 281. $25i$ 282. $2i\sqrt{2}$

283. $-5i\sqrt{2}$ 284. $-2i\sqrt{3}$ 285. $9+3i$ 286. $9+21i$

287. $12+7i$ 288. $37-35i$ 289. $30-30i$ 290. $14+2i$

291. $12-4i$ 292. $15-6i$ 293. $18-14i$ 294. $55+51i$

295. 73 296. 153 297. $2-6i$ 298. $4+5i$

299. $3+2i$ 300. $1+i$ 301. $7+6i$ 302. $5-i$

303. $-1+5i$, $-1-5i$ 304. $-4+i$, $-4-i$ 305. $\dfrac{4}{3}+\dfrac{7}{3}i$, $\dfrac{4}{3}-\dfrac{7}{3}i$

306. $\dfrac{3}{2}+\dfrac{3\sqrt{5}}{2}i$, $\dfrac{3}{2}-\dfrac{3\sqrt{5}}{2}i$ 307. $-\dfrac{1}{7}+\dfrac{3\sqrt{3}}{7}i$, $-\dfrac{1}{7}-\dfrac{3\sqrt{3}}{7}i$

308. $2+3i$, $2-3i$ 309. $3+5i$, $3-5i$ 310. $-2+i\sqrt{5}$, $-2-i\sqrt{5}$

311. $\dfrac{1}{8}+\dfrac{\sqrt{15}}{8}i$, $\dfrac{1}{8}-\dfrac{\sqrt{15}}{8}i$ 312. $\dfrac{1}{2}+\dfrac{\sqrt{15}}{10}i$, $\dfrac{1}{2}-\dfrac{\sqrt{15}}{10}i$

Writing/Conceptual Exercises

313. True

314. False; this equation is equivalent to $i^2 = 1$, rather than the true statement $i^2 = -1$.

315. False; the conjugate of $3-4i$ is $3+4i$.

316. True

317. False; the conjugate of $a+bi$ is $a-bi$, while the opposite of $a+bi$ is $-a-bi$.

318. False; multiply the numerator and denominator by $2-5i$. we need to use the conjugate of the denominator, not the conjugate of the numerator.

319. False; the complex numbers also include the real numbers.

320. False; although numbers like $i\sqrt{3}$ contains radicals, they are imaginary numbers, not irrational numbers. Irrational numbers are real numbers.

Section 10.5

Objectives 1 and 2

321.

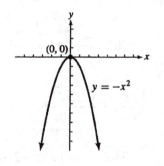

322.

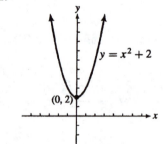

323.

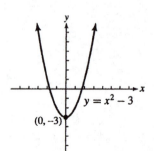

324.

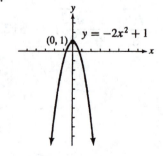

325.

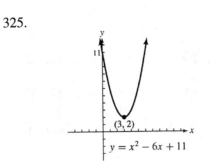

326.

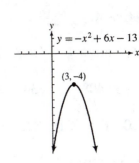

327.

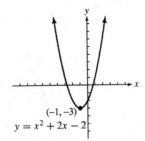

328.

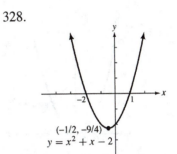

329.

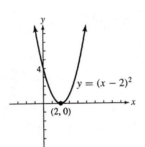

330.

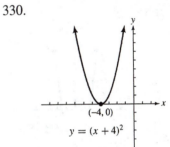

Objective 3

331. One solution: -3

332. One solution: 0

333. Two solutions: $-2, 2$

334. No real solution

335. Two solutions: $-5, -1$

336. No real solution

337. Two solutions: $-1, 3$

338. One solution: 2

339. No real solution

340. Two solutions: $-3, 3$

Objective 4

341. 63, 63   342. 195, 195   343. 1000, 1000   344. 25, 25

345. 408, 408   346. 239, 239   347. 12, 12   348. 39, 39

349. 204 ft × 408 ft; 83,232 sq ft   350. 400 ft × 800 ft; 320,000 sq ft

Mixed Exercises

351.

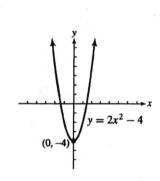

352.

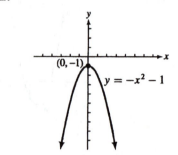

353.

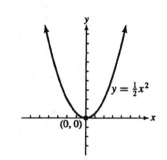

354.

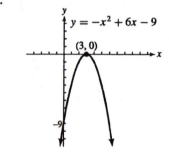

355.

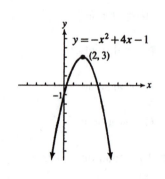

356.

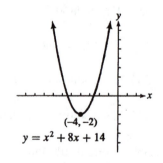

357.

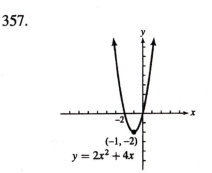

358.

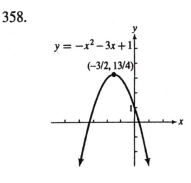

359.

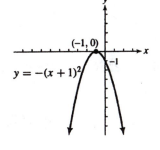

360.

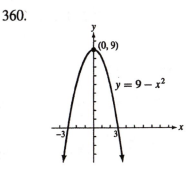

361. One solution: 1

362. One solution: 0

363. Two solutions: -2, 2

364. No real solution

365. Two solutions: 1, 3

366. Two solutions: -2, 2

367. Two solutions: -3, 1

368. One solution: -3

369. No real solution

370. Two solutions: 0, 4

371. 3.7320508 , .26794919

372. .30277563 , -3.3027756

373. .63745860 , -3.1374586

374. 2.5906672 , $-.25733395$

Appendix A

Objective 1

1. 65.51	2. 408.536	3. 640.125	4. 230.935	5. 424.002
6. 67.377	7. 522.9604	8. 349.096	9. 238.123	10. 12.255
11. 349.746	12. 226.661	13. 10.223	14. 36.314	15. 440.105
16. 1.203	17. 251.445	18. 1927.707	19. 93.855	20. 596.939

Objective 2

21. 475.8912	22. 164.19	23. 29.412	24. 32.2586	25. 66.72
26. .12138	27. .282	28. .0147	29. .0134	30. .833
31. 410	32. 20.1	33. .425	34. 19.654	35. 584
36. 6.57	37. .236	38. 525	39. 22.8	40. 25

Objective 3

41. .83	42. .45	43. .1	44. .28	45. 5
46. 3	47. 1.47	48. .02	49. .05	50. .08
51. .003	52. .059	53. .0025	54. .037	55. .0058
56. .32	57. .63	58. 7.21	59. .99	60. 1
61. 25%	62. 30%	63. 45%	64. 10%	65. 78%

66. .4%   67. 13.5%   68. 98.3%   69. 46%   70. 59%

71. 47.3%   72. 83.3%   73. .15%   74. .27%   75. 105%

76. 203%   77. 412%   78. 850%   79. 40.3%   80. 5.9%

Objective 4

81. 10   82. 9   83. 32   84. 1.5   85. 4.68

86. 21.2 87. 150   88. 702   89. .84   90. 2191.23

91. 52.94%   92. 47.06%   93. 500%   94. 86.67%   95. 6.25%

96. 20.74%   97. 31.11%   98. 2.03%   99. 161.9%   100. 28.53%

101. $3510   102. $13,770   103. $409.50   104. $50   105. 540 miles

106. $19.55   107. $3888   108. $10.92   109. $2436   110. $391